BIOLOGICAL PROCESSES

INDUSTRIAL WASTE TREATMENT PROCESS ENGINEERING

VOLUME II

Biological Processes

Gaetano Joseph Celenza

CRC Press
Taylor & Francis Group
Boca Raton London New York

CRC Press is an imprint of the
Taylor & Francis Group, an **informa** business

Biological Processes

First published 2000 by Technomic Publishing Company

Published 2019 by CRC Press
Taylor & Francis Group
6000 Broken Sound Parkway NW, Suite 300
Boca Raton, FL 33487-2742

First issued in paperback 2019

ISBN 13: 978-0-367-45557-6 (pbk)
ISBN 13: 978-1-56676-768-2 (hbk)

Visit the Taylor & Francis Web site at
http://www.taylorandfrancis.com

and the CRC Press Web site at
http://www.crcpress.com

Main entry under title:
    Industrial Waste Treatment Process Engineering: Biological Processes, Volume II

Bibliography: p.
Includes index p. 203

Library of Congress Catalog Card No. 99-66876

These books are dedicated to my parents and brother
who encouraged my efforts in engineering.

Table of Contents

Introduction

WHAT ARE THE BOOKS ABOUT?

THE title of the series defines the scope of the books. *Industrial,* indicating problems characteristic of, and solutions aimed at, manufacturing and processing plants. *Waste Treatment,* indicating treatment of wastes, solids, and residuals generated in manufacturing. *Process engineering,* indicating that design application and selection procedures are detailed. The tone of the technical discussions reflects industrial realities, requiring economic and site specific process engineering considerations to avoid noncompliance notoriety and fines, and to optimize capital and operating expenditures.

Waste control is treated as a single subject; sequencing process controls applied to (1) the raw materials entering the plant, (2) the generation source, (3) passage through a production plant, and (4) to a final central treatment. A facility treatment problem is addressed as two related subjects. One involving *manufacturing evaluation* to eliminate or minimize waste generation, and the other involving selection and application of appropriate *unit operations* to treat resulting effluents. The entire industrial problem is presented in a three volume set. *Volume I* details source correction, pollution prevention, development of a facility treatment system, and primary chemical treatment; *Volume II* details biological treatment unit operations; and *Volume III* details special treatment systems.

These books are written to guide experienced engineers through the various steps of industrial liquid and solid waste treatment. However, the structure of the text allows a wider application suitable to various experience levels; by beginning each chapter with a simplified explanation of applicable theory, expanding to practical design discussions, and finishing with system flowsheets and case study detail calculations. The reader can "enter or leave" a chapter according to their specific needs. As a result, it can serve as a primer for students engaged in environmental engineering studies, or a "one stop" source for experienced engineers. It includes

basic design principles which could be applied to municipal systems with significant industrial influents. In fact, substantial portions of the books are applicable to primary, secondary and tertiary municipal plant treatment. Conventional treatment methods are emphasized, allowing evaluation and implementation of emerging technologies akin to these methods, permitting enhancement to achieve improved performance. Innovative technologies are not downgraded, but encouraged as a choice based on understanding the basic operating principles; but not employed as a fad to replace applicable available technology.

HOW ARE THE BOOKS ORGANIZED?

Industrial Waste Treatment Process Engineering is prepared as a step-by-step implementation manual, detailing the selection and design of industrial liquid and solid waste treatment systems. Great emphasis is placed on identifying and compensating for highly variable and unpredictable manufacturing waste characteristics. The books are written as a *single source* industrial waste management text, consolidating all the process engineering principles required to evaluate a complex industrial waste problem; starting with pollution prevention and source correction, and finishing with "end-of-pipe" treatment. A complete treatment facility is discussed, not just an isolated unit process; including the ancillary as well as the major equipment. Emphasis is placed on the fate of the contaminants, warning of potential problems resulting from by-products formed; preventing a waste treatment problem from becoming an air emission or solid disposal problem.

What makes the books unique is the level of process engineering details included in both the facility evaluation and unit operations sections. The facility evaluation includes a step-by-step review of each major and support manufacturing operation; identifying probable contaminant discharges, practical prevention measures, and point source control procedures. This general plant review is followed by procedures to conduct a site specific pollution control program.

The unit operation chapters contain all the required details to complete a treatment process design, including:

(1) Basic concepts explaining the applicable design relations
(2) How adjustments are made for specific industrial waste characteristics
(3) How basic concepts are applied to equipment sizing
(4) Available equipment and system configurations
(5) Guidelines for selecting the appropriate equipment
(6) Process operating limits
(7) Reported industrial performance data
(8) Required design data
(9) Case studies with detailed calculations for the prominent applications
(10) Conceptual Engineering Flowsheets
(11) Commonly reported design and operating deficiencies
(12) A discussion of the fate of the contaminants; emphasizing total contaminant destruction
(13) General mechanical and associated engineering guidelines

PROCESS ENGINEERING PROCEDURES

The facility evaluation procedures discussed follow the process engineering considerations used in developing the manufacturing facilities. The manufacturing plant is subdivided into related production modules, which sequentially follow the production flow. This includes raw material storage, raw material preparation, the basic manufacturing process, the product separation process, product purification, the finishing operation, by-product recovery, packaging, and support facilities. The flow "in-and-out" of each of these segments is examined to identify potential pollution sources, and logically identify corrective action that would eliminate, minimize, or source correct any waste generation. Although a chemical plant is suggested, many of these separate operations are required in most product manufacturing, in one form or another. This breakdown into individual operations allows the Process Engineer to generically identify similar plant operations, and apply the suggested step-by-step analysis to implement applicable source correction actions.

The unit treatment operations discussed are those commonly applied for municipal waste or water treatment, those employed for industrial tertiary treatment, and those used for thermal destruction of hazardous wastes. Multiple references are included to identify sources of the design criteria cited, so that the interested reader can further review the concept details. Municipal and water treatment applications are frequently mentioned because they are the source of much of

the available operating data. Although some of the principles described originate from operating data from these facilities, the final design must be adjusted for industrial conditions to compensate for differences in treatment reactivity, waste volume variability, and the peculiar properties of industrial wastes. Not only do the properties of industrial wastes vary from municipal or water treatment influents, they are not consistent from industry to industry, or many times from plant to plant in the same industry.

In most cases, unit operation process design methodology is described in terms of models, with specific constants defining the range of applicability. The design method is further analyzed in terms of step-by-step procedures required to complete a system design. These models are supplemented by:

(1) Commonly applied capacity limits expressed in terms of hydraulic loadings, system residence time, or organic loadings
(2) Specific waste characteristics affecting capacity limits and process performance, and therefore the limits of the unit operation
(3) Practical guidelines for selecting and applying the unit operation

Industrial application of these models and associated process engineering procedures requires that an adjustment be made for the waste characteristics and variations, as well as scale-up from theoretical to actual operating conditions. Each chapter prescribes the following:

(1) The required design data that must be obtained from site specific testing
(2) Historical industrial performance data to gauge the applicability to meet required effluent quality
(3) Operating limits for the unit operation
(4) By-products produced from the process so that the fate of the contaminants can be evaluated
(5) General engineering limits to define the mechanical, control, or environmental limits of the system
(6) In some cases, safety factors to upgrade a system design from the theoretical to the operating level

The unit process engineering discussions culminate with Case Studies (calculations) and a Preliminary Concept Flowsheet.

The format in which the unit operations are presented could mislead the reader into ignoring the design criteria verification requirements, believing that a simple parameter can be used to size a system, leaving the details to the plant operator. The most important elements of Process Engineering are not, and cannot be, included in the discussions—common sense, engineering skills, and experience. All Process Engineering requires interpretation of available design methods to suit a site specific problem. Any shortcuts can only lead to a poor operating, or in extreme cases, a failed system; leading to noncompliance, fines, prohibitive operating costs, and endless capital to "patch" the system.

Finally, in reviewing the suggested steps in developing a treatment system the Process Engineer may not agree that a certain step is necessary, or reason that it could be scaled down considerably, or that it will require more attention than indicated. If this book has forced the engineer to consider all elements of a design, adjusting some elements from that indicated in the text to suit specific requirements, *it has served its primary purpose.*

PROCESS CALCULATIONS

Following the basic concepts, each unit operation chapter contains a complete discussion of the process design variables, detailing the individual elements influencing the equipment design and operation. The use of these variables are illustrated through process calculations, referred to as Case Studies. They represent a carefully sequenced procedure to size the system equipment within a range of acceptable operating parameters. They are called Case Studies because adequate examples are included to illustrate the various processing conditions and configurations, emphasizing operating limitations.

In some examples the results indicate (for the conditions evaluated) that the process has limited, or no, applicability because of the waste characteristics, the waste variability, or practical implementation. This is done to emphasize limitations encountered in full scale operations.

Finally, the calculations included in the texts were computer generated. The results were not always rounded out to whole numbers to avoid having to manually check each operation to the sensitivity of a preceding value, and because it could increase the chance of transferring an incorrect number to the text. Therefore, the number of decimals presented are a matter of convenience, and not intended to demonstrate the precision of an evaluation.

PROCESS FLOWSHEETS

Preliminary Conceptual Flowsheets are included in most unit operations to represent typical, not all, configurations employed; and to enable the Process Engineer to *initiate* the required Process and Engineering Diagrams. They will have to be altered and enhanced to meet site specific requirements. The instrumentation indicated is basic, and in many cases minimum, and must be upgraded to meet the specific plant control philosophy.

The instrumentation symbols are those commonly used in Chemical Process Engineering. The combined symbols, represented in a "balloon," are summarized and explained in Table I.1.

TABLE I.1. Instrumentation Symbols.

Symbol	Placement of Symbol		
	Primary	Second	Final
A	Analysis—pH, turbidity, etc.	Alarm	—
C	—	—	Control (ler)
E	Voltage	Element	—
F	Flow	—	—
G	—	—	Gauge
H	—	—	High (value)
I	—	Indicating	—
K	Time	—	—
L	Level	—	Low (value)
P	Pressure	—	—
R	—	Recording	—
S	Speed	—	Switch
T	Temperature	—	Transmit(ter)
V	—	—	Valve

A typical loop would include a sequence of elements, using a combination of symbols listed in Table I.1, to indicate a complete control system. As an example:

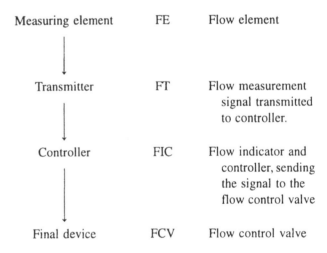

Measuring element	FE	Flow element
Transmitter	FT	Flow measurement signal transmitted to controller.
Controller	FIC	Flow indicator and controller, sending the signal to the flow control valve
Final device	FCV	Flow control valve

Complete discussions of instrumentation applied to waste treatment systems can be reviewed in environmental engineering texts [1].

HOW SHOULD THE BOOKS BE USED?

The three volume book set, and the chapters in the books, have been presented in the order that Process Engineering is commonly applied. Accordingly, the industrial waste treatment design can be initiated starting with Chapter I-1.

First, a pollution prevention evaluation should be conducted within the plant following the procedures outlined in Chapter I-1.

Next, a preliminary central treatment evaluation should be conducted to treat the remaining wastes generated, as discussed in Chapter I-2. This includes:

- a complete waste characterization
- a regulatory review
- preliminary selection of treatment system candidates
- laboratory tests to screen viable treatment options
- pilot studies to develop design criteria

Finally, a complete treatment facility concept should be developed, as discussed in Chapter I-3.

After the preliminary concept has been completed, detail design of the specific unit processes comprising the selected treatment system can be completed. The applicable unit operations are presented in the texts as follows:

- equalization in Volume I, Chapter I-4
- chemical treatment in Volume I, Chapters I-5 to I-9
- flotation in Volume I, Chapter I-10
- aerobic biological treatment in Volume II, Chapters II-2 to II-5
- anaerobic biological treatment in Volume II, Chapter II-7
- aerobic sludge treatment in Volume II, Chapter II-6
- anaerobic sludge treatment in Volume II, Chapter II-7
- sedimentation in Volume II, Chapter II-8
- industrial and hazardous waste incineration in Volume III, Chapter III-1
- adsorption in Volume III, Chapter III-2
- ion Exchange in Volume III, Chapter III-3
- wastewater stripping in Volume III, Chapter III-4
- filtration in Volume III, Chapter III-5
- membrane technology in Volume III, Chapter III-6
- dewatering in Volume III, Chapter III-7

The unit operations should be evaluated in a step-by-step procedure involving the following:

- A preliminary estimate of the equipment size should be made,using the common loadings cited, to establish site and economic feasibility.
- Determine the proper configuration for the technology selected.
- Review the operating limitations for the technology selected.
- Review the reported performance data.
- Do the detail process calculations required to establish the size of various pieces of equipment, according to the procedures described in the Process Engineering Criteria section and Case Study calculations.
- Review the General Engineering Criteria for engineering, construction, or operating limitations.
- Review the reported deficiencies as a check list to avoid commonly encountered process, mechanical, and general engineering pitfalls.
- Develop a Preliminary Process Flowsheet, as illustrated in the individual chapters, tailored to the *specific site and process conditions.*
- Work with the Instrument Engineer, and other engineering disciplines to develop the Process Flow Diagram, the Process and Instrument Diagram, and the plant layout.

This outlines the recommended use of the books' contents in the development of an industrial waste treatment system, with further details provided in the individual chapters.

REFERENCES

1. Water Environment Federation: *WPCF Manual of Practice No OM 6: Process Instrumentation and Control—Operation and Maintenance,* 1984.

Aeration

Aeration devices are employed to supply oxygen to suspended growth biological systems or to induce oxidation conditions for chemical treatment.

BASIC CONCEPTS

AERATION units are common components of chemical or biological oxidation systems, allowing a simple and inexpensive method of supplying oxygen by injecting air into the reactors. Their principal use is in suspended growth biological systems, where they supply oxygen, disperse biological solids, and mix the system reactants. Figure 1.1 illustrates the elements of any oxygen transfer process, with a device's oxygenation capacity defined by a rate coefficient (K_{la}), propelled by the driving force ($C_s - C$), delivering oxygen at the rate of dC/dt.

Chemical process oxygen requirements are stoichiometric quantities, easily established by material balance of simple reactions. Biological oxygen requirements are more difficult to estimate because they represent the stoichiometric sum of simultaneous reactions involving (1) biological carbonaceous oxygen demand, (2) nitrification, and (3) inorganic chemical oxygen demand. Chemical oxidation reactions are discussed in Chapter I-7, and biological systems are covered in Chapters II-2 through II-4.

OXYGEN TRANSFER RATE

Gas transfer rate in a reactor is defined by Equation (1.1) [7].

$$dC/dt = K \cdot A \, (C_s - C) \qquad (1.1)$$

where dC/dt is the change in oxygen concentration with time, K is the oxygen transfer coefficient, A is the gas transfer area, C_s is the oxygen saturation concentration in the waste at the process temperature and represents the maximum oxygen solubility, and C is the solution oxygen concentration in the waste and will always be lower than the saturation concentration.

Drawing an analogy between heat transfer and mass transfer, heat transfer rates (Q) are calculated on the basis of an overall heat transfer coefficient (U), a transfer area (A), and the temperature difference driving force ($T - t$):

$$(\text{BTU/hr}) = U \cdot A \cdot (T - t) \qquad (1.2)$$

As indicated by Equation (1.1), the basic oxygen transfer between two points is similar to the heat transfer relation, Equation (1.2).

Another analogy between heat transfer and mass transfer further illustrates the factors affecting oxygen transfer. Any action increasing transfer from the "film" transfer surface into the bulk volume will increase the heat transfer rate. High turbulent fluid conditions, measured by increased Reynold's number, increases the overall heat transfer coefficient and the heat transfer rate. Aeration devices essentially perform that function by dispersing oxygen to the liquid volume. This is accomplished by turbulent mixing of oxygen and waste in as much of the reactor volume as possible, increasing the contact area between the oxygen (air) and waste by forming small air or waste droplets. The net result is an increase in the mass transfer coefficient K_l and the area A. The contact area is difficult to establish because it is much greater than the basin surface area contributing to diffusion transfer; it is the total liquid droplet area for mechanical aeration or the total air bubble area for dispersed air systems. As a result the transfer capacity is represented by a new factor K_{la}, a product of the transfer coefficient and the area.

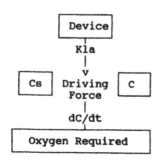

Figure 1.1 Oxygen system.

Using the combined coefficient, Equation 1.1 can be expressed as Equation (1.3) [6,8].

$$dC/dt = K_{la}(C_s - C) \qquad (1.3)$$

The relation defines the concentration change *between* points. The *mass* oxygen transfer rate for the entire basin can be calculated on the basis of an overall mass transfer rate (K_{la}), the aeration volume, and the bulk oxygen concentration difference $(C_s - C)$, as defined by Equation (1.4).

$$OTR = K_{la} \cdot V \cdot (C_s - C) \qquad (1.4)$$

where OTR is the overall mass transfer rate for the system, mass/time, V is the aeration volume, and K_{la} is the "apparent" overall mass transfer coefficient, 1/time.

The first process consideration is maximizing the driving force $(C_s - C)$. In process calculations, C_s is taken as the maximum possible value, the solubility of oxygen in the waste at the system temperature. C is taken as the minimum practical number to assure aerobic conditions, one to two parts per million (ppm), never below 0.5 ppm. For all practical purposes the driving force (the maximum concentration difference) is fixed in the process design, any change due to an uncontrollable variable, *temperature*. Therefore, oxygen transfer can only be improved by increasing the device capacity (K_{la}), the effected transfer area (the dispersion and mixing capability), or both. A device's energy output is considered a measure of its oxygen transfer and mixing capacity. As will be discussed in the next section there are many available devices capable of accomplishing the dual function required of aeration equipment, *oxygen transfer* and *mixing*.

PROCESS OXYGEN CAPACITY

The oxygen requirements of a treatment *system* are a function of the waste oxidation demand, which will be depleted at a rate defined by the reactor operating conditions and the waste reactivity. The specific process demand must be obtained from laboratory generated data or established from similar operating treatment facilities. This oxygen demand must be delivered by the aeration device at specific operating conditions, primarily defined by the release pressure and reactor temperature. Therefore, a rating system is required to compare equipment oxygen delivery rate on a common basis. Conventionally, the aeration capacity of a device is standardized to accepted reference conditions, reported as the standard oxygen transfer rate (SOTR) at the following conditions:

(1) 20°C
(2) One atmosphere oxygen release pressure and a specified relative humidity.
(3) Zero mg/L initial oxygen concentration
(4) Oxygen saturation concentration of tap water at 20°C

The SOTR capacity of specific devices is frequently reported as follows:

- Diffuser air system: The capacity is expressed as the diffuser standard air flow rate (L/S or SCFM) at SOTR conditions. Significantly, the standard oxygen transfer efficiency (SOTE) is used to estimate the portion of the diffuser oxygen transfer rate that is absorbed in the waste.
- Turbine system: The turbine diffuser capacity is expressed as the SOTR, similar to an air diffuser system. The horsepower ratio of the turbine to the diffuser compressor is often quoted to further define the turbine system effectiveness.
- Mechanical aerators: SOTR is expressed as the oxygen mass transfer rate per horsepower or kilowatt at 20°C, 1 atm, and an initial dissolved oxygen (DO) concentration of 0 mg/L.

Some device capacities may be defined by the supplier at conditions other than those mentioned. It is important that the units be understood, and that all process evaluations are based on consistent units.

The required or actual oxygen rate (AOR) and the standardized oxygen transfer rate (SOTR) must be equated to a common basis for design purposes. The relation used is a modification of Equation (1.4) [8]:

$$AOR = SOTR \cdot \frac{\beta \cdot C_{swc} - C_l}{C_s} \cdot 1.024^{T-20} \cdot \alpha \qquad (1.5)$$

where AOR is the actual oxygen requirements, kg (lb) per hour; SOTR is oxygen requirements in tap water at 20°C and 0 mg/l oxygen, kg (lb) per hour; β is correction of oxygen solubility for wastewater characteristics; C_{swc} is oxygen solubility at given temperature, altitude, and pressure at point of release; α is correction for oxygen transfer rate in wastewater; C_l is operating dissolved oxygen concentration, mg/L; C_s is oxygen solubility of tap water at 20°C, 9.17 mg/L; and T is wastewater temperature, °C.

Process calculations presented in design texts are sometimes confusing because of the method chosen to apply the correction. The application may be better understood if the previous equation is simplified to the expression designated by Equation (1.6), and actual oxygen rate and standard oxygen rate (SOR) used to define capacities.

$$AOR = SOTR \cdot F_c \qquad (1.6)$$

where F_c is a product of the individual correction factors. The relation can be applied in one of two ways:

(1) The correction factor can be applied to the actual required *process* AOR and the SOTR calculated, i.e.,

$$\text{Process SOTR} = AOR/F_c \qquad (1.7a)$$

The number of required *devices* can be calculated by

Number of devices
  = process SOTR/unit capacity SOTR   (1.7b)

(2) The correction factor can be applied to the SOTR rating of the device, corrected to the actual conditions (OTR), as follows:

$$\text{Device AOR} = SOTR \text{ rating}/F_c \qquad (1.8a)$$

The number of required *devices* calculated by,

Number of devices
  = required process rate/device AOR   (1.8b)

Application of these concepts are further explained in the Process Engineering Section and illustrated in Case Study 9.

OXYGEN-SATURATED SOLUBILITY VALUES

The saturated oxygen solubility (C_s) is defined by system conditions, set by the

(1) Partial pressure of oxygen above the waste, P_o
(2) Total system pressure
(3) System temperature
(4) Wastewater composition

Oxygen solubility in *clean water* can be estimated by Henry's law equation, relating gas solubility to the gas partial pressure above the solution and Henry's law constant H.

$$C_s' = P_o \cdot H \qquad (1.9)$$

However, oxygen solubilities are readily available; values at common operating temperatures are indicated in Table 1.1.

TABLE 1.1. **Oxygen Solubility.***

Temp, °C	°F	C_s, mg/L
4	39.2	13.1
10	50	11.3
16	60.8	9.9
20	68	9.2
21	69.8	8.9
27	80.6	8.0
32	89.6	7.3
38	100.4	6.6

*Constants in clean water, 1 atmosphere, 0 mg/L chloride.

Correction for Wastewater Characteristics

In most cases the wastewater solubility correction from reported tap water values to process conditions is obtained experimentally and expressed by a beta factor (β),

$$C_s = C_s' \beta \qquad (1.10)$$

where C_s' is the solubility of tap water, C_s is the solubility of the wastewater, and β is the oxygen saturation of wastewater divided by the saturation concentration of clean water, at the same temperature.

Pressure Correction

Solubility values of tap water at standard conditions of 1atm must be corrected for the site conditions and the depth of release in the vessel. The operating pressure will be equal to atmospheric pressure, *plus* the depth of water above the air release, *plus or minus* the elevation correction for the site.

Saturation values for other than 1 atmosphere can be estimated utilizing the following relation:

$$C_s = C @ 1 \, atm \cdot \frac{P - p}{P_s - p}$$

where C_s is the corrected surface oxygen saturation, mg/L, P is the prevailing barometric pressure, corrected for elevation, p is the saturated water pressure at the water temperature, and P_s is the standard atmospheric pressure, 1 atm or equivalent.

For below surface aerators, pressure correction from standard conditions can be estimated utilizing the following relation [1]:

$$C_b = C_s @ \text{surface} \cdot \frac{1}{2} \cdot \left[\frac{P_b}{P_s} \cdot \frac{O_t}{21} \right]$$

where C_b is the corrected oxygen saturation at the effective tank depth mg/L, P_b is the pressure at the bubble release, including the site elevation and bubble depth corrections, atm or equivalent units, and O_t is the oxygen concentration leaving the tank, %.

The use of these equations is illustrated in Case Study 9.

OVERALL TRANSFER COEFFICIENT (K_{la})

Overall transfer coefficents are usually lower for waste-waters than the published manufacturer's values for pure water. Published manufacturer's oxygen transfer values are specific for the aeration device, usually expressed on the basis of tap water at 20°C. Published values can be corrected to process conditions as follows:

$$(K_{la})_t = (K_{la})_{20} \cdot (1.024)^{(t-20)} \cdot \alpha$$

where $(K_{la})_t$ is the coefficient at t°C, $(K_{la})_{20}$ is the coefficient for tap water at 20°C, t is the wastewater temperature at °C, and α is the ratio of wastewater to tap water coefficients.

The overall transfer capacity is frequently reported as the standard capacity of the device in units previously defined.

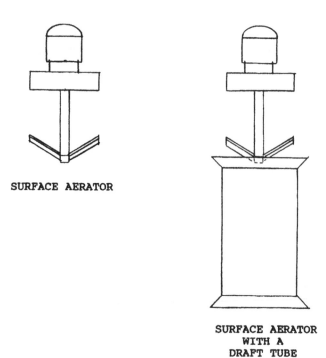

SURFACE AERATOR

SURFACE AERATOR WITH A DRAFT TUBE

Figure 1.2 Surface aerators.

AERATION DEVICES

Aeration devices can be categorized into two broad groups, *mechanical* and *diffused aeration*. *Mechanical aeration* can be described as *surface aerators, submerged aerators,* or *combined aerators*. *Diffused aerators* basically involve an external blower injecting air through some type of restricting orifice dispersing the air through the wastewater volume.

MECHANICAL AERATORS

The basic components of a mechanical aerator are a motor, a gear reducer (or variable speed motor), the impeller drive shaft, the shaft housing, the shaft support bearing, and the impeller. The motors are usually totally enclosed and fan cooled and equipped with a variable-speed motor or a gear reducer to decrease the speed to the operating range. The impeller is connected to a shaft, which in turn is connected to the shaft housing. Shaft bending is minimized by support bearings.

Figure 1.2 illustrates a typical surface aerator and a surface aerator with an extended draft tube. Some general *surface aerators* characteristics include [2,8,9]

(1) Large quantities of oxygen can be efficiently transferred using a few units.

(2) A high pumping action results in significant turbulence, producing good suspension of biological solids in dispersed growth reactors. In some cases, an extra lower impeller, or inclosing the primary impeller in an ex-

tended draft tube, increases the bottom pumping action in deep basins.

(3) Changing the speed or the impeller submergence provides some operating flexibility in adjusting the oxygen output.

(4) Major limitations include (a) an adequate surface area must be provided to suit the required spacing between individual aerators and aerator and walls, (b) they are prone to icing problems, (c) they must be serviced by boat or pulled from the basin, (d) high speed increases parts wear, and (e) low-speed units must be employed to achieve both high transfer efficiency and good mixing.

(5) Their major advantages include (a) high transfer efficiency, (b) the elimination of complex piping, (c) they can be float mounted, (d) standby equipment are not necessary for multiple unit system, and (e) relatively low unit cost.

Figure 1.3 illustrates a typical submerged turbine aerator. Some general *submerged turbine aerators* characteristics include [2,8,9]

(1) They provide an intermediate oxygen transfer efficiency, less than surface aerators but higher than diffused air.

(2) Mixing and oxygen transfer can be independently controlled.

(3) Because full power is applied near the basin bottom they are capable of dispersing high solids concentrations in tanks with working levels of 6 m (20 ft) or more.

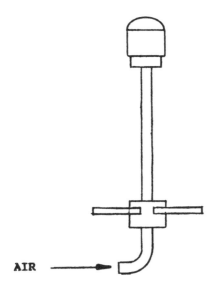

Figure 1.3 Forced air submerged turbine aerators.

(4) They allow a high degree of operating flexibility in controlling output oxygen capacity by varying the air flow rate, independently of the mixing power.

(5) Major limitations include (a) the required installed horsepower is generally higher than surface aerators because of lower transfer efficiencies, (b) the required submerged piping limits their use in shallow basins or lagoons, (c) high cost, and (d) potential reducer problems with the long shaft over turbulent sparged air.

(6) Their major advantages include (a) they are good for service in deep basins, (b) basin surface area is not as critical as with surface aeration and basins can be designed to optimize land use, (c) no significant winter icing problems, (d) they can be more economically applied to high industrial waste loadings because oxygen and mixing requirements can be separately controlled, and (e) they can be bridge mounted, making them convenient to service.

Vertical Axis Surface Aerators

Surface aerators are the least complex and most commonly employed aeration device in completely mixed industrial waste biological systems. As the name implies, they are basically surface action devices, with the applied energy resulting in a fine mist and a turbulent surface to assure high liquid to air contact for oxygen transfer. The unit flow pattern results in a waste pumping action upward and in a circular motion outward. They can be classified as upflow or downflow, depending on the impeller or turbine design, and as high or low speed.

The units are highly efficient oxygen transfer machines, and, although overall mixing is usually good, the effective mixing depth and radius of influence is limited. Because these units perform the dual function of oxygen transfer and mixing, improvement of one variable usually decreases the effectiveness of another. They are applied at depths less than 5 m (15 ft), extended to 6 m (20 ft) by using draft tubes or by adding a submerged lower impeller. The appropriate working depth depends on the specific manufacturer's design and the size of the unit. Generally, vendors quote maximum 5 m (15 ft) depth limits for units greater than 11 kW (15 hp), with recommended depths decreasing for smaller units.

Location of aerators in a basin is critical to assure an adequate mixing pattern and preventing "dead spots." The spacing depends on the unit's radius of mixing, which increases with increased horsepower. The minimum and maximum radius of influence defines the limits of the unit. Units spaced too closely result in interfering unit patterns, whereas those spaced too far apart result in areas of low or no mixing.

These units are highly efficient and when operated at design conditions usually require the least total installed horsepower. However, applied energy effectiveness depends on operating conditions, especially liquid to impeller contact level. Efficiency rapidly diminishes as the contact level changes from optimum conditions, which is no more than 2 to 5 cm (1 to 2 in.). Oxygen transfer rate can be altered by adjusting the basin liquid level, utilizing two-speed motors (generally 2:1 reduction) or variable-speed motors. The effective depth can be altered by using a draft tube.

Although very effective and commonly employed, surface units have some serious operating limitations. They present a special problem in optimizing oxygen transfer, mixing, and radius of influence because they cannot be independently controlled. When applied to high industrial organic loadings one of the operating variables will dominate the design criteria; the others are compromised to achieve the controlling objective. In addition, sprays generated from these units could create a serious problem in cold climates, causing icing of motors and working areas. Surface spraying also enhances heat loss from the system, potentially reducing basin temperatures to levels that diminish biological activity. Spraying can also cause odor problems from low-volatile waste components vaporizing. The exposed motor and spray action of the units can be noisy and undesirable in populated areas.

Vertical aerators are available as fixed or floating units. The fixed units are more stable but difficult to maintain without shutting them down and removing from service. Floating units, which are mounted on pontoons and held in place by guy wires, can be easily rearranged to (visibly) adjust mixing patterns and can be maintained without a system shutdown. However, continuous rotation can induce mechanical problems, and any break in the supporting guy wires can be disastrous.

Vertical *upflow aerators* are classified as low speed (centrifugal) or high speed (axial), with some manufacturers offering units with intermediate speeds (radial/axial). *Low-speed aerators* operate at a range of 20 to 100 rpm, up to 110 kW (150 hp) [8,9]. These units combine good oxygen transfer and mixing and are commonly used in activated sludge systems to achieve completely mixed characteristics. Gear box design is critical, generally requiring service factors of two or more to compensate for the severe service conditions and mechanical difficulties commonly encountered. Clean water standard capacities range from 1.2 to 3 kg O_2/kW · hr (2 to 5 lb O_2/hp · hr), reported field values range from 0.7 to 1.4 [8,9]. When used with extended draft tubes capacities are generally lowered to 1.2 to 2.8 kg O_2/kW · hr (2 to 4.6 lb O_2/hp · hr), reported field values range from 0.7 to 1.3. Unit performance is sensitive to impeller level, and therefore careful considerations must be given to location of fixed units in the aeration vessel.

High-speed axial aerators are normally directly motor driven at speeds ranging from 400 to 800 rpm or higher at ratings up to 110 kW (150 hp). They are more commonly used in aerated ponds, where sludge shearing is not a serious consideration and mixing and oxygen transfer efficiency is not critical. Clean water standard capacities range from 1.2 to 2.2 kg O_2/kW · hr (2 to 3.6 lb O_2/hp · hr); reported field values range from 0.7 to 1.2 [2,8].

Downdraft aerators can have a variety of proprietary designed configurations, two of which include the open and closed turbine. Liquid contact with ambient air is controlled, reducing potential cooling effects and providing high mixing forces in the aeration basin liquid. Clean water standard capacities range from 1.2 to 2.4 kg O_2/kW · hr (2 to 4 lb O_2/hp · hr); reported field values range from 0.6 to 1.3 [8].

Open-turbine aerators are constructed with a vertical shaft and motor, with an impeller on the shaft. The impeller induces downward liquid flow, entraining ambient air, resulting in bulk mixing in the effected aeration basin volume. The mass movement is downward toward the vessel bottom and upward toward the surface.

Closed-turbine aerators operate in a similar manner to the open turbine, the process differences a result of its physical configuration. An impeller is located in a confined casing, open at the bottom, and containing flow-directing vanes. The impeller action pumps liquid into the casing, causing a vacuum at the impeller, inducing ambient air into intake orifices located above the impeller at the top of the casing, forming a mixture consisting of fine bubbles and wastewater. The mixture is discharged into the aeration basin volume, causing a circulating pattern, and mixing of the effected vessel volume. An additional bottom turbine is frequently employed to enhance mixing in deep aeration basins. Because the transfer action in downflow devices occurs below the liquid, as opposed to mists produced by surface aerators, they are better suited for cold weather applications, reducing potential freezing problems.

Submerged Aeration

The major advantage of these units over surface aerators is that oxygen transfer can be controlled by controlling the air flow, whereas bottom mixing can be independently controlled by controlling a submerged impeller. Because oxygen capacity and mixing are independent operating variables, the basin can be more economically designed. The cross-sectional area can be optimized because the selected depth is limited only by the practical length of the shaft.

When employed in deep basins heat loss is reduced, external water sprays are eliminated, potential surface icing problems are prevented, and odor problems minimized. Noise problems can be eliminated by isolating the blower. Because air supply is independent of the aerator, these units are easier to automate. High pumping capacities allow excellent solids suspension in basins of all depths.

The disadvantages of these units over surface units are lower oxygen transfer efficiency and higher capital and operating costs. The higher capital costs are a result of both the requirement for more units and the additional costs of the blowers. Maintenance costs are usually also higher because mixers, blowers, and submerged piping must be maintained.

Mechanical, air-draft, turbine aerators use both surface and submerged turbines. The surface turbine contains air intake holes. When rotating, the surface turbine creates a vortex forcing air through the holes and downward to the bottom turbine. The bottom turbine shears the air bubbles, dispersing them into the liquid, and also serves as a mixer. The top turbine acts as a surface aerator. These units operate at reported clean water standard capacities ranging from 1.2 to 2.0 kg O_2/kW · hr (2 to 3.3 lb O_2/hp · hr); reported field values range from 0.7 to 1.1 [8,9].

Forced air draft turbine aerators employ an external source to deliver air to a sparger ring located below a mixing turbine beneath the liquid surface, usually near the bottom. The large bubbles formed are disintegrated into finer bubbles, which are injected into the aeration volume influenced by the turbine. A second turbine, located above the air injection area, adds additional mixing energy to the active region. These units operate at reported clean water standard capacities ranging from 1.2 to 2.4 kg O_2/kW · hr (2 to 4 lb O_2/hp · hr); field values range from 0.7 to 1.1 [8].

Horizontal Axis Aerators

Horizontal axis aerators are available for specially designed aeration basins, commonly long, narrow, and shallow. They operate as surface (and some specially designed submerged) aerators in a manner similar to their vertical counterpart, except on a horizontal axis. These units operate at reported clean water standard capacities ranging from 0.9 to 2.2 kg O_2/kW · hr (1.5 to 3.6 lb O_2/hp · hr); field values range from 0.5 to 1.1 [8].

DIFFUSED AIR SYSTEMS

Diffused air systems consist of a blower, an air filter, air transfer piping to the aeration basin, aeration basin distribution piping, and diffusers. The diffusers are attached to headers that can be either fixed or retractable. Blowers supply air to each of the diffusers at a discharge head adequate to overcome piping, diffuser, and liquid head losses. The critical component is the diffuser, which establishes oxygen transfer and mixing efficiency. Manufacturers offer a variety of diffuser designs, broadly classified as *porous* or *nonporous*. Porous diffusers are usually ceramic plates or tubes designed with tiny restricting orifices to develop fine or coarse bubbles. Nonporous diffusers use nozzles, orifice, valves or shear forces to develop contact between the wastewater and air.

Diffuser oxygen transfer depends on effective contact between the wastewater and oxygen belched from the diffuser. Contact area is a direct function of the bubble size, whereas the transfer rate depends on effective contact between the bubbles formed and the waste. Contact efficiency increases with decreasing bubble size down to an optimum 2.2 mm, below which the additional required power is economically ineffective [8]. In general, the bubble size is determined by the pore size in porous diffusers or shearing energy in nonporous diffusers. Air flow, waste properties, and diffuser material characteristics affect air release and the bubble size.

Process factors affecting SOTE include orientation and depth. Diffuser characteristics affecting efficiency include size and shape, especially those features directly influencing bubble release and total bubble area. One of the critical operating parameters is diffuser depth, which when increased increases SOTE as a result of greater contact time and higher oxygen partial pressure.

Diffuser *operating characteristics* are affected by air flow rate sensitivity and fouling potential. Fouling can result from particles transferred with the air, particles generated from piping erosion and corrosion, or chemical and biological action on critical diffuser surfaces. Diffuser effectiveness depends on cleaning to retain full working capacity. Cleaning is facilitated by ease of access, removal, and cleaning. The individual unit's sensitivity to air flow in the required operating range establishes its stability.

The *total aeration effectiveness* depends on the combined effects of the individual diffusers, influenced by the aeration basin geometry, as well as location and pattern within the basin. Theoretically, a basin geometry (its length and width) is designed to promote either a plug flow or completely mixed configuration. Although diffusers can be designed for plug flow or complete mix systems, they have traditionally been associated with plug flow municipal systems. Total aeration system stability depends on the ability to adjust total or individual air flow to meet dissolved oxygen requirements or mixing requirements, without radically affecting transfer efficiency.

Diffuser *physical characteristics* are determined by mechanical strength and the ability to withstand physical and thermal stresses encountered during installation, operation, and maintenance. This starts with the diffuser's material of construction and assembly and includes the diffuser, attachments to the aeration basin, and support and connections to the air piping. It also includes its susceptibility to mechanical shock during installation, operation, or cleaning and its resistance to corrosion or biological induced fouling.

The number of diffusers and their location in completely mixed systems is primarily defined by mixing requirements; the aeration requirements are usually met under these conditions. In plug flow configurations oxygen demand is the controlling criteria, and the diffuser system design is driven by an aeration demand varying with linear distance, balancing heavy inlet organic demand and lower exiting demand. Ideally, this can be accomplished without excessive total basin aeration or creating local anaerobic conditions. Varying oxygen demand makes diffuser location and pattern a major design consideration. Process economics favor low diffuser density, whereas some investigative data suggest increased density improves oxygen transfer efficiency [8]. The selected installation pattern must provide adequate spacing to check, service, and clean the units.

In summary, diffuser effectiveness depends on its oxygen transfer characteristics, operating characteristics, turndown limits, physical limits, and contribution to the total basin performance. Total aeration effectiveness depends on the specific diffuser properties and turndown limits, balanced with its location, depth, and basin density, all of which must be consistent with the varying oxygen demands within the basin, maintaining the configuration mixing requirements and minimizing energy consumption.

Some general characteristics of *diffused aeration* can be summarized as follows [2,6,8,9]:

(1) Oxygen transfer efficiencies are generally at the low end of the aeration devices, with fine bubble diffusers more efficient than large bubble diffusers.
(2) They are commonly employed in plug flow aeration basin configurations.
(3) Careful attention to basin design and diffuser placement is required to insure solids dispersion.
(4) The ability to vary diffuser air flow provides complete operational flexibility.
(5) Major limitations include (a) diffusers, especially fine bubble devices, are prone to plugging problems, (b) they are generally employed in more expensive (long-narrow) basins, and (c) they could potentially require higher installed horsepower if they cannot be designed to accommodate the varying load consistent with a plug flow configuration.

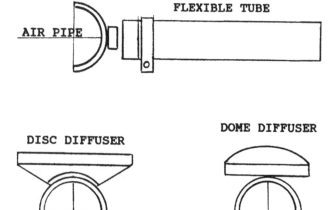

Figure 1.4 Porous diffusers.

(6) Their major advantages include (a) the flexibility to tailor individual diffusers to area oxygen demands, (b) a relatively quiet aeration basin operation is achieved by isolating the blowers, and (c) no icing problems.

Porous Diffusers

Porous diffusers, illustrated in Figure 1.4, are generally manufactured as tubes constructed of flexible materials such as cloth or plastic sheaths or as rigid ceramic or plastic plates, domes, or discs. Plates are not commonly employed, being replaced by the other porous diffusion devices.

Dome diffusers are circular discs constructed of aluminum compounds, commonly 3.8 cm (1.5 in.) high and 18 cm (7 in.) in diameter. They are mounted on base plates and factory connected to air piping. Aeration basin designs include total floor coverage or grid patterns installed in the basin floor. Air capacity when ceramic units are used in a grid pattern range from 0.25 to 1.2 L/s (0.5 to 2.5 scfm) per diffuser, with a 25 to 37% SOTE at a 5 m (15 ft) submergence, and a 15 to 64 cm (6 to 25 in.) headloss for clean units [8].

Disc diffusers are similar to domes, except that they are relatively flat, and available in different sizes, shapes, materials, and methods of attachment. They are constructed in 18 to 24 cm (7 to 9.5 in.) diameters and 13 to 19 mm (0.5 to 0.75 in.) thickness. They can be factory or field mounted to distribution piping. Air capacity when ceramic units are used in a grid pattern range from 0.3 to 1.5 L/s (0.6 to 3.0 scfm) per diffuser, with a 30 to 34% SOTE at 5 m (15 ft) submergence and a 13 to 49 cm (5 to 19 in.) headloss for clean units [6,8]. Diffusers constructed of flexible membranes have diameters up to 520 mm (20.5 in.) and capacities ranging from 1.5 to 10 L/s (3 to 20 scfm) [8].

Tube diffusers are typically 50 to 60 cm (20 to 24 in.) long, of 6.4 to 7.6 cm (2.5 to 3.0 in.) outside diameters, and constructed of rubber or plastics [8]. They are commonly installed at one or both of the basin walls but can be installed in uniform grind patterns. The air rates for porous plastic tubes in a grid pattern are 1.1 to 1.9 L/s (2.4 to 4.0 scfm) per unit, with a 28 to 32% SOTE at a 5 m (15 ft) submergence; 1.4 to 4.6 L/s (3.0 to 9.7 scfm) per unit, with a 18 to 28% SOTE in a dual-spiral roll pattern; and 1 to 5.7 L/s (2.1 to 12.0 scfm) per unit, at 13 to 35% SOTE in a single-spiral roll pattern. The air rate for flexible sheath tubes in a grid pattern is 0.5 to 1.9 L/s (1 to 4 scfm) per unit, with a 22 to 29% SOTE at a 5 m (15 ft) submergence; 0.9 to 2.8 L/s (2 to 6 scfm), at 19 to 24% SOTE in a quarter points pattern; and 0.9 to 2.8 L/s (2 to 6 scfm), at 15 to 19% SOTE in a single-spiral roll pattern [8].

Nonporous Diffusers

Nonporous diffusers can be broadly classified as fixed or valved orifice, which encompass a wide variety of devices. Fixed orifice diffusers include devices constructed in a variety of forms, such as perforated piping, spargers, and slotted tubing. Nonperforated diffusers include valved orifices, static tubes, and perforated hose. Other devices include jet aeration, aspirating devices, and U-tube aerators. Characteristics of these devices depend on their specific construction and their location in the aeration basin—with air rates ranging from 1.5 to 21 L/s (3–45 scfm) at a 5 m (15 ft) submergence, at a 9–13% SOTE [2–8].

PROCESS ENGINEERING DESIGN

Aeration design requires a definition of the process oxygen requirements, selecting an aeration device, and tailoring the basin design to optimize the aeration performance.

PRACTICAL LIMITS

Aeration devices are stable oxygen-producing machines with performance dependent on reactor liquid level, waste temperature, and waste characteristics. The reactor oxygen concentration is a valid measurement of aerator performance, with inadequate aeration device capacity denoted by depleted basin oxygen levels. With adequate capacity the operator has complete control of the aeration or oxidation process. Specific operating characteristics are cited in the Aeration Devices section. Aerator device capacity is a characteristic of the specific device and reactor conditions. Generally, the capacities are in the range cited for the individual class of devices.

REQUIRED PROCESS DESIGN DATA

Oxygen uptake rates are obtained in biological treatment system studies utilizing procedures detailed in Environmental Engineering handbooks [2,9]. Aeration devices are rated

TABLE 1.2. **Required Design Data.**

(1) Design temperature
(2) Summer temperature effects
(3) Winter temperature effects
(4) Device oxygen capacity at standard conditions.
(5) Alpha factor
(6) Beta factor
(7) Mixing efficiency
(8) Appropriate reactor geometry

on the basis of clean water, expressed in terms of air capacity or oxygen output, depending on the specific device. The rated capacity must be corrected to the waste water oxygen saturation value (β ratio) and the transfer rate (α ratio), both defined in the Basic Concepts section. The value of these constants are obtained in reactor studies using common laboratory procedures (Standard Methods). Regardless of the manner of obtaining data, the design criteria listed in Table 2 are required.

SELECTION

The devices discussed in the Aeration Devices section have specific characteristics that must be compared to the service intended. There are no rigid rules dictating aeration equipment, the selection being primarily governed by definitive process requirements, basin geometry, site-specific climatic conditions, and *economics*. However, some general considerations can be summarized as follows:

(1) Industrial completely mixed systems generally utilize mechanical aerators to dampen high influent loadings and optimize total oxygen distribution, although diffused air systems can be adapted for most services encountered. Plug flow kinetics, utilizing long narrow basin configurations, are best adapted to diffused air systems.

(2) Industrial systems frequently involve small highly concentrated waste volumes, with large aeration basin volumes. In such cases, mechanical aerators often provide a convenient combined oxygen supply and mixing capability.

(3) Where high influent concentrations are encountered, mechanical aerators, employed in a completely mixed system, often offer the best protection against shock loading by providing instant dilution and a better use of the aerator oxygen output.

(4) Diffused aerators can be installed in deep basins, allowing for minimum land use.

(5) When employed for deep basins, diffused aerators provide minimum exposure of waste droplets to climatic conditions and maximum cold weather protection.

(6) Submerged aerators are an attempt to optimize the characteristics of a diffused air and mechanical aerator.

These considerations must be balanced with the device characteristics, such as fouling potential, capacity range, operating flexibility, transfer efficiency, installation costs, operating costs, and required maintenance. Specific aerator device performance characteristics, limitations, and disadvantages are discussed in the Aerator Devices section.

PROCESS DESIGN VARIABLES

Optimizing aeration equipment performance to specific process demand requires evaluating the following criteria:

(1) Oxygen demand
(2) Basin/tank configuration
(3) Number of units
(4) Mixing requirements
(5) Winterization
(6) Installed horsepower

Oxygen Demand

The most critical process decision made in an aeration system assessment is selecting the waste flow and corresponding influent concentration design basis. Aeration capacity can be affected by significant influent flow and concentration variations from the design basis. Volume changes could result in changing aeration device oxygen output, whereas influent substrate changes affects required oxygen demand. Successful treatment requires that process demands be met over the range of expected influent conditions.

The effect of volume variations depend on the specific device. In diffused air systems, a changing level alters the hydraulic head above the diffuser, changing its SOTE, which for any specific device is clearly defined by manufacturer characteristic charts. In mechanical aerators, level effects are more subtle. As an example, surface aerators are commonly mounted in fixed positions, with level changes resulting in basin discharge points above the device and reduced device capacity; or below the device so that the aerator is exposed and nonproductive. In many cases, surface aerator efficiency is affected by minute level changes. These units must be installed with provisions for variable outlet weirs to allow proper initial setting to optimize performance and investigated to assure that flow variations do not result in levels critically affecting performance. Submerged aerators with external air sources are also pressure sensitive, following the general characteristics of diffused air systems.

The substrate feed rate establishes the process oxygen demand, which will vary with changing influent biological or chemical oxygen demand. Unless elaborately controlled, aerators are constant production machines, supplying the same oxygen output regardless of process demand. Based on design conditions the aerator output will be deficient at excess oxygen demands and wasteful at lower demand loads.

The Process Engineer, extremely sensitive to system deficiency often overdesigns the aeration system, thereby unintentionally introducing two operating problems. First, excess aerators are installed in the reactor as "operating" spares, to be activated when principal units are inoperable or in repair, higher loads are encountered, or as a design contingency. Unfortunately, these units are many times operated "because they are there," resulting in excessive dissolved oxygen levels, which besides being costly, has been associated with poor sludge settling in some activated sludge systems. Next, the Process Engineer has probably included the nonoperating units in evaluating the mixing patterns or ignored that basin section completely. In either case, dead spots and poor mixing result when these units are not operating. The best solution is influent equalization and preparation to control the waste loads to the plant, designing for an average (equalized) condition and avoiding large influent variability. The system should also include capabilities to monitor and control the reactor dissolved oxygen level to effectively adjust aerator production.

Basin/Tank Configuration

As previously mentioned, a primary consideration in designing an aeration basin is controlling liquid level variations to stabilize aeration device output. An aeration basin design is primarily based on required retention time, which establishes the basin volume. The vessel geometry is established by process kinetics and mixing requirements, which set the length to width proportion. The aeration device limitations and basin construction costs set the depth. Generally, industrial reactors operate at long retention times and corresponding large volumes so that the installed aerator capacity may be controlled by mixing rather than the oxygen production.

Two important considerations that define a basin or tank configuration are the available (land) area and the required effective reactor volume. The available area plays a significant, but not necessarily an obvious, factor in aeration selection. Given a required process volume, the basin configuration will have either a dominant cross-sectional area and a modest or shallow depth or a deeper basin with a smaller cross-sectional area. A deeper basin is usually selected because of limited land area or to minimize heat losses in cold climates. These configurations present different criteria in selecting aeration equipment.

A large surface area could require more units than that required to satisfy oxygen demand or mixing as defined by the total volume, the additional units being necessary for complete surface coverage, as governed by the aerator radius of influence. If the number of units installed is not adequate, reactor dead spots will occur, and the reactants will not be uniformly dispersed throughout the basin. In suspended growth biological systems, biological solids will deposit, minimizing reaction efficiency. Reduced mixing, dead spots, or solids deposition result in reaction times less than the

residence time, and a corresponding reduced effective reactor volume, and in some cases ineffective performance. However, a larger surface area usually results in a corresponding shallower depth and better mixing patterns in the reactor bottom layers. A smaller surface area frequently results in a deeper basin, requiring fewer units to cover the surface. However, in such cases mixing at the bottom levels may be reduced. In fact, deeper basins limit the aeration devices types that can be effectively used.

The reactor geometry affects the mixing pattern. In a completely mixed system a turbulent pattern is desired, with mixing occurring from bottom to the top as well as from the side to side. As an example, although circular basins are economical to construct, they tend to induce a high-velocity circular motion, swirling the contents. An irregular shape disrupts mixing patterns, interrupting continuous patterns, resulting in more effective mixing. Where practical, baffles should be included to induce turbulent irregular patterns, especially in circular configurations. Another consideration is to have multiple aerators circulate in opposite directions to avoid a continuous pattern.

Design guidelines for completely mixed systems suggest maintaining the length-to-width ratio less than 3:1, no greater than 5:1, and ideally at 1:1. They are commonly designed at a length-to-width ratio from 1.5 to 2 and depths less than 5 m (15 ft) for mechanical units less than 22 kW (30 hp) and no more than 5 to 6 m (15 to 20 ft) for larger units [2,9]. Where these criteria cannot be met, multiple aerators can be installed, each influencing a square segment, and the total basin acting as a series of completely mixed units. Where extremely large basins are required, multiple components may be warranted, each conforming to the suggested configuration. Basin construction and capital costs must be balanced against aerator capital and operating costs and the resulting reactor efficiency.

In a plug flow reactor configuration a substrate concentration gradient is maintained throughout the basin length, the concentration diminishing with time (length). A plug flow configuration is usually achieved by maintaining a length-to-width ratio of at least 5:1, and in many cases over 10:1 [2,9]. The width-to-depth ratio can vary from 1 to 2.2:1, commonly 1.5:1 [2]. Plug flow conditions can usually be more readily achieved with a diffused air system. The dilemma with designing an aeration system for a plug flow reactor is that any aeration creates some turbulent condition counter to the required process kinetics. Therefore, to minimize this turbulence, mixing is minimized. However, the device capacity cannot be radically decreased because within any sector mixing must be adequate to assure uniform oxygen concentration, adequate reactant contact, and minimum solids deposition. Balancing mixing and varying oxygen demand, aerators must be strategically distributed to economically meet process requirements.

Where deep reactors are necessary, the selected system components must be specifically designed to influence the bottom layers. Because under the best of conditions complete solids suspension is difficult at the lower levels, the design must include the ability to remove deposits. Total deposited solids reduce the effective reactor volume, with the deposited organic material anaerobically degrading, resulting in what to the operator is unexplained poor effluent quality.

Number of Units

The number of units required is calculated on the basis of two quantities:

(1) The required oxygen demand expressed in pounds per hour corrected to SOR conditions.
(2) The rated device oxygen capacity converted to equivalent pounds per hour at SOR conditions.

Based on these two definitions the number of units required can be estimated by dividing the required oxygen demand by the device unit oxygen capacity, *at the same SOR conditions*. The required correction factors are discussed in the Basic Theory section, and a step-by-step calculation for each of the devices is detailed in Case Study 9.

Mixing Requirements

Aeration equipment performance, although reported as oxygen transfer capacity, is complex, involving (1) contacting wastewater and oxygen, (2) mixing the basin contents to obtain near homogeneous oxygen levels, (3) dispersing the reactants, and (4) providing and energy level consistent with the system configuration (plug flow or complete mix). An aeration device's mixing capacity is extremely important and should not be neglected when evaluating the device's oxygen output capacity. In fact, maintaining high surface and bottom basin velocities to assure uniform composition and minimize solids deposition is critical. The Process Engineer, along with the manufacturer, must develop the required number of aeration units to achieve adequate mixing, whether than be in excess or in agreement with the quantity required for oxygen production. In addition, these units must be installed in a basin pattern that assures optimum performance. Some common guidelines are cited subsequently; specific design criteria must be developed for each treatment system.

Mixing Guidelines

In general, aeration devices should maintain liquid velocities of 0.3–0.6 ms (1–2 ft/s) to assure adequate mixing and prevent solids deposition. As a guide, the required mixing velocity can be achieved with a mechanical aeration energy input of 19 to 39 kW per 1000 m³ basin volume (0.75 to 1.50 hp per 1000 ft³) for the broad spectrum of available mechanical aeration devices, and from 15 to 30 kW per 1000 m³ basin volume (0.6 to 1.15 hp per 1000 ft³) with units generating a significant vertical mixing regime [2,9]. Similar mixing patterns can be achieved with a diffuser air input of 20 to 30 m³/min per 1000 m³ tank volume (20 to 30 cfm per 1000 ft³) for a spiral roll pattern and 10 to 15 m³/min per 1000 m³ tank volume (10 to 15 cfm per 1000 ft³) for a uniformly installed grid pattern [2,9].

The number of units required for mixing depends on the tank design, the device chosen, and the installed pattern. Specific recommendations should be obtained from the aeration device supplier, developed for the selected basin geometry. The mixing capability is critical to good performance because adequate oxygen capacity is meaningless if it cannot be distributed uniformly throughout the reactor. Inadequate mixing reduces the effective aeration basin capacity and therefore the treatment plant performance.

Basin Pattern

Adequate overall mixing energy is pointless if the devices are not located in an effective distribution pattern, avoiding "dead spots" or overactivity in isolated areas. Again the Process Engineer should utilize the supplier's experience in applying specific aerator characteristics. Some common guidelines are cited as a *frame of reference*.

- Mechanical aerators: The radius of influence and unit spacing is directly related to the aerator power output. The diameter of influence of an aerator can be from 12 to 18 m (40 to 60 ft), for units ranging from 15 to 40 kW (20 to 50 hp), influenced by the number of units and their location [3]. A study of some operating facilities utilizing multiple units suggests that surface aerator spacing can range from 9 to 19 m (28 to 62 ft) apart, generally in the 11- to 14-m (35 to 45 ft) range for units ranging from 45 to 75 kW (60 to 100) [4]. It should be noted that the diameter of influence could be greater than the spacing employed in the study cited.
- Diffusers: Tube diffusers headers are installed along one side or along both sides and if required in parallel or a total floor configuration to achieve a more turbulent mixing pattern. Increasing the diffuser headers from a simple single or dual-header system complicates not only the construction but also maintenance and removal of the units. The headers can be placed from 3 to 9 m (10 to 30 ft) apart, whereas the diffusers on the headers should be at least 15 cm (6 in.) apart, usually from 30 to 91 cm (12 to 36 in.) [6,7]. The allowable spacing of nonporous devices depends on its capacity. Spargers are frequently spaced 30–60 cm (12–24 in.) apart, whereas static tubes are spaced 120–300 cm (48–120 in.) apart [8]. The specific pattern employed is best

left to the aeration equipment supplier, tailored to the specific device characteristics.

- Submerged turbines: Turbine pattern is based on the area of influence of the unit, approximately one unit being used for every 84 to 232 m² (900 to 2500 sf) of aeration *area*; the number of units and placement are specific to proprietary designs [3].

Winterization

Aerators present unique winterization problems because good mass transfer devices make good heat transfer devices. First, surface devices are designed to create a mist to achieve effective air waste contact. This transfer mechanism results in "snow," which if dispersed in the motor drive area could freeze the unit and cause mechanical failure. In addition, large heat transfers could lower basin temperatures to levels where the reaction is minimal. The degree to which these problems occur depends on specific device characteristics and the depth of the aeration basin. The Process Engineer should be aware of these potential problems when treatment plants are to be operated in cold climates, and the design should reflect worst weather conditions.

Installed Horsepower

Effective aerator design involves achieving required performance by installing the minimum required horsepower and operating the installed horsepower at optimum conditions. This becomes extremely complicated with high-strength wastes (high-rate activated sludge systems) because the required mixing horsepower is out of proportion to the oxygen transfer horsepower. This can be further complicated in reactors with large areas when the unit's effective radius of influence results in inadequate basin coverage, necessitating more units than that required for (volume) mixing or oxygen transfer.

In such cases the Design Engineer is inclined to establish the "worst case" design for that condition and install excessive capacity. In suspended growth biological systems, an extremely conservative design could present a problem if the units are installed with minimum flexibility, so that the units are operating at maximum capacity regardless of process requirements. In such cases, the reactors operate at high dissolved oxygen concentrations, affecting both operating costs and biological activity. In cases where the mixing horsepower requirements are radically different from oxygen demand, the Process Engineer should either install separate units for mixing and oxygen transfer or combined units which allow the oxygen level to be separately controlled.

An important point specific to completely mixed activated sludge systems is where "complete" may be overstressed. In such systems, mixing must be accomplished without excessive floc shear, which could result in either poor clarifier settling or excessive polymer requirements. Generally, this can be accomplished with velocities of 0.3 m/s (1 fps), achieved with the mixing conditions previously cited.

GENERAL ENGINEERING CRITERIA

The number of units installed in an aeration basin depends on process conditions, but single units should seldom be installed. Single units eliminate any operating flexibility because mixing patterns cannot be adjusted, and unit failure results in a complete system shutdown. Units must be installed so that they can be readily inspected and serviced. Mechanical aerators should be platformed to insure they can be inspected, isolated, and serviced without a system shutdown. Diffusers should be installed so that they can be readily removed, inspected, and cleaned without shutting and draining the system.

CRITICAL EQUIPMENT SELECTION

Aeration equipment is an integral part of the aeration basin design, selected to achieve specific reactor characteristics and produce required reactor performance. Some common design requirements are presented as a guideline for the Process Engineer.

Diffused System Components

Blowers

Generally, two types of air movers are considered for generating and transferring air in a diffused air system: a blower, with discharge pressures up to 103 kPa (15 psi) and a compressor, which can deliver pressures above 103 kPa [2,9]. Blowers commonly employed include rotary displacement blowers having a capacity range from 2 to 24,000 L/s (5 to 50,000 acfm) or centrifugal blowers with capacities ranging from 240 to 70,000 L/s (500 to 150,000 acfm). The rotary blower is a positive displacement device capable of delivering a fixed volume "against" a varying diffuser system pressure and therefore varying basin depth and diffuser fouling. However, its capacity is more difficult to adjust than a centrifugal unit because the outlet cannot be effectively throttled; instead the machine speed is commonly adjusted to vary the flow. The centrifugal blower operates similar to a centrifugal pump, delivering a varying flow dependent on the "throttled" or restricted discharge pressure. The fan blade characteristic establishes its sensitivity to downstream pressure change. The selection of a blower depends on the volume required, the discharge pressure, the expected variation in the aeration basin depth, the turndown range, and the selected control method.

Blower motor capacities should be adequately sized for the range of system demands. The required discharge head is equal to the sum total of the distribution piping loss, the aeration bed liquid head, and the diffuser operating pressure drop. The diffuser loss is not for a clean unit but that required prior to removing the unit for cleaning, usually 3.5 to 7 kPa (0.5 to 1.0 psi). Pressure loss calculations should be based on maximum expected air humidity conditions. The required diffuser loss is one of the considerations in evaluating diffusers. Multiple units, including an operating spare, should be considered to provide maximum process flexibility. Blowers should be provided with intake and discharge silencers as well as air filters to prevent diffuser clogging.

Finally, blower system installation design should include (1) sufficient space around the unit for maintenance and monitoring, (2) provisions for removing the blowers from the pad for maintenance, (3) selection of suitable intake filters, and (4) locating the intake filters for easy inspection and replacement.

Air Distribution

Air piping design involves appropriate distribution of clean air, with provision for controlling and monitoring the distribution. This includes the following:

(1) Air flow monitoring should be provided for all major headers, with provision to monitor air flow to individual drop legs.

(2) The inlet pressure into the distribution system should be monitored and automatically controlled as well as the pressure to each major header. Pressure gages should be provided at all headers and at critical elements. Provision must be included to permit balancing the transmission legs. Pressure variations to the system should be minimal.

(3) Shut-off valves should be provided for the inlet to the distribution system, each major header, and each drop leg.

(4) Each diffuser should be provided with restrictions to prevent excess flow in case of tube breakage.

(5) Dry air or freeze protection should be provided for cold climate air transmission systems.

(6) Piping low points which could collect moisture should be avoided. If low points cannot be avoided, manual or automatic drains should be provided.

(7) The piping system should be designed with provisions for contraction and expansion.

(8) All submerged piping and devices should be constructed and installed to prevent leakage and contamination of the air lines. Special consideration should be given to the condition when the blower is not operating, preventing hydraulic conditions resulting in siphoning into the air piping system.

(9) Provision should be made to prevent destruction of air support lines, including the diffuser, within the aeration basin.

(10) Pipe headers should be hydraulically designed to assure equal air distribution and air flow. This should be included regardless of any automation provided to control air flow.

(11) Special care must be taken during construction to prevent debris collecting or being trapped in the pipe, causing damage to the air transmission system or diffusers.

Piping Materials

Generally, diffuser piping systems should include exterior piping (surface) materials suitable for anticipated site corrosion conditions. Inside piping surfaces must be kept clean and resistent to any corrosion or erosion conditions resulting in particles transmitted to the diffusers. Typical transmission piping materials include carbon steel, ductile iron, galvanized and stainless steel, reinforced fiberglass plastic, PVC, and HDPE [2,9]. Basin submerged piping should be adequate for corrosion conditions and should be easy to remove and replace. Typical submerged piping materials include stainless steel, PVC, or coated steel [2,9].

Mechanical Aerator Motors

Surface aerator drives can be horizontally or vertically mounted, and either integral or coupled. Most aerators use a drive operating at 900 to 1800 rpm, with a gear reducer to achieve the desired aerator speed. The physical aspects of surface aerators has been evaluated in an extensive field study, and the major contributors to mechanical failure identified [4]. The study concluded that major mechanical failures were a result of gear reducer failures, themselves a result of manufacturing defects, operating loads, and improper maintenance. Manufacturing defects resulted in failure of (1) intermediate gears and ball bearings, (2) intermediate pinions, and (3) the keyway on intermediate pinions. Gear reducer failures attributed to operating load were a result of (1) final gears pitting, (2) shafts cracking, (3) intermediate bearing failure, and (4) pinion failure. Failure as a result of improper maintenance was evident at the lower output bearings.

Bearing span ratio was cited as the most significant factor affecting gear reducer service life. Reducers with span ratios greater than 4.3:1 experienced high failure potential, suggesting that this ratio should be maintained at 4:1 or less. The bearing span ratio is defined as the ratio of the aerator shaft length to the distance beween the bearings on the gear reducer. The next major contributor to gear reducer stability was the service factor, with units having factors less than two prone to failure.

Because of the importance of mechanical stability the specifications for mechanical aerators should insure that the gear reducers are (1) suitable for continuous operation under moderate shock loads, (2) have a life span of at least 100,000 hours, (3) designed for a life factor equivalent to at least 100 million cycles, (4) have a service factor of at least 2, and (5) have a bearing span ratio of 4:1 or less [4]. The aerator output should be within the required system range, with the ability to easily adjust the machine output with system requirements, and the unit's level sensitivity compatible with the process volume changes. Finally, the motor output should be optimized for the required unit size with motor losses around 5% of its rating, and no greater than 15%.

DISSOLVED OXYGEN CONTROL

Dissolved oxygen control, whether automatic or manual, is necessary to minimize operating costs and to avoid excessive oxygen levels that could promote filamentous growth. The size of most industrial biological systems usually does not warrant large capital expenditures for automation to manage energy costs. With proper influent control the aerator output can be easily maintained manually, without influent control the biological process will be difficult to manage, and the aerator output of secondary importance. However, the size of a facility, or plant practice, may warrant automatic aeration control. The complexity of the required control depends on the system configuration, the influent variability, and the accuracy of the element measuring the dissolved oxygen.

In completely mixed systems theoretically only one dissolved oxygen reading is required, usually at the outfall, although strategically located readings can serve as a verification that homogeneous conditions are approached. Plug flow systems require control of operating zones, each zone maintained at a specified range and individually controlled.

On the assumption that upstream equalization is not provided, the degree of the influent variability will determine the complexity of the required controls. With a minimum of influent variation, a feedback control system can be used to monitor the outfall (or zone effluent) and smoothly modulate the aeration system output. If the influent is highly variable but definable by some measurable variable such as flow, the influent variable can be monitored and a feedforward system used to adjust the aerator output as determined by feedback controls. When the influent is highly variable and unpredictable, the savings in upstream equalization can be used as partial payment for complex aeration control using a combined feedback, feedforward system, or an even more complex automation system. In designing an automatic dissolved oxygen control system the Process Engineer should use an instrument engineer to establish a suitable system and specify the mechanical elements. A critical part of automatic aeration control is selection of a DO measuring probe, location of the probe, assuring a representative basin or area reading, and maintaining the probe under extremely nonideal wastewater conditions.

COMMON AERATION SYSTEM DESIGN DEFICIENCIES

Some design deficiencies reported for municipal system aeration systems have been investigated by the U.S. Environmental Protection Agency and are identified for the reader as a checklist of process and mechanical design considerations [5]. Reference is made to the cited literature for a more detailed discussion of these potential problems.

General

(1) Aerator spacing between units and walls is not adequate. Generally, diffusers should be spaced more than 15 cm (6 in.) and less than 60 cm (24 in.) on center. Mechanical aerators spacing can vary from 1.5 to 9 m (5 to 30 ft).
(2) Aerator mixing capacity is not adequate to provide uniform suspended solids and DO concentrations.

Dissolved Oxygen Monitoring and Control

(1) No means is provided to monitor and manually adjust DO in the aeration basin or a means of automatically controlling blower output or aerator speed to optimize aeration effectiveness in large systems.
(2) DO monitoring systems cannot be easily inspected or maintained. Probes are difficult to access, remove, clean, and calibrate.

Diffused Air Systems

(1) Insufficient mixing, as previously discussed.
(2) Insufficient aeration capacity for aeration, mixing, or both.
(3) Diffuser plugging results from waste constituents because upstream screening, or similar primary treatment, is not employed.
(4) No air filters are installed in the blower intake to minimize diffuser plugging.
(5) No provision was made for removing air diffuser drop pipes for inspection and maintenance.
(6) No provision was made for controlling diffuser air flows.
(7) Air drop pipe supports are not visible when the aeration basin is at the working level, making maintenance difficult.

Blowers

(1) Inadequate capacity provided.
(2) Inadequate or no air cleaner provided.
(3) No blower silencer provided.

Air Piping

(1) The air flow cannot be monitored. Flow meters on all main air headers and graduated valves in drop legs were not provided.

(2) The wrong air piping material was selected. Scaling results because of interior deterioration, a result of poor piping or coating selection.

(3) No provision was made for removing drop pipes from the aeration basin, as previously discussed.

Mechanical Aerators

(1) Inadequate consideration of basin level change affects the surface aerator performance. No variable weir was provided to adjust the basin level to a suitable operating level. If frequent level changes are required, diffused or submerged aeration should be considered.

(2) Inadequate aerator spacing, as previously discussed.

(3) Mixing is inadequate.

(4) Aeration capacity is inadequate.

(5) Wastewater debris affects the aerators; no upstream pretreatment was provided.

(6) Oil drains from the gear box into the aeration basin.

(7) Amp meters were not provided at the motor control center to monitor amperage drawn, to establish that proper aeration is being obtained or that too much amperage is drawn that could damage motor during start-up.

(8) No provision was made to limit aerator gear shock when shifting from high to low speed. Time delay relays to limit stress shock were not provided.

(9) Hazardous conditions result from aerator spray onto the walkway because aerator splash shields were not provided.

(10) Floating aerators are not properly installed:
   a. Low-voltage starters were not provided to minimize initial torque levels and resulting excessive lateral movement.
   b. "Snubbers" were not provided to minimize cable slack.
   c. Pontoons were not properly aligned.
   d. Aerators were not located a safe distance from the nearest walls.

(11) Quick disconnect plugs in the aerator splashing range were not weatherproof and shorted when wet.

CASE STUDY NUMBER 9

Evaluate aeration systems for the activated sludge system described in Case Study 10, based on 6140 lbs per day of oxygen requirements in a 920,000-gal basin. Specific design criteria include

(1) Oxygen requirements: 6140 lb/day = 255.8 lb/hr
(2) Aeration basin volume, gal: 920,000
(3) Design temperature, °F: 68
(4) Maximum operating temperature, °F: 80
(5) Alpha value: 0.8
(6) Beta value: 0.9
(7) Elevation above sea level, ft: 200
(8) Minimum aeration basin DO, mg/L: 2

PROCESS CALCULATIONS

General Process Conditions

(1) Select the tank depth: 15 ft for diffused air and 10 ft for mechanical aeration.

(2) Aeration basin cross-section for diffused air
920,000 gal · 0.1337 cu ft/gal/15 ft = 8200 sq ft
Select *two* parallel units, each 4100 ft^2.

(3) Select basin physical dimensions for diffused air
Width twice the depth = 2·15 = 30 ft
Length = 4100/30 = 137, select 140 ft

(4) Aeration basin cross-section for mechanical aerators
920,000 gal · 0.1337 cu ft/gal/10 ft = 12,300 ft^2
Select *two* parallel units, each 6150 ft^2.

(5) Select basin physical dimensions for mechanical aerators
Length twice the width, width 55 ft
Length = 112 ft
Overall "footprint" 110 · 112 ft, two basins.

All estimates based on worst (summer) conditions!

Diffuser System

(6) Establish diffuser characteristics
Nonporous diffusers used SOTE: 10%
Unit capacity at 68°F: 10 scfm
Mixing energy: 20 scf per 1000 cu ft

(7) C_{sw} at operating conditions

(8) Basic conditions
Design/limiting temperature, °F: 68 / 80
C_s, tap water at 68°F and 1 atm: 9.17 mg/L
C_s, tap water at 80°F and 1 atm: 7.95 mg/L, 7.89 mg/L at the site elevation

(9) Pressure at bubble point
Atmospheric pressure = 14.7 psi
15 ft of water depth = 6.52 psi
Elevation, 200 ft, = −0.11 psi
NET = 21.1 psi

(10) Percent oxygen exiting basin
% O$_2$ leaving
= 21 · [(100 − % diffuser efficiency)/100]
= 21 · [(100 − 10)/100)] = 18.9%

(11) C_{sw} calculation
$C_{sw} = C_s · [P_b / 29.4 + O_t/42]$

$$C_{sw} = 7.89 \cdot [21.1 / 29.4 + 18.9/42]$$
$$= 9.21 \text{ mg/L at } 80°F.$$

(12) Correction for standard conditions

$$AOR = SOR \cdot \frac{\beta \cdot C_{swc} - C_l}{C_s} \cdot 1.024^{T-20} \cdot \sigma$$

See Equation (5) for definitions used.

(13) Correction for adjusted solubility:

$$\frac{\beta \cdot C_{swc} - C_l}{C_s} = \frac{0.9 \cdot 9.21 - 2}{9.17} = 0.686 @ 80°F$$

(14) Temperature correction:
$$1.024^{T-20} = 1.024^{20-20} = 1.000 @ 68°F (20°C)$$
$$= 1.171 @ 80°F (27°C)$$

(15) Correction for transfer rate: $\sigma = 0.8$

(16) Maximum SOR requirement:
AOR = 255.8 pounds oxygen per hour
SOR = 255.8 · 1/0.686 · 1/1.171 · 1/0.8 =
398 lb/hr corrected to 80°F (27°C), @ 20°C, 1 atm,
0 mg/L.

(17) SCFM air (20°C, 68°F) required
Basis 1 mol of air
1 mol air = 359 cu ft · 528/492 = 385.2 ft³ @ 20°C
1 mol air = 0.21 · 32 = 6.72 lb oxygen
6.72/385.2 = 0.0174 lb oxygen/ft³ of air
0.0174 · 0.1 (10% SOTE) = 0.00174 lbs O_2 @ 20°C,
1 atm, 0 mg/L
Worst case @ 398 lb/hr oxygen demand.
(398/60)/0.00174 =
3812 SCFM (20°C) at the maximum demand condition.

(18) Number of diffusers
Diffuser capacity of 10 SCFM
3812 scfm air /10 scfm ≡ 380 diffusers

(19) Air required for mixing
Mixing energy level: 20 scf per 1000 ft³
Total air = 920,000 gal · 0.1337 ft³/gal · 20/1000 =
2460 SCFM
2460 scfm/10 scfm per unit = 246 diffusers

(20) Basin pattern
No. of diffusers used: 380 for maximum oxygen demand.
Each basin 30 x 140 ft, requiring 190 diffusers each!
No. rows = (width/max distance between) + 1
= 30/10 + 1 = 4 rows
No. units per row = (length/max distance between) + 1
= 140/3 + 1 = 47, use 48
Number of diffusers = 4 · 48 = 192
Final basin diffuser pattern must be checked with diffuser manufacturer specifications!

(21) Preliminary selection
Select a total of 192 diffusers per basin, a total of 384
diffusers, each at 10 scfm, for total air flow of 3840 scfm.
Horsepower = scfm · depth/[529 · efficiency]
= 3840 · 15 ft/529 · 0.75 = 145 hp

Select two 75 hp motors
The motor size was estimated assuming a simple centrifugal compressor, and the horsepower rated at standard inlet conditions of 20°C, 1 atmosphere. The final mechanical design will include a detailed piping design, an estimate of the line losses at the operating flow and temperature, and the motor horse power adjusted to expected operating conditions.
System oxygen efficiency = 256 *AOR* lb/hr oxygen/150 hp
= 1.7 lb *AOR*/hr/hp

Turbine System

(22) Establish turbine characteristics
SOTE: 10%
Unit capacity at 68°F: 500 scfm
Mixing energy: 0.2 hp per 1000 gal (1.5 hp/1000 cu ft)
Optimum turbine to compressor hp ratio: 1
Area of influence: 1000 sq ft/unit
Blower efficiency: 75%
Turbine efficiency: 75%

(23) SCFM air required
Worst case is 3812 scfm (see item 17)

(24) Number of turbines
Turbine capacity of 500 scfm
3812 scfm air/500 scfm ≡ 8 turbine units
≡ 4000 scfm

(25) Mixing energy level
Mixing energy level: 0.2 hp per 1000 gal
Total hp = 920,000 gal · 0.2/1000 gals ≡ 200 hp

(26) "Unit" basin pattern
Each basin 30 × 140 ft; two basins
30 · 140 (area)/4 = 1000 sq ft/unit
Install 4 units per basin, each @ 25 hp

(27) Final selection

(28) Turbine
Select 4 turbine units, each at 25 hp, per basin, a total of 8 25 hp turbines or a total of 200 hp.

(29) Compressor
Horsepower = scfm · depth/[529 · efficiency]
= 4000 · 15 ft/[529 · 0.75] ≡ 150 hp
Select two 75 hp units, total 150 hp

(30) Turbine:compressor ratio
$P_d = 200/150 = 1.3$, split is good

(31) Oxygenation efficiency
256 (*AOR*) lb oxygen/hr/200 hp = 1.3 lb/hr/turbine hp
256 (*AOR*) lb oxygen/hr/150 hp = 1.7 lb/hr/ compressor hp
256 (*AOR*) lb oxygen/hr/350 hp = 0.7 lb/hr/total hp

Surface Aerators

(32) Establish aerator characteristics
Unit capacity (SOR): 2 lb/hr/hp

Mixing energy: 0.1 hp/1000 gal (0.75 hp/1000 cu ft)
Area of influence: 40 ft between aerators

(33) C_{sw} at operating conditions

(34) Basic conditions
Design/limiting temperature, °F: 68/80
$\alpha = 0.8$
$\beta = 0.9$
C_s, tap water at 80°F and 1 atm: 7.95 mg/L
Elevation correction, Fe = 0.992

(35) Correction for elevation:
$C_{sw} = C_s \cdot$ Fe $= 7.95 \cdot 0.992 = 7.89$ mg/L @ 80°F

(36) Correction for standard conditions:

$$AOR = SOR \cdot \frac{\beta \cdot C_{swc} - C_l}{C_s} \cdot 1.024^{T-20} \cdot \sigma$$

$$\frac{\beta \cdot C_{swc} - C_l}{C_s} = \frac{0.9 \cdot 7.89 - 2}{9.17} = 0.556 \text{ @ } 80°F$$

(37) Correction for temperature:
$1.024^{T-20} = 1.024^{20-20} = 1.17$ @ 80°F(26.7°C)

(38) Maximum SOR requirements (80°F)
SOR $= 255.8 \cdot 1/0.556 \cdot 1/1.17 \cdot 1/0.8 = 491$ lbs/hr

(39) Required aerator capacity
Basis: 2 lb oxygen/hr/HP
491 lb/hr/2 = 246 HP @ 80°F

(40) Aerator mixing requirements
Basis: 0.1 hp/1000 gal
(920,000 Gal/1000) $\cdot$ 0.1 = 92 hp
Basis: 40 ft between units, 40 ft from walls.
175-ft length would require about three along length.
Each basin would have one row. Approximately 6 units

would be required for two basins. *Check* with supplier when final selection made!

(41) Preliminary selection
Since 60 hp units would not be adequate for required oxygen demand,

Select four 75-hp units, two each basin, total 300 hp
Oxygen efficiency = 256/300 = 0.85 *(AOR)* lb/hr/hp installed.

REFERENCES

1. Benefield, L.D., Randall, C.W.: *Biological Process Design for Wastewater Treatment,* Prentice-Hall, 1980.

2. Medcalf & Eddy, Inc.: *Wastewater Engineering-Treatment, Disposal, Reuse,* McGraw-Hill, 1991, Third Edition.

3. Ramalho, R.S.: *Introduction to Wastewater Treatment Processes,* Academic Press, 1977.

4. Stukenberg, J.R: "Physical Aspects of Surface Aeration Design," *Journal WPCF,* V 56, No 9, Pg 1014, September, 1984.

5. U.S. Environmental Protection Agency: *Handbook for Identification and Correction of Typical Design Deficiencies at Municipal Wastewater Treatment Facilities,* EPA-625/6-82-007, 1982.

6. U.S. Environmental Protection Agency: *Design Manual, Fine Pore Aeration Systems,* EPA-625/1-89-023, September, 1989.

7. WEF Technical Committee: "Aeration in Wastewater Treatment-Manual of Practice No5," *Journal WPCF,* V 41, No 11, Pg 1863, November, 1969 (MOP-5).

8. WEF Manual of Practice FD-13: *Aeration, Water Environment Federation,* 1988.

9. WEF Manual of Practice: *Design of Municipal Wastewater Treatment Plants,* Water Environment Federation, 1992.

Aerobic Biological Oxidation

This chapter serves as a primer for the biological processes discussed in Chapters II-12 through II-17.

THEORETICAL concepts for *aerobic biological oxidation* discussed in this section are applicable to all biological treatment system configurations and therefore not repeated in the biological processes discussed in Chapters II-2 to II-7, except to advance applicable process design criteria.

GENERAL BIOLOGICAL CONCEPTS

Biological systems are effective methods to economically remove large quantities of biodegradable wastewater organics, converting them to water and carbon dioxide. The reactions are complex but can be compared to combustion oxidation where the required process energy is released as part of the reaction, and the biological floc can be compared to the "pilot flame" to sustain the process. The achievable oxidation of any compound depends on the reaction efficiency, defined in terms of

(1) Theoretical or complete oxidation in which all components are combined with oxygen to form carbon dioxide, water, and the inorganic oxide.

(2) Chemical oxidation in which less than complete oxidation occurs.

(3) Biological oxidation in which oxidation is limited by the ability of the biological floc to assimilate the organic.

Many waste water organics can accommodate both chemical and biological oxidation, but none will be completely oxidized. Stoichiometrically, theoretical oxidation of a wastewater organics can be expressed as indicated in Figure 2.1. The final form of any influent nitrogen compounds depends on the available oxygen and the reactor oxidizing conditions.

BIOLOGICAL OXIDATION MEASUREMENTS

The environmental effects of organic pollutants are usually not expressed as direct concentration measurements but as the *oxygen-depleting effect* on a receiving stream. This is defined as the biological (BOD) or chemical oxygen demand (COD), which in essence is the oxygen required to satisfy the chemical oxidation reaction indicated in Figure 2.1 or the biological oxidation reactions indicated in Figure 2.2. Biological treatment reduces the waste's affinity for oxygen. As an example, the oxidation of methanol can be represented by the equation:

$$CH_3OH + 1.5O_2 = 2H_2O + CO_2$$
$$32 \text{ g} \quad\quad 48 \text{ g} \quad\quad 36 \text{ g} \quad\quad 44 \text{ g}$$

Theoretically, 1.5 g of oxygen is required to completely oxidize 1 g of methanol. Measured values could result in a COD ranging from 1 to 1.3 and a BOD value of 0.8 to 0.9. The measured COD value is always less than the theoretical oxygen demand value of a compound, and the BOD is always less than the COD. The more biodegradable the compound, the less the difference between the COD and BOD value.

BIOLOGICAL OXIDATION REACTIONS

Biological processes differ from chemical oxidation systems in at least two ways:

(1) Biological systems are selective to biodegradable organics, requiring microorganisms to act as a catalyst to promote the reaction.

(2) Many chemical oxidation systems reduce the reactants to final products that are environmentally inert. All bio-

$$Ca \cdot Hb \cdot Oc + (a + 0.25b - 0.5c) \cdot O_2 = a \cdot CO_2 + 0.5b \cdot H_2O$$

OR

$$Ca \cdot Hb \cdot Oc + (a + 0.25b - 0.5c) \cdot O_2 =$$

$$a \cdot CO_2 + 0.5b \cdot H_2O$$

NOTE: 1. Any nitrogen will be converted to the nitrate
        2. Any sulfur to the phosphate
        3. Any phosphorus to the phosphate
        4. COD, oxygen demand, will be increased accordingly.

Figure 2.1 Complete oxidation equation.

logical processes produce excess sludge that may have to be further reduced and stabilized before disposed.

Stoichiometrically, as illustrated in Figure 2.2, the process is commonly represented as two simple reactions, indicating (1) oxidation of an organic to carbon dioxide, water, and new cells and (2) the destruction of excess cell population. The sludge composition is commonly expressed as $C_5H_7NO_2$.

Substrate utilization, based on available reactants and reaction time, is illustrated in Figure 2.3.

Actually, biological reactions are *complex*, embodying a chain of activities involving a substrate capable of being biologically oxidized, serving as a nutrient for cell growth, releasing energy, and being metabolized by cells. In addition, to sustain the biological process the cell growth must be select and controlled. The process is sustained by maintaining adequate substrate, nutrients, oxygen, and a biological population and stabilized by producing a settleable floc and removing (and processing) excess sludge.

The *complex* biological reactions involve the occurrence of many simultaneous activities to sustain a viable biological process, such as [4]

(1) The formation of a competitive microorganism population.

(2) The use of some substrate to synthesize new cells, whereas some are converted to final products by direct oxidation. The oxidation process is an exothermic reaction, supplying the required energy to perpetuate the process.

(3) Because part of the reactants (substrate) is oxidized, an equivalent amount of reactants must be reduced.

Oxidation and Synthesis Reaction

Organic Matter + O_2 + P + NH_3 + Bacteria
=
New Cells + CO_2 + H_2O

Endogenous Respiration Reaction

Cells + O_2
=
CO_2 + H_2O + NH_3 + Excess Sludge

Figure 2.2 Biological process reactions.

The electrons lost in the oxidation must be "accepted" by reactants being reduced. In aerobic reactions oxygen acts as the "acceptor," or the reduction component of the (energy releasing oxidation) reaction, receiving the electrons released from the oxidation components.

Cell production (maintenance) must be perpetuated in some direct relationship to the available substrate to balance the "food" available to the microorganisms. Figure 2.4a illustrates a typical cell growth curve, with microorganism increasing proportionally to the available substrate to a maximum value, the population decreasing until an equilibrium concentration is reached. The relationship (batch system) among cell growth, oxygen consumed, oxygen uptake rate, and available food is indicated in this diagram.

A similar continuous system relation is illustrated in Figure 2.4b, the common operating ranges for suspended growth systems being designated as dispersed growth, high rate, conventional, and extended aeration systems. These systems are further defined in Chapter II-3.

BIOLOGICAL SYSTEM MICROORGANISMS

As illustrated in Figure 2.4, a significant characteristic of the log phase cell growth curve is that there is a proportional relationship between cell growth and substrate removal, with oxygen utilization proportional to substrate removal. Initially, cell growth *rate* increases significantly until a maximum is reached.

The cell growth is commonly expressed by a constant (μ), a characteristic specific to the individual microorganism defining the time related cell growth rate. Generally, relationships developed for individual microorganisms are not directly applicable to multicomponent mixed cultures, where competition and environmental conditions govern the proportional survival of the individual cultural in an integrated population. However, the general principals have been applied to biological treatment and the batch growth patterns to continuous systems.

Understanding the batch cell growth mechanism, the next step is to investigate biological kinetics in a continuous reactor, illustrated in the Figure 2.5 simple once-through vessel.

As indicated, Q is the influent and effluent flow rate, V is the reactor volume, μ is the cell growth rate (mass rate per unit time per mass of cells), and X is the cell concentration in the reactor. Assuming a completely mixed system, the cell balance for the system can be written as follows:

Rate of change of = cell growth − cell loss from
   cell solids   increase   system

$$V \cdot dX/dt = \mu \cdot X \cdot V - Q \cdot Xe$$

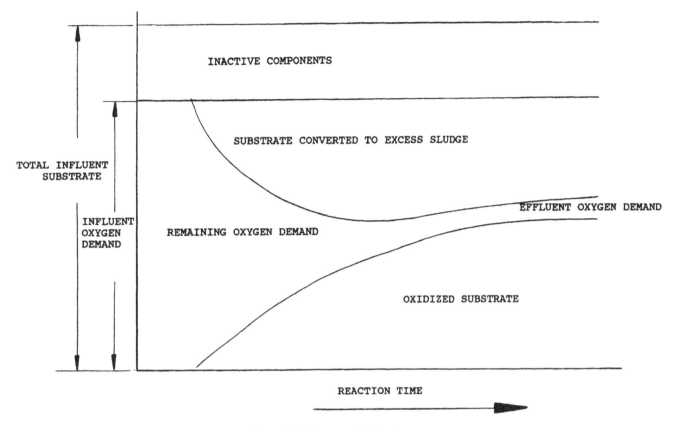

Figure 2.3 Substrate utilization diagram.

Because at steady state $dX/dt = 0$,

$$\mu = Q \cdot Xe/X \cdot V = \text{(mass cells/unit time)/mass cells}$$

As will be discussed, XV/QX is the cell retention time in the system, referred to as the solids retention time (SRT), so that,

$$\mu = 1/\text{SRT} \qquad (2.1)$$

or

$$\text{SRT} = 1/\mu \qquad (2.2)$$

Significantly, this statement indicates that the sludge retention time, or the average mass retention time, is inversely equal to the specific growth rate of the cell(s) in the system. If adequate retention time is not available the required microorganisms will not grow in the reactor, and there will be cell washout. As will be discussed for each biological system, this applies for all processes.

In the case where sludge is not recycled, the solids concentration in the effluent and reactor are equal (concentration cancels out in the equation), resulting in

$$\mu = Q/V = \text{dilution rate} = 1/\text{hydraulic retention time}$$

Process viability is directly related to sustaining the population specific to the substrate that must be degraded (utilized by the microorganisms). SRT or HRT is the basic criteria establishing cell and enzyme balance and process stability required to sustain the *microorganism food chain.*

Microorganism Food Chain

In a biological treatment system specific biological species must be cultivated to assure both a *consumption of the substrate and developing a settleable mass.* The organisms found in a biological system fall into three major groups: bacteria, protozoa, and rotifers. The functions of these various species are best understood by describing them in the food chain in the biological environment, as illustrated in simplified form in Figure 2.6.

Biodegradable organics are consumed by bacteria, the fastest-growing population, with the rate proportional to the available food. They can exist in an aerobic, anaerobic, or facultative environments. The bulk of bacteria is facultative, permitting them to exist in either aerobic or anaerobic conditions, allowing them to survive during oxygen-deficiency conditions. Substrate is metabolized by an enzyme produced and released by the bacteria cells, specific to the organic substrate, which accelerates the biological reaction. The abil-

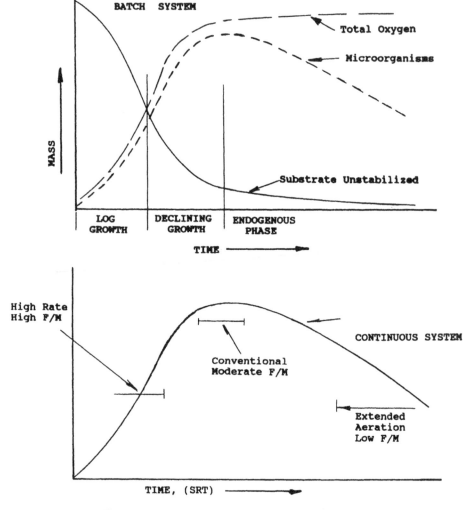

Figure 2.4 Ideal growth curves (adapted from Reference [58]).

ity of the cell to produce enzymes capable of reacting with the substrate is a measure of the substrate's "biodegradability." The bacteria are poor settling and (along with the enzyme) sensitive to environmental conditions such as temperature, pH, available oxygen, and nutrient level. Fungi are indicators of upset reactor conditions, evident by low pH, lack of oxygen, and nutrient deficiency. A dominant fungi population will compete with bacteria for soluble organics, promoting filament growth that results in poor sludge settling.

Flagellate and amoeboid presence are indicating signs of the relative health of the biological population, indicative of low microorganism (floc) concentration and high BOD concentrations. As other organisms form, and a balanced

population achieved, the flagellate concentrations diminishes. The predominance of these organisms would be evident as the system is starting up or recovering from an upset.

The next link in the food chain are the protozoa, which are aerobic, depending on bacteria for their survival. The most significant are the free-swimming and stalked ciliates. The more active free-swimming ciliates require and consume much of the available bacteria and are initially formed at a rate proportional to the bacteria growth. The inactive stalked ciliates form after the swimming ciliates, at a lower growth rate and resulting population, consuming smaller amounts of bacteria. The two protozoa combine to perform three important operational functions. First, because they are strictly aerobic their presence is an indication of the system viability. Next, they exist to control the bacteria population. Finally, they form a floc, and flocculating conditions, which improve sludge settling and result in high effluent quality.

The final link in the food chain is the rotifers, which feed on bacteria and are present when the stalked ciliate population has diminished. Rotifers are present when the

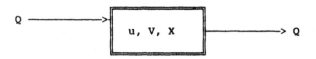

Figure 2.5 Once through reactor.

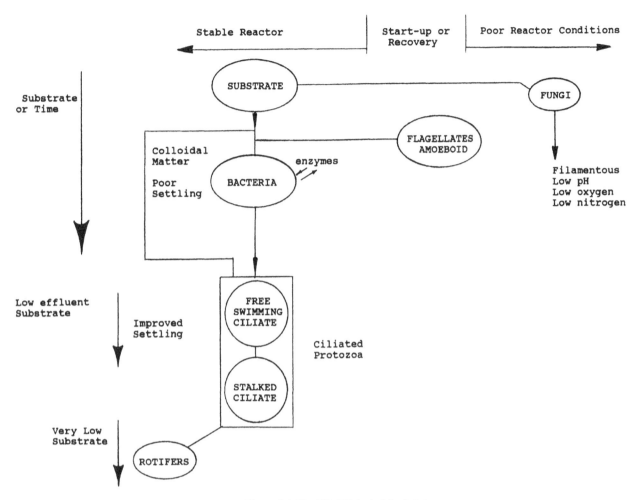

Figure 2.6 Simplified biological food chain.

food supply is low and diminishing. Their presence is usually indicative of high organic removal efficiency and long retention times.

It is important that the aerobic microorganism population be balanced to assure an active environment, resulting in both high substrate removal efficiency and sludge quality. This can be accomplished by maintaining stable operating conditions and optimizing cell growth rate. As demonstrated previously, cell growth rate is related to SRT in reactors with recycle capabilities and hydraulic retention time (HRT) in once-through reactors. The relative growth rates of microorganism species are illustrated in Figure 2.7. As illustrated, reaction time (or equivalent cell growth rate) is significant in assuring a viable flocculating sludge.

A stable biological reactor results in

(1) Select enzyme generation specific to the substrate characteristics
(2) Bacteria cell growth consistent with the available substrate
(3) Balanced microorganism population
(4) Settleable floc formation

When satisfactory reactor conditions are not sustained, and the system is upset, investigators have observed a major shift occurring from zoogloea bacterium, the basic substrate removal microorganisms, to filamentous bacterium [39]. Common causes cited for this shift are low food-to-microorganism (F/M) ratio, low dissolved oxygen (DO), presence of sulfides, low pH, and nutrient deficiency.

It is important to note that a balanced biological population can perform poorly if the food source or rate is changed. This is because the bacteria have acclimated to a specific food source, releasing specific enzymes, and must acclimate to any new conditions. The other organisms in the food chain must also adjust to a "new menu," when the food supply to the bacteria is altered.

Finally, the microbiology of a biological system was simplified for the purposes of emphasizing its importance in treatment system operation and design. Understanding that biological systems are impacted by microbiology has always been considered essential in municipal system operation [19,55]. In fact, advanced design methods give careful consideration to SRT evaluation, "enhanced" and "selector" design considerations to influence biological population, and

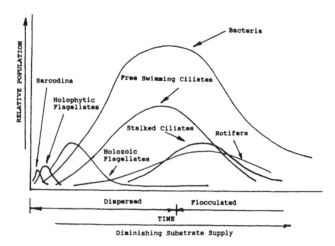

Figure 2.7 Relative microorganism growth rates (adapted from Reference [58]).

stabilizing influent variations by equalization or similar control. For a more detailed, and probably a more accurate, accounting of the microbiology encountered in biological systems the reader should review the many texts devoted to this subject [17,22,30,39]. Its importance in this section is to serve as a basic primer for the various times that cell growth rate will be encountered in developing design procedures and discussing operating parameters.

Process Design Significance

Reactor microbiology, SRT, and cell growth rate were discussed so that the design engineer understands that biological system design does not involve a "black box" but a complex reactor design developed to assure the operator a range of conditions to control process stability and optimize treatment efficiency. This begins with a representative and convenient measure of total microorganism concentration as an operating and design indicator. Regardless of the level of design sophistication, the system mixed liquor volatile suspended solids (MLVSS) is commonly considered a direct measure of the biological microorganisms, its mass quantity adjusted to the feed BOD (or COD) to achieve the required treatment effectiveness. However, when suspended solids become the primary, or the sole, reactor sizing consideration, other design objectives may be neglected; such as

(1) Selection of an SRT, along with sludge management criteria, to fix the cell growth rate(s) governing specific viable reactor microorganisms

(2) Selection not only of a single reactor microorganism concentration to balance feed substrate, but an operating range to assure operating flexibility

(3) Avoiding the introduction of high feed nonbiodegradable components, which reduces the reactor *active* microorganism concentration

Finally, it should be emphasized that poor settleable floc is not always an indication of poor clarifier operating conditions or low polymer addition. In some cases poor clarifier performance can be attributed to an unstable upstream treatment system, either by *design* or because of poor operating procedures. What is important is that the Process Engineer consider all factors that affect the influent quality and variability, designing to a suitable operating range defined by the process limits. For that reason, related process components will be discussed in each of the biological process chapters.

BIOLOGICAL KINETICS

To define the specific kinetics of biological treatment appropriate reaction paths must be assumed, and experimentation must be developed to verify the assumed cell growth and substrate removal mechanisms. Because substrate removal is the natural function of biological treatment, it is commonly assumed that the treatment process follows log phase kinetics and is not substrate limiting.

Exponential first-order growth rate (dX/dt) of single microorganisms can be expressed mathematically by the expression [4,46,48]:

$$dX/dt = \mu X$$

$$X_t = X_o \, exp \, \mu t$$

where μ is the specific growth rate, cell growth per unit of biomass. The time for the growth to double (td) is commonly represented by the relationship:

$$\mu = 0.693/td$$

Monod expanded the kinetic expression for exponential growth, utilizing the Michaelis-Menten enzymatic kinetic expression, to include declining growth, expressed as [4]

$$dX/dt/X = [K_o \cdot S \cdot /(K_m + S)] \qquad (2.3)$$

where $dX/dt/X$ is the specific cell growth rate, 1/time, K_o is the maximum substrate utilization rate, 1/time, S is the reactor substrate concentration, mass/volume, and K_m is the substrate concentration when dX/dt is 0.5 K_o.

This equation in modified form has been used by Lawrence and McCarty to define biological reaction kinetics, as follows [21]:

$$dS/dt = [k \cdot S \cdot X/(K_s + S)] \qquad (2.4)$$

$$dX/dt = Y \cdot dS/dt - k_d \cdot X \qquad (2.4a)$$

$$dX/dt = [Y \cdot k \cdot S \cdot X/(K_s + S)] - k_d \cdot X \qquad (2.4b)$$

where dX/dt is the cell growth rate, mass/volume/time,

dS/dt is the substrate depletion rate, mass/volume/time, k is the maximum substrate utilization rate, 1/time, S is the reactor substrate concentration, mass/volume, K_s is the substrate concentration when dX/dt is $0.5\ K_o$, X is the microbial mass concentration, mass/volume, k_d is the microorganism decay coefficient, 1/time, and Y is the yield ratio, mass cells produced/mass substrate utilized.

In this relation the feed substrate concentration is buried in the definition and can be included if Equation (2.4) for substrate utilization is rearranged, as follows:

$$dS/dt = (S_o - S_e)/t \qquad (2.4c)$$

So that,

$$(S_o - S)/S = [k \cdot X \cdot t/(K_s + S)] \qquad (2.4d)$$

or

$$S_o/S = [K' \cdot X \cdot t/(K_s + S)] + 1 \qquad (2.4e)$$

which for a first-order reaction $(K_s \gg S)$ reduces to

$$S_e/S_o = 1/[1 + (K'' \cdot X \cdot T)] \qquad (2.4f)$$

The Monod equation is sometimes further simplified to reflect the concentration of the substrate S relative to the substrate concentration coefficient K_s.

As an example, when S is $\gg K_s$

$$dS/dt = k'X \text{ (a zero-order reaction)}$$

and when S is $\ll K_s$:

$$dS/dt = k''X\,S \text{ (a first order in reaction)}$$

In some circumstances organic substrate is not the limiting growth constituent, but the reaction is controlled by dominant factors such as oxygen availability, temperature, pH, or the presence of toxic components. In such cases data are often gathered and kinetic design procedures developed to include these limitations as part of the Monod or similar kinetic expressions [47]. As an example, if A is a limiting growth factor to a system, its influence can be represented by the expression:

$$F_a = A / [K_a + A] \qquad (2.5)$$

So that if a process has a number of limiting factors controlling the process, the overall kinetic expression can be expressed:

$$dX/dt = [kSX/(K_s + S)][Fa\ Fb\ Fc] \qquad (2.6)$$

Researchers have, and will continue to, collect data to develop enhancements to these and other models to better establish a universal model to predict design and operating data for industrial waste systems, especially to define the inhibitory effects of specific regulated organic compounds. Discussion of the merits of various models is beyond the scope of this book. It is the author's opinion that the cost of obtaining pilot data specific to the waste to be treated is insignificant when compared to the system capital costs, the operating costs, and the potential noncompliance penalty costs. Once this approach is accepted site-specific pilot data can be fit to most models, and the model will reasonably predict operating performance within the data range investigated.

BIOLOGICAL OXIDATION PROCESS VARIABLES

Biological reaction effectiveness depends on physical factors important to all biological oxidation systems, as illustrated in Figure 2.8. First, the reactants must be transported to the system at concentrations which sustain the process. Next adequate mixing is required to assure contact of the reactants and biomass and assure material diffusion into the active cells. Energy release is necessary to perpetuate the process. Finally, the resulting slurry must be separated so that treated effluent can be discharged and the cell containing solids discarded or recycled. Optimization of these various physical elements, along with the related reaction factors, is essential to sustaining a viable biological operating *system*.

Biological reaction performance depends on environmental factors sustaining or inhibiting a biological reaction. Factors that must be controlled to sustain a biological reaction include

(1) Biodegradable organics
(2) Oxygen
(3) Nutrients
(4) Biological population
(5) Temperature
(6) pH

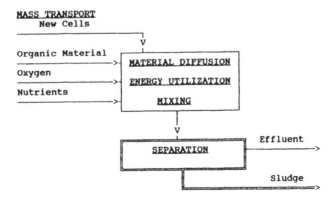

Figure 2.8 Physical processes in biological systems.

Factors that could affect a biological reaction and therefore must be considered in the design of biological systems include

(1) Feed characteristics
(2) Pretreatment
(3) Solids management
(4) Mixing

Finally, a system's biological activity is controlled by sustaining the organic-to-microorganism mass contact (F/M) ratio, which also establishes the sludge quality produced (sludge age).

FACTORS SUSTAINING A BIOLOGICAL REACTION

BIODEGRADABLE SUBSTRATE

The primary component required to sustain a biological system is a biodegradable substrate, amenable to the microorganism population commonly encountered in treatment systems. In municipal treatment systems domestic waste is degradable, any reaction deterioration in a well-operated plant easily attributed to industrial wastes. In industrial systems the waste strength and characteristic is site specific and biological treatment effectiveness established by pilot testing. The need for industrial waste treatment design data has resulted in considerable laboratory investigations to evaluate specific organic biodegradability [1,7,16,24,26–28,36,37]. More recent evaluations have been aimed at establishing biological treatability of toxic, priority pollutants, or RCRA classified compounds [35,38,42,43,50,51]. Data from these type studies are too voluminous to summarize, sometimes contradictory, and the results are specific to the testing methods and concentrations considered. The reader should review the references cited to evaluate test results and cautiously consider the suitability of these pilot studies for full-scale design or to the specific wastes under consideration. In many cases stripping performance and biodegradability are not isolated, sludge quality is not recorded, and the ability to sustain cell growth under varying influent conditions is not clearly reported. Such studies should be used as a process guide but should not replace specific site evaluations for permanent industrial treatment facilities.

Laboratory test data have been utilized to develop "rule-of-thumb" guidelines to correlate a compound's physical structural to its biodegradable potential [25]. Generally, the affect of compound structure to biodegradability can be summarized as follows:

(1) Biodegradability decreases with compound branching.
(2) Biodegradability increases with compound chain length.
(3) Unsaturated aliphatics are more biodegradable than the corresponding saturated hydrocarbon.

(4) Substitutions on simple organic compounds affect biodegradability. Increased substitution generally decreasing degrading properties, whereas some substitutes that form alcohols, aldehydes, acids, esters, amides, or amino acids improve biodegradability.
(5) The position and type of a substitution on a simple compound can significantly affect degrading properties.

More details on the noted effects of compound structure can be reviewed in the cited references.

As a result of an accumulation of test data, primarily from static testing, other attempts have been made to develop generalized biodegradability *screening* guidelines based on physical or chemical properties such as solubility, BOD/COD ratio, or COD decrease. Some of these methods are summarized subsequently. The reader should review the cited references to examine the data or the basis for the *screening method.*

Solubility

Generally, water-insoluble compounds are thought to be more resistant to biodegradability because of the mass transfer limitations governing reaction rate, the ability of the compound to reach the cell, and removal of the compound by adsorption or trapping in inert solids [25].

BOD/COD Ratio

The waste BOD has traditionally been used to estimate biodegradability, especially when compared to the COD value. Lyman and Rosenblatt used the BOD/COD ratio to classify the biodegradability of pure compounds, suggesting the following rules-of-thumb [25]:

(1) Compounds with a ratio of less than 0.01 are relatively nonbiodegradable.
(2) Those with a ratio between 0.01 and 0.1 are moderately degradable.
(3) Those with ratios greater than 0.1 are degradable.

The use of BOD has the advantage of being relatively simple but must be employed with some caution because the BOD test itself is highly sensitive and subject to (1) the viability of the seed used, (2) adherence to test procedures, (3) the effects of waste impurities inhibiting biodegradability, which can be removed by pretreatment, (4) introduction of contaminants in the test procedure, (5) seed acclimation procedures, (6) dilution of the test sample to the test limits, thereby diluting any toxicity, (7) the relatively small quantity of sample utilized, and (8) any other intrinsic test limitations.

COD Decrease

Another method of estimating waste biodegradability was developed by Pitter (as reported in Reference [25]) using activated sludge inoculum with 20 days of adaptation to the substrate and measuring the percent reduction of the total COD. Pitter defined removal of at least 90% of the COD as readily biodegradable, equivalent to a removal rate of at least 15 mg COD/g hr.

General Discussion

The methods discussed, as well as those cited, are helpful in screening waste characteristics for potential biological treatment, estimating the effects of specific components and in developing a specific pilot test program. Regrettably, these methods are sometimes used to establish "absolute" criteria by arbitrarily selecting rate constants for (1) pollutant fate evaluations, (2) establishing definitive design criteria, (3) establishing economic impact of treatment schemes, or (4) determining treatment feasibility. For all practical purposes the methods discussed are *not recommended* for predicting oxidation rates and *definitely not recommended* for establishing biological treatment design criteria.

OXYGEN REQUIREMENTS

A critical element in maintaining aerobic systems is supplying sufficient oxygen for *cell maintenance* and sustaining the *oxidation* process. The total oxygen supplied must reflect the substrate and cell concentrations. Theoretically, the system oxygen demand can be estimated by subtracting the (*remaining*) ultimate oxygen demand of the system products (*effluent* and *waste sludge*) from that of the feed. Based on a stoichiometric requirement of 1.42 g of oxygen per gram of cell ($C_5H_7NO_2$), the oxygen demand can be estimated as follows [29]:

$$\text{kg O}_2 \text{ per day} = \frac{Q(So - Se)}{\tau} - 1.42\, W_x \quad (2.8)$$

where Q is the waste flow in cubic meters per day, So is the influent BOD_5, kg/m^3, Se is the effluent BOD_5, kg/m^3, W_x is the mass organisms wasted, VSS, kg/day, and τ is a factor converting BOD_5 to BOD_{ult} (commonly 0.68). Where nitrification is significant, the added oxygen requirements can be estimated using the expression

$$\text{NOD} = d\text{NH}_4\text{-N} \cdot 4.57 \quad (2.9)$$

or a conservative estimate calculated using the expression

$$\text{NOD} = \text{TKN}_0 \cdot 4.57 \quad (2.10)$$

where NOD is the nitrogen oxygen demand, mass per unit time, $d\text{NH}_4$-N is the mass of ammonia nitrogen removed in

a unit of time, and TKN_0 is the total Kjeldahl nitrogen in the influent, mass per unit of time.

In industrial waste treatment application, the oxygen required for a specific system is usually measured in pilot studies as the uptake rate, defined as the weight of oxygen consumed per unit of time, expressed as mg/L per unit time. This is related to the system solids concentration as the respiration rate, expressed as the weight of oxygen consumed per unit time per unit weight of MLVSS. Commonly, the respiration rate is between 8 to 20 mg/hr/g MLVSS. Lower rates indicate inadequate substrate for the available microorganisms, resulting in poor sludge settling quality. Higher rates are an indication of system overloading, too high a substrate for the available microorganisms.

The respiration rates for the system operating range are used in estimating the required oxygen requirements according to the relation

$$[\text{Respiration rate}] = a \cdot \text{F/M ratio} + b$$

$$[(dO_2/dt)/X] = a \cdot [(dS/dt)/X] + b \quad (2.11)$$

where dO_2/dt is the oxygen uptake as mass per hour, dS/dt is the BOD removal rate as mass per hour, X is the mass of MLVSS under aeration, a is the oxidation requirement factor, mass oxygen mass substrate removed, and b is the endogenous oxygen factor, mass of oxygen per mass of mixed liquor solids under aeration.

Actually, the respiration rate is measured as mg/L of oxygen utilized per unit time per mg/L of MLVSS under aeration, which is equivalent to mg of oxygen per unit of time per mg of MLVSS. Respiration test data are plotted as shown in Figure 2.9 to obtain specific biological treatment constants.

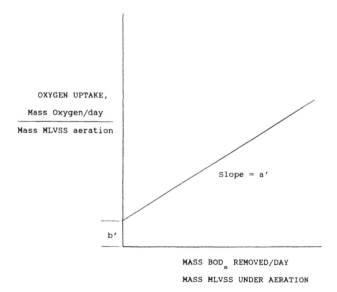

Figure 2.9 Oxygen respiration data plot.

The slope of the line is equivalent to a, the oxidation factor, and the ordinate intercept is equivalent to b, the endogenous oxygen factor. Specific oxygen limitations for each of the biological system will be discussed in the appropriate chapters.

NUTRIENT REQUIREMENTS

Primary nutrients required for biological processes are nitrogen and phosphorus. Any supplements added to deficient industrial wastes must balance the need for the biological process and potential effluent compliance problems if present in excess. In addition, excess nitrogen participates in the biological reaction as a nitrification reaction, which if not anticipated in estimating design oxidation requirements could limit the system efficiency. If massive quantities of nitrogen or phosphorus are present, alternative chemical or physical removal processes must be evaluated.

Based on the biomass production, and a biomass composition of $C_{60} \cdot H_{87} \cdot O_{23} \cdot N_{12} \cdot P$, the nutrient requirements can be estimated on the basis of net cell production as follows [4]:

$$\text{Nitrogen (mass/day)} = 0.122 \cdot \text{net cell produced/day}$$

$$\text{Phosphorus (mass/day)} = 0.023 \cdot \text{net cell produced/day}$$

Based on influent BOD_5, nutrient requirements can be approximated using the ratio of influent BOD_5 to N to P as 100:5:1. The nutrient estimate has to be accurate enough to establish whether nitrogen or phosphorus has to be supplemented and to determine the size of storage, transfer, and feed equipment. The plant operator can adjust the nutrient feed rate to the point that traces are detected in the effluent to minimize costs and meet effluent (nutrient control) requirements.

Although the requirement for nitrogen and phosphorus is accepted, the need for other trace elements is not well understood. Wood and Tchobanoglous have studied and documented the role of trace elements in biological processes [53]. They conclude that trace elements are required for metallic enzyme activators and for the electron transfer in the oxidation/reduction reaction. The trace elements deemed essential are listed as potassium, magnesium, calcium, sulfur, iron, manganese, molybdenum, boron, chloride, sodium, and vanadium.

Nutrient deficiency could present a real dilemma for a Process Engineer because the lack of the essential elements could result in poor treatment performance, whereas at the same time any excess could present a problem in meeting discharge limits. "Trace" quantities required for biological activity should be present as a result of the domestic water used in the process. However, some manufacturing processes may require highly treated process feed waters where these elements would be purposely removed for manufacturing reasons. What should be established in the early pilot study

stages, especially in manufacturing facilities using purified water, is the quantity of these elements in (or not present in) the plant waste streams. Any observed pilot plant inefficiencies should be first corrected by adding any deficient essential elements.

VIABLE BIOLOGICAL POPULATION

As discussed previously, the natural food chain will inherently result in a balanced biological population. Therefore, what is critical is cultivating a population that will optimize treatment performance. This requires providing a suitable and consistent feed quantity and quality with low variability and eliminating conditions that promote poor sludge quality. It is obvious that an inconsistent feed supply affects the biological process; what is not as obvious is that changes in feed species could be detrimental to the system. As discussed previously, a balanced system is one in which the bacteria have adjusted to the food supply, emitting the enzymes required for feed degradation, and the system microorganism growth rate is in balance with the feed supply. When the food supply changes, the bacteria population and the organisms balance change accordingly. When the food type is constantly changing the system will never be balanced, and the treatment operation becomes unpredictable.

In addition, a viable biological population requires avoiding conditions promoting filamentous growth, resulting in bulking sludge. In the case of suspended growth, where settling is critical to maintaining a viable system, researchers have studied process conditions promoting filamentous growth [8–13,15,57]. Design considerations to minimize this condition are discussed as *selector* design enhancements in Chapter II-3. Finally, a suitable biological population is a result of what should not enter the system as well as conditions that should be present, the major detriment being the presence of toxic constituents, as discussed subsequently.

TEMPERATURE

Temperature effects are often neglected in evaluating biological systems, commonly limited to its affect on the reaction rate. Actually, the wastewater temperature has multiple effects on a biological treatment system, influencing (1) the biological population that will survive, (2) sludge growth rate, (3) substrate utilization rate, and (4) oxygen transfer capacity.

Although some biological activity can be established in a wide range of temperatures, the practical design range is quite narrow. Biological treatment is commonly classified as processes effectively operating in (1) a mesophilic temperature range of 25 to 40°C (77 to 104°F), (2) a thermophilic range of 55 to 65°C (131 to 150°F), or (3) a cryophilic range of 12 to 18°C (54 to 65°F) [29]. For all practical purposes aerobic treatment systems operate at ambient temperatures within the mesophilic temperature range.

The temperature effect on sludge growth and substrate utilization is estimated by correcting the appropriate reaction rate coefficients. Coefficients, such as the endogenous decay rate (b), the maximum rate constant (k), or the first-order constant (k_i), are most frequently corrected using the Arrhenius relation:

$$d(\ln k)/dt = Ea/RT \qquad (2.12a)$$

where k is the reaction rate constant, Ea is the activation energy, cal/mol, R is the ideal gas constant, 1.98 cal/g/mol, and T is the temperature in Kelvin.

The integrated form of the previous equation is commonly expressed as

$$K_2/K_1 = \Phi(T_2 - T_1) \qquad (2.12b)$$

The value of Φ depends on the specific biological system, the specific configuration, aeration type, and loading levels, as discussed in each of the biological treatment chapters.

The effect of temperature on the cell yield coefficient (Y) and the half-velocity constant (K_3) is not well documented and could depend on the system configuration.

The final consideration involving temperature is maintaining aerobic conditions, necessitating oxygen transfer to the microorganisms at the utilization rate. The driving force required to maintain aerobic condition depends on the difference between the aeration basin oxygen concentration, usually greater than 1 mg/L, and DO saturation concentration (C_s), which is temperature related. The saturation concentration decreases with increasing temperature, thereby decreasing the driving force. The Process Engineer must consider the varying climatic conditions affecting oxygen availability, supplying enough flexibility to adjust to system or environmental changes.

pH

Optimum biological activity depends on the process pH, which affects both biological population and activity. Industrial waste systems should be maintained above 5 and below 9, with optimum conditions within the 6 to 8 range. This can be easily achieved by pretreating the feed to that range. Although pH adjustment is normally straightforward, nitrification processes alter the system alkalinity, requiring either a compensating feed adjustment or direct reactor monitoring and control.

FACTORS AFFECTING OR INHIBITING A BIOLOGICAL REACTION

FEED CHARACTERISTICS

Achievable treatment performance depends on the collective biodegradability of substrate components, as discussed previously. Of equal importance is avoidance of excessive inert, nonbiodegradable, and toxic components. Although the waste generated by a manufacturing facility cannot always be controlled, how individual streams are treated is a design decision. Relevant feed properties include (1) *waste content,* (2) *toxicity,* (3) *solids content,* and (4) *influent variability.* Those characteristics detrimental to biological processes can be corrected by *pretreatment.*

Waste Content

The performance of a biological stream will be significantly influenced by the waste components fed to the system. There are three criteria that should govern the selection of waste streams to *any* biological treatment system:

(1) Only biodegradable organic wastes should be fed to a biological treatment system. All other wastes will reduce process effectiveness.
(2) Unless required as nutrients, inorganic wastes to the system should be avoided or minimized.
(3) Dissolved inorganic solids adversely affect a biological treatment system, especially the sludge settling characteristics.

The waste's organic content is not the only important criterion; its concentration may restrict applicable configurations. Theoretically, in a suspended growth system any feed concentration can be tolerated if the reactor volume is increased, the microorganism concentration increased, or both. However, at some point the total pounds of feed substrate, the product of concentration and *waste volume,* will result in a prohibitively large system or required oxygen demand. In fact, as the influent BOD_5 concentration exceeds 500 mg/L the Process Engineer should carefully evaluate the biological system limits, especially for fixed-film processes.

For any configuration, abnormally high organic concentrations cannot be tolerated over extended time without depleting the available oxygen or producing an environment unstable to aerobic biological treatment. Under extreme overloading, anaerobic conditions can develop at the microorganism surfaces, and the organic itself results in toxic effects from "overactivity." *In some configurations* exceptionally strong influents may have to be diluted to avoid anaerobic conditions at the reactor inlet. Although dilution may be effective in adjusting an influent to tolerable inlet limits, the larger flow may increase the reactor size; in some cases the content of the selected diluting streams could adversely affect the system.

In many plants, the biological system is a "plant toilet," where everything and anything from the plant is discharged. Any plant waste stream considered for dilution, or simply deposited into the general waste sewer as a "perfect" treatment alternative, should be analyzed to establish its composition and effect on the biological process. Typical plant wastes that should be excluded from a biological process include:

(1) Boiler blowdown containing dissolved solids and alkalinity
(2) Air pollution acid scrubber system discharges
(3) Water treatment discharges
(4) Cooling tower discharges
(5) Air pollution (particulate) wet scrubber discharges containing large quantities of solids

These streams could contain excessive solid or inorganic concentrations that reduce the "active" aeration solids capacity, or add toxic inorganic constituents, and may have to be pretreated before they are acceptable for a biological system. Before any "dilution" plant waste streams are mixed with high strength wastes, they should be analyzed for inorganics, biologically toxic materials, total solids, and total dissolved solids. No stream should be excluded for consideration from a treatment facility, but every stream should be evaluated on the basis of composition, variability, and availability.

In addition, effluent from sludge-dewatering equipment should not be recycled to the clarifier or biological system without pretreatment because they contain fines difficult to settle. Frequently, Process Engineers think they find a good way of disposing of these discharges by "drawing an arrow on the flowsheet back to the front of the system," resulting in scum accumulation in treatment plant components or higher-effluent suspended solids.

One further word of caution, "clean streams" from research, testing laboratories, or pilot facilities should be "suspect," collected separately, and tested before discharged to the treatment facility. As an example, if a treatment facility has low permitted chromium limits, care should be taken in discharging environmental laboratory analytical wastes.

Toxicity

Toxic materials are an unstabilizing factor in a biological system and of particular concern in industrial waste treatment plants. The most common materials found in industrial wastes include acids, alkalies, arsenic, chromium, copper, cyanide, iron, lead, manganese, mercury, nickel, silver, and zinc. Because alkalinity or acidity can be easily adjusted their presence is not as much a concern as some of the other contaminants mentioned. Specific feed limits are difficult to define because the significant parameter is the ratio of the toxic ion to the biomass, the sensitivity of which is specific to the reactor configuration and the biomass level. *As a guide,* and only a guide, "alert" wastewater values historically cited are summarized in Table 2.1. Continuous feeding influents at the concentration indicated or higher could result in corresponding high reactor concentrations, which could affect biomass viability. If these values are approached or exceeded specific wastewater laboratory in-

TABLE 2.1. Alert Toxic Concentrations (adapted from Reference [55]).

Constituent	Alert Concentrations, mg/L*
Total dissolved solids	16,000**
Total chlorides	15,000**
Oil, grease or floatables	50**
Aluminum	15
Ammonia	480
Arsenic	0.1
Borate (boron)	0.05 to 100
Cadmium	10
Calcium	2500
Chromium (hex)	1
Chromium (tri)	50
Copper	1
Cyanide	0.1
Iron	1000
Lead	0.1
Manganese	10
Mercury	0.1
Nickel	1
Silver	5
Zinc	0.08 to 10
Phenol	200

*Reference [55] *lower* values are cited as threshold influent concentrations, which, if accumulated in the sludge, could result in reduced biological activity, or generate hazardous classified sludge difficult to dispose of. When these "alert" concentrations are evident, laboratory investigations are required to define process limits and assure process stability.
**The Process Engineer should monitor pilot reactor performance for deteriorating effluent quality, a result of excessive suspended solids or scum concentrations, caused by excessive influent oils, grease, or dissolved solids.

vestigations should be conducted to establish any inhibitory effects.

In the final analysis, the evidence of an unstable process will be *poor sludge quality, poor sludge control,* and *variable effluent quality.* In these cases polymer addition replaces sludge and process control. Exact risk levels for most potentially toxic compounds are not well documented, but the Process Engineer should review manufacturing procedures to identify routine process "dumps" that could contain large quantities of constituents identified in Table 2.1. In the author's experience with industrial systems the two problems all to often encountered include (1) the unintentional dumping of mass quantities of toxic materials as a result of a process adjustment or vessel cleanout and (2) large variations in the influent toxic component concentration, making it difficult for the biological system to stabilize. The sensitivity of an activated sludge system to specific toxic compounds may require their removal, equalization to minimize their affect, or both. Source correction or pretreatment should be evaluated to remove constituents that are identified as adversely affecting the biological system. In addition, slug amounts of these or other constituents over a short period could be detrimental to the system. In some cases, waste dumps may have to be isolated and stored, and a hold and bleed feed system may have to be considered for pretreatment. Finally, toxic affects of many compounds can be stabilized, or the affects minimized, by increasing the reactor cell concentration level. For that reason, the design engineer must carefully develop the sludge recycle capabilities of a system with that in mind. Where applicable, sludge recycle capabilities must be carefully evaluated to allow for increased MLVSS levels in response to toxic reactor conditions.

Process Adaptation

Limiting criteria indicated in Table 2.1, commonly cited in early design manuals, were based on threshold concentrations believed to inhibit biological processes and were therefore imposed as influent limitations [56]. Indeed, a sudden influx of toxic material to a biological treatment system may reduce the system efficiency or, in extreme cases, terminate the process. However, with gradual introduction of a toxic substance(s) the microorganisms may be able to adapt themselves to the materials and continue the process. At some point toxicity occurs because of acute quantities accumulated *in the biomass,* and a critical ratio of toxic-to-biomass is reached. In such cases, the influent or reactor toxin concentration is frequently higher than the values indicated in Table 2.1.

Although toxic materials injected into a biological system will affect the *overall* treatment efficiency, the extent of the problem is defined by the path of the toxic substance, that is, the quantity partitioned to the sludge or the effluent. Past studies have often focused on affects of concentration on the substrate reaction rate, which under ever more stringent regulatory constraints is only part of the problem, for example,

(1) If a toxic substance is not adsorbed by the sludge, its affects on the reaction rate will probably be minimal because all or most of the compound passes to the effluent. However, most materials classified as "potentially biologically toxic" are also regulated by effluent standards and could result in specific effluent limits being exceeded.

(2) If the substance is significantly adsorbed by the sludge, the *initial problem* may not be reduced reaction rate but deteriorating sludge settling qualities, resulting in increased effluent suspended solids. It has been the author's experience that low-level toxicity is initially evident in microscopic examination of the sludge and in deterioration of the clarifier performance, not the aeration basin performance.

(3) *Any* material adsorbed by the sludge may impose considerable difficulty (and expense) in sludge disposal, because its characteristics may make landfill disposal impossible and require transfer to a controlled hazardous waste facility.

(4) At some point, the toxic compound concentration in the sludge may reach a critical ratio, resulting in deterioration in reactor performance as well as the three other conditions mentioned.

That a system may adjust to higher concentrations than those commonly cited as influent limitations, with no apparent affect on removal efficiency, complicates evaluating the toxic affects of specific compounds. Taking into account contributing substrate removal factors such as stripping and microorganism adaption, it is possible that high removals can be achieved, yet the system may be operating under unstable, potentially toxic conditions. When a toxic component is present in low levels, and consistently fed to a reactor, the biological process may adjust (or be made adjust) to the component. With a gradual introduction of toxic substance(s) (1) the microorganisms may be able to adapt themselves to the toxicity and continue to perform satisfactory, (2) the operator can increase the microorganism content of the reactor to reduce its impact, or (3) both.

Heavy Metals

The affinity of a biomass for heavy metals has been researched, with results generally indicating that biosludge affinity was greatest for lead, decreasing for cadmium, mercury, chromium III, chromium VI, zinc, and nickel (in that order) [34]. Continuous activated sludge pilot studies by Neufield et al. produced results suggesting that (1) it was possible to maintain a viable biological culture at levels of mercury, cadmium, or zinc at much higher influent concentrations than previously reported, (2) the weight ratio of metal in the biofloc to that in the acqueous phase for the three metals could range from 4000 to 10,000, and (3) all these metals were rapidly adsorbed to *almost* equilibrium values in approximately 3 hr, although it took about 2 weeks to stabilize to the equilibrium level [33].

Organics

Some waste organics may themselves exhibit inhibitory effects. Studies conducted with influents containing xylenols, cresols, and phenols show that, although a system can acclimate to these compounds, there was great difficulty in sustaining cell growth, with activated sludge MLVSS reaching only 1500 mg/L and the biofloc exhibiting poor settling characteristics [38].

Ammonia

Ammonia concentrations in the influent imposes a design dilemma. Nitrogen-deficient wastes require nutrient addition, whereas excessive levels could exceed effluent limits, increase oxygen requirements, and at excessive levels affect the biological population by encouraging nitrifier production. Ammonia partitioned to the sludge is usually not a problem because of its nutrient value, although high-influent ammonia concentrations could result in a significant air emission problem.

Dissolved Solids

Dissolved solids, especially salts, could potentially be inhibitory to biological systems. Studies have indicated that systems can generally tolerate salt concentrations less than 1500 mg/L and may acclimate to levels as high as 5000 mg/L [20,52]. However, at high concentrations the settling characteristics of the biofloc deteriorates. A study evaluating high-salt injections concluded that low-salt content reactors adjust to high-salt dosage shocks better than systems acclimated at high-salt levels (30,000 mg/L) can withstand freshwater surges [20]. Low-salt systems withstood heavy-salt dosages with a reduction in treatment efficiency of 30%, whereas high-salt system efficiency was reduced 70% when purged with freshwater. It has been the author's experience that indeed a high-salt environment biological system is unstable and easy to upset.

Solids Content

A complete influent analysis is necessary to determine dissolved and suspended solids composition, active and inactive distribution, and the biodegradable portion of the active solids. Unlike municipal waste streams, industrial waste solids are not always compatible with a biological process, sometimes not being biodegradable. Their affects cannot be generalized because their characteristics are specific to the generating manufacturing process. As a result they can potentially reduce the reactor *volatile suspended solids* (VSS) loading, add to the excess biological solids mass generated, sometimes be detrimental to the biological reaction efficiency, and sometimes produce sludge legally defined as hazardous or toxic. Their path through an aerobic treatment process is illustrated in Figure 2.10 [4].

Influent Variability

Selection of a biological design basis more often than not presupposes a uniform feed with minimum variation. In reality, variations can be tolerated within the system's hydraulic capacity to dilute and dampen to acceptable process limits, above which the plant performance is impacted. A highly variable effluent is the product of either an extremely variable feed or constantly changing manufacturing conditions. Because the operating response of a biological system is limited, an inconsistent effluent quality is generally a result of excessive "leakage," a result of frequent influent changes.

Flow changes are generally directly related to planned changes in manufacturing procedures or unintentional upsets, both of which must be considered in the treatment design.

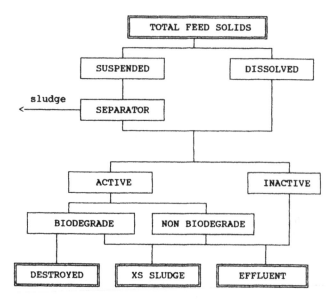

Figure 2.10 Feed solids classification.

Large and abrupt volume fluctuations could influence reactor performance due to changes in the resulting residence time, directly related to the reaction kinetics. Extreme variations could result in reduction of MLVSS or reactor wash out conditions. In some cases, the flow may exceed the system capacity, and the treatment plant will be completely out of control for an element of time. Manufacturing factors affecting effluent discharge are discussed in Chapter I-1.

Varying organic load conditions can result in unstable reactor conditions because of a resulting imbalance between population growth and the available organic food, the system striving to stabilize but unable to reach equilibrium. In addition, the organic load affects the *aeration* requirements, *sludge* production, *clarification, pumping, recycle* requirements, and *final solids processing* capacity. Because biological systems are not fine-tuned processes that quickly adjust to variation, constantly varying feed conditions will result in unstable conditions that may critically affect some process component or exhibit a domino effect on the entire system. Where varying waste discharges is inherent in the manufacturing process, equalization has to be considered as a pretreatment for the biological system.

MIXING

Mixing is imparted in a biological system as a result of energy input (dispersed system) or turbulence created by the waste flow (fixed-film system) through the system. The resulting turbulence influences reactants mass transfer through the system phases, could affect cell maintenance, and establishes the reactor concentration homogeneity. Mass transfer within a system under steady-state conditions is governed by physical factors, the most significant being the turbulence caused by mixing. The system efficiency will be influenced by the effective transfer of components through single and multiple phases. Substrate will be transferred from the bulk of the solution to the microorganism cell interface. Oxygen must be transferred from the atmosphere (gas phase) to the waste (liquid) and to the microorganism cell (solid). All other things equal, the transfer of constituents through the biological phases increases with increased energy input. Studies have demonstrated that, although oxygen system efficiency is generally considered to be optimized at a bulk concentration of 1 to 2 mg/L, oxygen uptake rates could be increased with increased mixing, resulting in an overall reaction rate increase [18].

PROCESS VARIABLES TO CONTROL A BIOLOGICAL SYSTEM

Two significant variables defining biological treatment system capacity, allowing control and operating flexibility, are (1) the quantity of microorganisms available to biodegrade the substrate and (2) the system sludge management capabilities. Specifics for controlling these variables are different for each of the reactor configurations, but in general terms can be explained in terms of the F/M ratio or sludge age.

F/M RATIO

A common factor used to relate process variables in a biological treatment plant is the F/M, expressed as the mass BOD fed per day per mass of activated sludge under aeration. Process variables such as oxygen requirements, effluent quality, substrate reduction, cell growth, endogenous rate, and sludge quality can be related to this ratio. A common variation is the applied F/M ratio, expressed as the mass BOD removed per day per mass of activated sludge under aeration. The F/M ratio is a directly controllable variable in dispersed growth biological system but an indirect variable in fixed growth systems. In either case, system performance can be related to the available biodegradable substrate and the available microorganism contact ratio.

Mathematically, the F/M ratio (U) can be expressed as

$$U = \frac{Q \cdot S_o}{V \cdot X_v} = \frac{S_o}{t \cdot X_v} \qquad (2.13)$$

The applied F/M ratio U_a can be expressed as

$$U_a = \frac{Q \cdot (S_o - S_e)}{V \cdot X_v} = \frac{S_o - S_e}{t \cdot X_v} \qquad (2.14)$$

where S_o is the influent concentration, mg/L, S_e is the effluent concentration, mg/L, V is the reactor volume, X_v is the reactor

mixed liquor volatile solids concentration, mg/L, and Q is the influent flow rate, volume/day.

EXCESS SLUDGE

The quantity of sludge produced by the biological oxidation can be expressed by a reactor material balance:

Net sludge = sludge produced − sludge digested

$$Q \cdot (S_o - S_e) \cdot Y_o = Q \cdot Y \cdot (S_o - S_e) \\ - K_d \cdot V \cdot X_v \quad (2.15)$$

where Y_o is the apparent synthesis of soluble BOD converted to solids, net mg of solids produced/mg BOD removed, Y is the actual synthesis of soluble BOD converted to solids, total mg of solids/mg BOD, k_d is the endogenous decay coefficient, 1/time, and $K_d \cdot V \cdot X_v$ is the endogenous cell destruction rate, mg/day.

Based on the previous material balance and the applied F/M (Ua) definition, a relation between the apparent sludge coefficient (Y_o) and the actual sludge coefficient (Y) can be developed:

$$Y_o = \frac{Y \cdot Ua - K_d}{Ua} \qquad (2.16)$$

Y and K_d constants are obtained from pilot studies for the specific waste to be treated. Typical values are given in Table 3.2 through 3.5 in Chapter II-3.

Besides the biomass (net X_v), the total *dry* sludge generated includes the nonbiodegradable volatile solids, the portion of the biodegradable solids that is not treated, and the fixed influent suspended solids, as discussed under Feed Solids. The *wet* sludge generated includes its water content.

SLUDGE AGE

Sludge age or solids retention time (SRT) is a common parameter, comparable to F/M, used to design and control biological treatment systems. SRT is directly related to cell growth, *being equal to the reciprocal of the cell growth rate*, as discussed under Biological System Microorganisms. It is defined as the average time that activated solids are retained in the process. SRT is equal to the total mass of activated solids present in the process divided by the mass of organisms wasted daily, calculated as follows:

$$SRT = M / (dM/dT) \\ = XV / [X_rW + X_e(Q - W)] \quad (2.17)$$

where M is the total solids in the system, kg (lb), dM/DT is the solids wasted in kg (lb) per day, X is the system solids concentration, V is the mass system capacity, kg (lb), X_r is the concentration of the recycle stream, W is the wasted stream in kg (lb) per day, X_e is the effluent concentration, and Q is the forward feed rate in kg (lb) per day. Note: Because concentration appears in all the terms, any consistent units can be used.

For a once-through system (no recycle) the sludge age (SRT) equals the hydraulic retention time. A relation between SRT and F/M can be derived by combining Equations 2.15 and 2.17, resulting in the equation

$$1/\text{SRT} = Y \cdot E \, (\text{F/M})_{\text{fed}} - K_d \qquad (2.18)$$

where E is the removal efficiency, $(S_o - S_e)/S_o$ and Y and K_d are defined previously.

Based on the definition of Y_o,

$$Y_o = \frac{Y}{1 + K_d \cdot \text{SRT}}$$

SYSTEM PERFORMANCE

SLUDGE QUALITY

Sludge quality is a major consideration in evaluating biological treatment performance, tracing its fate in the reactor, clarifier, sludge-dewatering systems, and its suitability for final disposal. Generally, sludge quality is dependent on reactor conditions and the resulting cell solids size distribution, which is directly dependent on the predominant microorganism population. A common sludge quality measure is solids settling, measured by the sludge volume index (SVI). A sludge with an SVI of 100 to 150 is indicative of quality sludge, an indicator that the sludge will settle in the clarifier, and can be concentrated and returned to the aeration basin for solids control (in suspended growth systems). Another significant sludge characteristic is dewatering quality, greatly influencing dewatering performance, final disposal quantity, and costs. Generally, a good settling rate will have good dewatering characteristics.

Finally, sludge composition establishes the ease of disposal, with or without final stabilization. Generally, any sludge containing regulated toxic constituents, which could leach into a landfill underground water table, will be expensive to dispose. Sludge metal content represents a special disposal problem, restricting direct land disposal, and affecting sludge incineration because of limits imposed on metal air emissions and ash disposal. Sludge composition is affected by reactor conditions as well as the quantity of feed components not biologically oxidized, adsorbed, and retained by the biofloc. The Process Engineer must consider waste sludge quality as part of the design criteria.

DESTRUCTION EFFICIENCY

Historically, biological treatment system efficiency is measured by overall system substrate removal, generally measured as the quantity of substrate (BOD or COD) removed from the system, as follows:

$$\frac{\text{BOD}_{\text{in}} - \text{BOD}_{\text{out}}}{\text{BOD}_{\text{out}}} \times 100 \qquad (2.20)$$

However, other criteria play a significant part in evaluating performance. The quantity of suspended solids in the effluent must be considered in the system performance. There is generally a limit on the allowable suspended solids, from 5 to 20 mg/L, and each mg/L of degradable suspended solid is equivalent to 1.42 mg/L of effluent *ultimate BOD*. A poor performing biological system can achieve high organic removal efficiency by combined degradation, volatilization, and partitioning to the sludge and still produce a poor and unacceptable effluent quality as a result of high suspended solid content.

Many regulatory agencies impose limitations on specifically defined priority pollutants, which individually must be controlled to water quality limits. The required treatment of these individual components imposes an even stricter treatment performance requirement, whether that be for the biological system alone or the biological system combined with tertiary treatment. In some cases ammonia nitrogen destruction to a specified nitrate or nitrogen level is imposed on the system, further complicating biological system performance requirements.

Finally, after all substrate treatment requirements are satisfied, the *fate* of the waste components must be considered. Will the waste component(s) be biodegraded? Adsorbed by the biomass and discharged as waste sludge resulting in the treatment of hazardous sludge? Is a large portion of the component(s) stripped or volatilized from the waste as an air stream, possibly presenting an air emission violation or an odorous nuisance? *Total* acceptable system performance is usually not measured by substrate destruction efficiency alone but effluent control requirements for nutrients, priority pollutants, and suspended solids as well as fate of contaminant considerations.

FATE OF CONTAMINANTS

Currently biological systems have undergone a more stringent scrutiny to evaluate the *fate of the substrate contaminants,* more specifically the pathways by which the substrate contaminants are separated and their final deposition. In general terms the route the contaminants can take will be one, a combination, or all of the following:

- Stripping from the bulk of the waste
- Volatilization from the waste surface

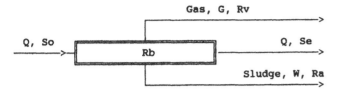

Figure 2.11 Potential substrate paths in aerobic systems.

- Sorption by the biological solids
- Biodegradation in the treatment system

Of the four mechanisms, biodegradation is the only mechanism designed into the system. The other paths are a result of the system configuration and equipment selection. The type biological system selected, and the configuration within that system, will affect the substrate removal mechanism. Possible substrate contaminant paths in a steady-state process indicated in Figure 2.11, can be expressed by the equation:

$$Q \times (C_{in} - C_{out}) = R_v + R_a + R_b$$

where Q is the waste flow in m³/day, V is the aeration volume in m³, S is the waste composition in and out in g/m³, R_v is the volatile removal rate in g/day, R_a is the adsorption removal rate in g/day, and R_b is the biodegradation removal rate in g/day.

Considerable effort has been undertaken to develop estimating methods to determine partition paths of priority pollutants in municipal systems [2,3,5,6,23,32,44]. The U.S. Environmental Protection Agency (EPA) has developed a model applicable to municipal systems, detailed in a manual entitled "CERCLA Site Discharges to POTWS: Treatability Manual" [45]. The same concern about priority pollutants from industrial facilities can be expected, although the same assumptions made for municipal system models may not be applicable. Where industrial waste systems are piloted component partitioning information can be established, and those data can be used in lieu of estimated results. The methods discussed in this section are generally those outlined by Namkung and Rittman [32]. They are basic and general in nature and can be easily upgraded. The reader should refer to the cited references for alternative methods and more details about the procedures discussed.

Biodegradation

Based on Monod kinetics, the biological rate can be expressed as

$$R_b = \frac{k \cdot X_a \cdot S}{K_s + S} \cdot V \qquad (2.21)$$

where k is the maximum specific substrate utilization rate,

g/g VSS · day, X_a is the active cell concentration, gVSS/m³, S is the aeration tank (effluent) substrate concentration, g/m³, K_s is the half-velocity concentration, g/m³, and R_b is biodegradation removal rate, g/day.

Because many constituents of concern are present in low concentrations, the Monod kinetic is frequently expressed as a first-order reaction in fate models, such that

$$R_b = -k_i \cdot X \cdot S \cdot V \qquad (2.21a)$$

Stripping and Volatilization

Stripping refers to the quantity of influent contaminant removed from the waste as a result of the aeration device, mechanical or diffused. This quantity can be explained in terms of its volatility, measured by Henry's law, and the aerator power output. Volatilization is directly related to the volatile compound diffusivity and wind velocity inducing volatile contaminant transport from the surface.

One method of estimating the volatilization and stripping is based on ideal gas law considerations, assuming that all volatile materials are carried from the system with the escaping gas (air) to its maximum capacity:

$$R_v = -\frac{G(P_e - P_i)}{RT} \cdot MW \qquad (2.22)$$

where R_v is the combined volatilization and stripping effect in g/day, G is the total gas volumetric flow rate, m³/day, P_e is the partial pressure of the component in the exit gas, atm, P_i is the partial pressure of the component in the inlet gas, atm, R is the universal gas constant, 0.00008206 m³ atm/K mole, T is the reactor temperature in K, and MW is the molecular weight of the component.

Assuming the exit gas concentration is the equilibrium concentration, based on the liquid phase solute concentration,

$$P_e = H \, S/MW$$

where H is the Henry's law constant, atm·m³/mol and S is the component liquid concentration, g/m³.

In addition, the entrance P_i is normally zero. Based on these assumptions,

$$R_v = -\frac{G \cdot H \cdot S}{R \cdot T} \qquad (2.22b)$$

This method may be an oversimplification and does not include the effects of the aeration device or process characteristics. Other methods of estimating volatilization and stripping rates are discussed in the references cited in the introduction, which the interested reader should review.

Adsorption

Organic accumulation in the biofloc is the first biodegradation step; the amount remaining in the floc depends on biodegradability and reaction time. In most proposed models the amount of a component adsorbed and removed with the biofloc can be estimated as follows:

$$R_a = -W \cdot X_v \cdot q \qquad (2.23)$$

where W is the waste sludge flow, m^3/day, X_v is the concentration of the wasted cells, g VSS/m^3, and q is the compound sorption ratio, g/g VSS.

It is generally assumed that the sorption ratio is proportional to component concentration in the aeration basin, so that

$$q = k_p S \qquad (2.23a)$$

where S is the component concentration in the aeration tank and k_p is the partition coefficient, m^3/g, VSS.

The partitioning ratio is commonly related to the octanol/water coefficient (K_{ow}) and is defined as the ratio of a specific compound concentration in the octanol phase to its concentration in water. A high K_{ow} is an indication of a high adsorption potential from the water to organic (biological solids) phase. Namkung and Rittman [32] propose that K_p can be defined as

$$K_p = -0.00000063 \cdot \int oc \cdot K_{ow} \qquad (2.23b)$$

where $\int oc$ equals 0.531 when biological cells can be represented by the formula $C_5H_7O_2N$.

The resulting adsorption equation reduces to

$$R_a = -(0.0000003345) \cdot W \cdot X_v \cdot K_{ow} \qquad (2.23c)$$

The final form of the equation will vary with individual models, depending on the relationship selected for the partition coefficient (k_p) [2,3,5,6,23,32,44]. Based on experimental data, many correlations have been proposed for the partition coefficient for municipal sludge, generally in the form expressed by Equation (2.23d) [45].

$$k_p = a K_{ow}^b \qquad (2.23d)$$

where a and b are constants specific to the sludge type and specific organics adsorbed.

Specific Fate Models

The relations specified previously are included in all fate models, with different (and sometimes more complex) expressions substituted for the individual rates. The specific implication to the Process Engineer designing a biological system is that an air emissions or solids waste disposal problem *not be substituted* for a liquid waste problem.

NITRIFICATION AND DENITRIFICATION

Biological systems are sometimes utilized for ammonia or complete nitrogen removal as part of advanced municipal treatment. Similarly, biological systems are commonly evaluated for industrial wastes containing significant quantities of ammonia and related nitrogen components. The major difference in industrial systems is that nitrogen is not an inherent waste component, it must be introduced as a deliberate manufacturing discharge, and the species may not be a simple inorganic ammonia or related derivative. As a result, all nitrogen waste components may not be readily amenable to biological oxidation, and indeed the nitrogen species or concentration may be toxic to biological systems. Therefore, destruction of nitrogen-containing components in industrial wastes must include evaluation of not only biological but also other competitive systems.

Biological nitrogen removal chemistry includes a series of sequential steps involving

(1) Aerobic oxidation of carbonaceous materials
(2) Aerobic oxidation of nitrogenous materials to nitrates
(3) Anaerobic reduction of nitrates to nitrogen

Specific equations for nitrification and denitrification are discussed in the proceeding paragraphs. Unless otherwise noted, information cited can be reviewed in the EPA Nitrogen Control Process Manual [47].

NITRIFICATION [47]

Nitrification reactions in a very weak carbonic acid aqueous system can be stoichiometrically represented as follows:

Step 1: $NH_4^+ + 1.5 O_2 + 2 HCO_3^- = NO_2^- + 2 H_2CO_3 + H_2O$
Step 2: $NO_2^- + 0.5O_2 = NO_3^-$
Overall: $NH_4^+ + 2.2 O_2 + 2 HCO_3^- = NO_3^- + 2 H_2CO_3 + H_2O$

The combined oxidation synthesis reaction, in a weak carbonic acid aqueous system, can be stoichiometrically represented as

(1) For the nitrosomonas synthesis

$$55 NH_4^+ + 109 HCO_3^- + 76 O_2 = + 54 NO_2^- + C_5H_7NO_2 + 57 H_2O + 104 H_2CO_3$$

where $C_5H_7NO_2$ is representative of the *Nitrosomonas* bacteria generated and required for the step 1 nitrite production equation. The nitrosomonas growth rate is limited by the concentration of the ammonia nitrogen.

(2) For nitrobacter synthesis

$$NH_4^+ + 400\ NO_2^- + 4\ H_2CO_3 + HCO_3^- + 195\ O_2$$
$$= 400\ NO_3^- + C_5H_7NO_2 + 3\ H_2O$$

$C_5H_7NO_2$ is representative of the *Nitrobacter* bacteria required for the step 2 nitrate production equation. The nitobacter's growth rate is limited by the nitrite concentration.

Aerobic oxidation of nitrogen ammonia in the two steps indicated is accomplished under specific conditions encouraging both *Nitrosomonas* and *Nicrobacter* nitrifier cells. However, even under the best of conditions cell growth is minimal, with stoichiometric yields of 0.15 mg of nitrosomonas cell produced per mg of ammonia nitrogen oxidized and 0.02 mg of *Nitrobacter* cells per mg of nitrite nitrogen. Experimental yields of 0.04 to 0.13 mg of VSS (associated with the *Nitrosomonas* cells) and 0.02 to 0.07 mg of VSS (associated with the *Nitrobacter* cells) have been reported. Significantly, the maximum *Nitrobacter* growth rate exceeds that of the nitrosomonas, so that nitrite will not accumulate, and the ammonia conversion step is the controlling reaction.

These reactions occur in a weak aqueous carbonic acid environment, resulting in alkaline destruction of ammonia nitrogen and bicarbonate reactants. This results in a pH depressing effect on the system according to the expression:

$$pH = \log K_{eq} - \log [H_2CO_3/HCO_3^-]$$

System configuration conditions encouraging carbon dioxide stripping, compensating for the effect of increasing pH, may alleviate this situation.

Generally, the kinetics of a nitrification system are represented by the following Monod kinetic expressions:

Microorganism growth rate	Maximum growth rate	Ammonia concentration	
μn	$=$	$\mu n \cdot \dfrac{N}{[K_n + N]}$	(2.24)
dN/dt	$=$	$\mu n/Yn$	(2.25)
dN/dt	$=$	$q'n\ (N/[K_n + N])$	(2.26)
$q'n$	$=$	$\mu n/Y_n$	(2.27)

where N is the ammonia concentration and the other variables are as defined in the Biological Kinetics section.

Temperature Effects

Nitrification is temperature sensitive, significantly affecting the cell growth and nitrification rates. Corrections for municipal systems includes adjustments to the peak growth rate (k_n) and the half-saturation constant (K_n) include [47]

$$\mu n = 0.47\ e^{0.098(T - 15)},\ 1/\text{day} \tag{2.28}$$

$$K_n = 10^{0.051T - 1.158},\ \text{mg/L as } N \tag{2.29}$$

pH Effects

Biological nitrification has been reported to be effective at a wide pH range, decreasing as the pH moves to the acid side. Acclimation tends to moderate the effect of pH changes to levels as low as 5.5, although sudden decreases from the 7.2 to 5.5 inhibit the reaction until the pH increases. Control is complicated because the process inherently results in pH depression, with an estimated 7.14 mg of alkalinity (as calcium carbonate) reduced per mg of nitrogen nitrified, 7.07 when the synthesis reactions are included. The reaction of CO_2 and water (HCO_3^-) buffers the reaction, whereas CO_2 stripping increases pH. However, even under ideal conditions, the system alkalinity is depleted by the acid nitrification process unless the waste contains adequate alkalinity or alkalinity is added.

Generally, the system kinetics favor a pH range of 7.2 to 8.0 for optimum treatment efficiency; the effect on cell growth outside this range can be estimated using the Downing expression:

$$F_p = [1 - 0.833\ (7.2 - pH)] \tag{2.30}$$

Oxygen Requirements

Oxygen requirements are readily estimated from stoichiometric balances, with theoretical oxygen demand being 4.57 mg oxygen per gram of ammonia nitrogen oxidized or 4.19 mg if the synthesis reaction is considered.

DO Level

As in carbonaceous oxidation, oxygen is an active process reactant, which if deficient could result in limited cell growth. Oxygen effects can be represented by a Monod growth kinetic expression:

$$F_o = k\ Do/[K_o + DO] \tag{2.31}$$

where K_o can range from 0.15 to 2 and taken as 1.3 for municipal design purposes.

Toxins and Nitrogen Level

The nitrification process is adversely affected by toxins; some compounds identified as adversely affecting nitrification are indicated in Table 2.2 [47].

TABLE 2.2. Nitrification Toxins (adapted from Reference [47]).

Organics	Inorganics
Thiourea	Zinc
Allyl-thiourea	OCN ion
8-Hydroxyquinoline	ClO$_4$ ion
Salicyladoxine	Copper
Histidine	Mercury
Amino acids	Chromium
Mercaptobenzthiazole	Nickel
Perchloroethylene	Silver
Trichloroethylene	
Abietec acid	

Ammonia concentration affects the nitrification process up to approximately 2.5 mg/L, above which the process approaches zero-order reaction kinetics.

Combined Kinetic Expression

Based on Equation 2.26, the complete kinetics expression includes adjustments for ammonia nitrogen limiting conditions, corrected for temperature, pH, and oxygen limiting factors as follows:

$$\mu n = \mu'n \left[N/[K_n + N] \right][F_p\, F_o\, F_t] \qquad (2.32)$$

Nitrifier Growth Rate

Sustaining the nitrification oxidation reaction involves a deliberate control of the nitrifying growth cell, especially in the environment of organic carbon substrate, where the corresponding heterotrophic bacteria growth is favored over the slower nitrifying population. That being the case, the reactor conditions must favor nitrifier growth. As discussed previously, the specific cell growth rate is reciprocal of the SRT, so that the SRT for the nitrifying environment must be *less than that required for the heterotrophic* (carbonaceous) growth rate.

For nitrification to occur, the maximum nitrifier growth rate (μn) must be greater than the heterotrophic (carbonaceous design μb) growth rate. Neglecting nitrification effects from nitrogen concentration ($K_n >>> N$), the population growth requirements can be represented as follows:

$$\mu n > \mu b$$

or

$$1/SRT_n \text{ (maximum growth)} > 1/SRT \text{ design}$$

When these conditions are not met nitrifier washout occurs because either the nitrifying growth is too slow or the heterotrophic growth rate is too high.

Conditions depressing nitrifier growth rate can be evaluated from the appropriate kinetic expressions. First, the maximum growth rate μn *may* be depressed to a value $\acute{o}n$ as a result of available oxygen, pH condition, or temperature. The impact of these variables can be estimated using Equation 2.33, assuming $N >>> K_n$ and the EPA-cited constant values.

$$\acute{o}n = 0.47 \cdot (e^{0.098\,(T-15)}) \cdot DO/(KO_2 + DO)] \\ \cdot [1 - 0.883 \cdot (7.2 - pH)] \quad (2.33)$$

Based on these environmental adjustments, the nitrification can be further corrected for nitrogen concentration using the expression

$$\mu n = \acute{o}n \cdot [N/(K_n + N)]$$

Vigorous conditions to assure nitrification can be established by meeting the criteria

$$\acute{o}n \geq \mu b$$

or

$$SRT \text{ design} \geq SRT \text{ min for nitrification}$$

that is, the design SRT must be greater than the minimum nitrification SRT at the given nitrogen concentration, pH, temperature, and DO.

Heterotrophic Growth Rate

The heterotrophic growth rate can be defined by appropriate kinetic expressions and to assure nitrification must be adjusted to appropriate levels. The growth can be defined by the relationship

$$\mu b = 1/SRT_c \text{ design} = Y_b \cdot qb - K_d \qquad (2.34)$$

where qb is defined as

$$qb = \frac{S_o - S_e}{X_i \cdot HRT} \qquad (2.35)$$

where S_o is the feed substrate concentration, S_e is the effluent substrate concentration, and X_i is the MLVSS.

Assuming Y_b and K_d to be constant, carbonaceous growth rate can be controlled by controlling qb as follows:

(1) In a *combined treatment* system the MLVSS (X_i) can be increased; or the HRT could be increased, but would require increased aeration volume; or the influent flow reduced, which is not a controllable variable.

(2) The other method of controlling qb is to reduce the organic concentration (S_o) in the nitrification reactor,

by *separate* stage nitrification, employing an upstream carbonaceous oxidation stage.

This general discussion is applicable to any biological treatment system in which combined or separate stage nitrification is a design consideration. Although a suspended growth system is implied, the same theoretical limitations apply to fixed growth nitrification systems.

DENITRIFICATION [47]

Denitrification occurs under anoxic conditions, described as nitrate dissimilation, where nitrate replaces oxygen as the electron donor in a carbonaceous environment, resulting in oxidation of the organic and reduction of the nitrate to nitrogen. The reaction, which commonly takes place in a weak carbonic acid system, can be represented as follows:

Step 1: $NO_3^- + 0.33\ CH_3OH = NO_2^- + 0.33\ H_2CO_3$
$+ 0.33\ H_2O$

Step 2: $NO_2^- + 0.5\ CH_3OH + 0.5\ H_2CO_3 = 0.5\ N_2$
$+ HCO_3^- + H_2O$

Overall: $NO_3^- + 0.833\ CH_3OH + 0.167\ H_2CO_3$
$= 0.5\ N_2 + 1.33\ H_2O + HCO_3^-$

Denitrifier synthesis equations in a weak carbonic acid aqueous system can be stoichiometrically represented as follows:

$14\ CH_3OH + 3\ NO_3^- + 4\ H_2CO_3$
$= 3\ C_5H_7O_2N + 20\ H_2O + 3\ HCO_3^-$

where $C_5H_7O_2N$ is representative of the denitrifying bacteria generated.

Because biological process conditions favor oxygen over the nitrate as an electron acceptor, denitrification must be conducted in an anoxic environment to assure nitrate destruction. Electron donors considered in denitrification include municipal waste substrates, volatile acids, brewery wastes, molasses, and methanol. The most frequently used is methanol, generally considered the organic of choice based on treatment effectiveness, sludge production, and cost. Stoichiometrically, the methanol requirements can be estimated as follows:

Methanol, mg/L $= 2.47\ NO_3^-N$ mg/L
$+ 1.53\ NO_2^-N$ mg/L $+ 0.87\ DO$ mg/L   (2.36)

Generally, it is expected that a methanol to nitrate nitrogen ratio of 3 assures a complete reaction and a 95% nitrate removal.

Biomass production from a denitrification reaction can be estimated using the following relation:

Biomass, mg/L $= 0.53\ NO_3^-N$ mg/L
$+ 0.32\ NO_2^-N$ mg/L $+ 0.19\ DO$ mg/L   (2.37)

Denitrification results in bicarbonate production, carbonic acid concentration reduction, and a resulting pH system elevation. Generally, 3.57 mg of alkalinity as $CaCO_3$ is produced per mg nitrate or nitrite nitrogen reduced to nitrogen gas. This increased pH (alkalinity) partially offsets the pH reduction resulting from nitrification reactions, when the processes are combined.

Denitrification growth kinetics can be represented by the expression:

$$\mu d = \acute{\mu} d \cdot [D/(K_d + D)] \qquad (2.38)$$

where μd is the growth rate, 1/day, $\acute{\mu} d$ is the maximum denitrifier growth rate, 1/day, D is the concentration of the nitrate nitrogen, mg/L, and K_d is the half-saturation constant, mg/L NO_3^-N

K_d values have been reported as 0.16 mg/L NO_3^-N with solids recycle and 0.08 for no recycle for suspended growth systems at 20°C, and 0.06 for attached growth systems at 25°C [47]. These values imply that the effects of nitrate concentrations above 1 to 2 mg/L nitrate results in zero-order reaction rate, i.e., concentration does not affect denitrification.

As with the other biological processes, the denitrification rate can be related to the cell growth rate by the equation

$$qd = \mu d/Y_d \qquad (2.39)$$

where qd is the nitrate removal rate, expressed as kg (lb) nitrate nitrogen removed per kg (lb) VSS per day, Y_d is the denitrifier gross yield, expressed as kg (lb).

The growth rate is affected by temperature, carbon concentration, pH, and DO.

Temperature

Growth rate and denitrification rate are both affected by reduced temperatures below 20°C, with reported data indicating anywhere from 20 to 40% reduction of the 20°C rates at 5°C, some data suggesting less temperature effects by attached growth than suspended growth systems [47].

Carbon Concentration Effects

Carbon concentration effects, where applicable, are included in a Monod-type expression:

$$\mu d = \acute{\mu} d\ \frac{M}{K_m + M} \qquad (2.40)$$

where M is the methanol concentration, mg/L, and K_m is the half-saturation constant for methanol, mg/L methanol.

K_m is generally in the order of 0.1, implying that (1) small excess concentrations are required and (2) this correction, within practical estimating limits, is insignificant unless the methanol concentration is well above 1 mg/L.

pH Effects

Data indicate depression of denitrification rates outside a 6 to 8 pH range and optimum rates occurring within 7 to 7.5.

DO Effects

DO depresses denitrification as a result of the higher growth rate of carbonaceous oxidation cells.

Combined Relationship

Theoretically, a combined relationship similar to nitrification could be developed for denitrification, except that a mathematical expression for temperature, DO, and pH effects is not readily available. Therefore, based on a specific cell growth rate (μd), which has been adjusted for temperature, pH, and DO level, the combined effect relationship reduces to

$$\mu d = \hat{\mu} d \, \frac{D}{K_d + D} \cdot \frac{M}{K_m + M} \qquad (2.41)$$

where $\hat{\mu} d$ is the peak rate of denitrifier growth at the adjusted temperature T and pH and μd is the rate of denitrifier growth adjusted for nitrate and methanol concentration, as well as T, pH, and DO.

In addition, because methanol concentration is not a significant factor at concentrations less than 1 mg/L, the correction is usually neglected, and the basic process is defined by Equation (2.38).

Denitrification evaluation follows the same kinetic evaluation proposed for carbonaceous oxidation and nitrification, such that

$$1/\mathrm{SRT}_d = Y_d q_d - b \qquad (2.42)$$

where SRTd is the denitrification solids retention time, b is the decay coefficient, day^{-1}, Y_d can range from 0.6 to 1.2, and b is commonly taken as 0.04 day^{-1} for engineering evaluations.

$$1/\mathrm{SRT}_d \text{ minimum} = Y_d \cdot q_d - b_d \qquad (2.43)$$

where q_d is the nitrate removal rate and $\underline{q}_d$ is the peak nitrate removal rate.

These equations are further developed in the suspended growth sections, and comparable procedures are suggested for attached growth systems.

BIOLOGICAL PROCESSES

As illustrated in Figure 2.12, a variety of treatment systems is available to biologically treat industrial wastewaters. These systems can be operated aerobically or anaerobically and as dispersed or fixed growth systems. Dispersed growth systems employ a biological environment in which suspended microorganisms react with the substrate. Fixed growth involves contact of the substrate with microorganisms attached and accumulating on a media. Dispersed systems can be operated on a batch or continuous basis, whereas fixed-film systems are commonly operated continuously or in some cases continuously dosed with waste at regular intervals. The reactors for these different systems are relatively simple, but all perform extremely complicated biological reactions, attempting to balance substrate removal with cell production and excess cell wasting.

Dispersed batch reactors, commonly referred to as sequential batch reactors (SBRs), conduct the various treatment steps in a time sequence, involving full, reaction, settling, and withdrawal. Continuous dispersed systems conduct the same operations in single or multiple flow-through vessels. The basic parameter in a batch system is *cycle time;* in a continuous system it is *flow rate.* Continuous dispersed systems include activated sludge, lagoons, and special systems such as oxidation ditches. Fixed-film systems include packed beds, trickling filters, and rotating biological contractors (RBC). Each of these systems will be discussed in subsequent chapters. These systems can be enhanced by carbon addition, selector design, nitrification, or denitrification.

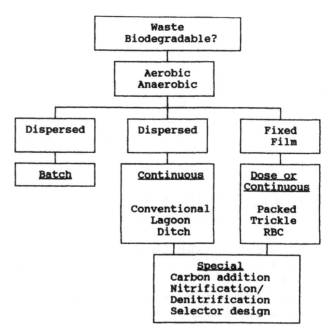

Figure 2.12 Biological treatment systems.

TABLE 2.3. Biological Systems Effluent Characteristics (adapted from Reference [54]).

	BOD, mg/L		SS, mg/L	
	Range	Median	Range	Median
Extended aeration	3–28	7	2–33	10
Oxidation ditch	3–29	8	4–32	12
Contact stabilization	4–45	13	4–47	14
Activated sludge	4–41	15	5–52	14
RBC	5–32	17	7–23	15
Plastic TF	7–55	19	6–43	19
Stabilization pond	9–21	23	13–133	40
Rock TF	8–58	26	9–101	25

HISTORICAL DATA

As a frame of reference reported municipal effluent quality offers an indication of potential achievable industrial effluent quality. Performance of 416 *well-operated* municipal treatment plants reported in the WPCF Municipal Design Manual are summarized in Table 2.3 [54]. The cited reference should be reviewed for more complete details of the facilities reported.

PROCESS SELECTION

Which system is applicable to what wastes? What are the criteria for matching a waste to a system configuration or process? A step-by-step evaluation of process alternatives can be applied to compare waste characteristics with suitable configuration alternatives.

- Step 1: Is the waste biodegradable or capable of being pretreated to improve biodegradability?

The first consideration is that the waste must be biodegradable or capable of being pretreated so that it can be made biodegradable. Biodegradable criteria and acceptable feed characteristics were discussed previously.

- Step 2: Is the organic concentration of the waste greater than 100 ppm or less than 3000 ppm?

The influent concentration of any biodegradable waste should be screened for extreme (high or low) conditions resulting in oxygen limitations or extreme reactor sizes. When substrate concentrations exceed 3000 mg/L the feeds volatility or caloric value should be investigated for alternative distillation or incineration treatments. Low substrate concentrations present a special problem that could limit adequate cell growth to sustain the system or adequate treatment efficiencies to reliably achieve water quality levels of 10 ppm or less. In such cases effluent quality may be more reliably achieved by selective adsorption methods or chemical treatment.

- Step 3: Is the organic loading dissolved or suspended solids?

The quantity of waste organic suspended solids will affect the biological system configuration selection. When the influent suspended solids loading exceeds 100 mg/L, primary clarification must be evaluated to optimize the "effective" secondary treatment volatile solids capacity. In such cases, if influent solids are not removed the secondary treatment system will assume the characteristics of an aerobic digester. As the waste stream biodegradable solids exceeds 10,000 mg/L anaerobic digestion is a viable economic consideration. Wastes containing primarily dissolved biodegradable organics are treated in conventional treatment systems.

- Step 4: Does organic loading warrant aerobic or anaerobic waste treatment?

In evaluating the viability of aerobic or anaerobic biological treatment the primary considerations are the oxygen requirements, the physical limitations of oxygen transfer equipment, required energy, and the energy cost to sustain an aerobic system. In general, an aerobic reaction is simple to operate, able to sustain itself over a wide range of operating parameters including some inhibitory conditions, and capable of correcting abnormal conditions over a reasonable time span. Basically, the aerobic microorganisms are "fighters" and "survivors." However, the waste organic concentration, the corresponding oxygen requirements, and the practical economical limitations will limit the use of aerobic systems. Theoretically, enough aeration capacity can be included to assure aerobic conditions for a wide range of feed concentrations, but from an energy and operating cost aspect aerobic suspended growth systems are limited to treatment of wastes at concentration of 3000 BOD_5 mg/L or less, most often in the 500-mg/L range for large waste volumes. Above 3000 BOD_5 mg/L an anaerobic system, with the possible fuel recovery advantage, becomes a viable alternative. See Chapter II-7 for anaerobic treatment selection considerations; aerobic treatment selection logic continues subsequently.

- Step 5: Dispersed or fixed growth system?

The significant factors in considering dispersed versus fixed film growth are organic loading, oxygen rate, excess sludge generation, and operating flexibility. Dispersed growth systems can be more effectively controlled because aeration capacity, degree of mixing, sludge concentration, and recycle can be independently controlled and adjusted to changing influent conditions. Fixed-film growth rates are oxygen limited, based on the natural oxygen diffusion rate into the waste, defined by the equilibrium concentration. Dispersed growth rates can be supplied with enough oxygen through a wide range of oxygen demand rates. In addition, excess sludge can be easily disposed in dispersed systems but can present a plugging problem in some forms of fixed-film systems. As a result fixed-film biological systems are generally applied at organic concentrations less than 400 to 500 mg/L.

- Step 6: Dispersed growth alternative: batch or continuous treatment?

Dispersed growth batch systems offer a high degree of operating flexibility, and the opportunity to treat until an acceptable treatment level is achieved, as long as the treatment can be completed within an acceptable cycle time. The restrictive parameter is the flow volume and the installation of reasonable size and manageable batch reactors. Based on an economical treatment volume, dispersed growth batch reactors are generally restricted to treatment flows of less than 3785 m^3/day (1 MGD).

- Step 7: Dispersed growth alternative: continuous activated sludge system.

Continuous activated sludge treatment offers the highest degree of process control for a wide range of influent concentrations. Theoretically, acceptable reactor loadings (F/M of 0.2 to 0.5) can be maintained at any substrate concentration by increasing reactor volume. However, practical MLVSS limitations of less than 5000 mg/L and economical reactor size considerations limit influent biodegradable organic concentrations to less than 1000 mg/L for large waste volumes. At flow rates less than 1 MGD substrate concentrations up to 3000 mg/L can be considered, with large reactor volumes and correspondingly high sludge handling facilities required. Acceptable influent loading decreases rapidly with increasing influent flow volume.

The ability to individually control process variables makes the system applicable to a wide variety of industrial wastes and allows implementing control measures during upset conditions. It is also the most complex and expensive system to design, construct, and operate. A wide variation of available configurations allows tailoring the system design to expected waste conditions. However, design criteria must be carefully selected to allow an acceptable range of operating conditions. Unlike fixed-film systems, where the selection of loading rates and tower height (or number of discs) can be used to estimate the equipment size, correctly or incorrectly, each process variable must be individually selected in an activated sludge system. This is because each variable is controlled by a separate piece of equipment.

In addition, the configurations selected must be carefully "tailored" to the waste characteristics and the reaction kinetics desired. As an example, the reactor can be designed for plug flow or completely mixed, contact stabilization or conventional processing, step aeration or conventional, atmospheric air or pure oxygen operation, diffused air or mechanical aeration, and enhanced with carbon addition.

- Step 8: Dispersed growth alternative: oxidation ponds or carousels

Oxidation ponds cover a wide group of once-through suspended growth processes. The specific configuration depends on the natural and mechanical oxygen capacity and the lagoon size. Aerated lagoons allow some control of oxygen input by controlling the mechanical device output, allowing the operator some performance control. Aerobic stabilization pond performance is less predictable, and considerably less dependable, because of the dependence on natural aeration. Lagoons are designed on the basis of two predominant physical characteristics, depth and area. The selection of these parameters fixes the SRT and HRT.

(1) Aerated lagoons allow for deep basin design, within the limits of the aeration device. Stabilization ponds are shallow to maximize energy (sunlight) penetration to the basin bottom.

(2) All lagoons operate at SRTs equal to the HRT and therefore at limited and uncontrollable F/M contact.

(3) Because a lagoon is designed based on the HRT it requires a large land area and multiple units to achieve required performance.

Carousels are suspended growth systems operated as extended aeration processes and are generally employed for small wastewater flows. Some available operating data indicate that, as long as BOD loadings of less than 160 g/m^3 (10 lb BOD_5 per 1000 ft^3) of reactor volume per day are maintained, effluent BOD and suspended solids levels of less than 10 and 25 mg/L can be achieved [4].

- Step 9: Fixed-film growth treatment

Three reactor configurations are generally employed as fixed-film treatment systems: trickling filters, rotating biological discs, and fixed bed reactors. Trickling filters are the most common fixed film systems, applied to a wide range of waste characteristics. Space considerations generally limit the practical number of units that can be applied, and sludge control can sometimes be a problem because of restricted media passages.

Rotating biological contractors offer a wide flexibility because they can be installed in module form. In addition, RBC systems offer the opportunity to evenly distribute the organic loading, and the corresponding oxygen requirement, by tailoring the number of discs per stage. Oxygen to RBC units can be supplemented by introduction of air into the vessel contents.

Packed beds are an enhancement of the trickling filter, designed to operate with supplementary air, allowing the system to more effectively operate with varying loads and seasonal temperature changes. In addition, packed beds are commonly operated as sparged air systems, resulting in fluidized bed conditions, allowing for media self-cleaning. Packed beds can also be adapted to anaerobic conditions by operating in an upflow condition, adaptable to anoxic systems such as denitrification. Considerable research is being conducted to enhance the state-of-the-art of packed beds and develop reliable continuous performing systems. The Process Engineer must carefully establish required pilot data to obtain specific design and performance for industrial wastes.

- Step 10: General engineering criteria

Viable alternatives selected on the basis of meeting process criteria for biological oxidation must be subjected to related engineering considerations, which may limit or eliminate some of the choices. Important considerations include

(1) Process reliability
(2) Process efficiency
(3) State-of-the-art, as defined by the number of full-scale operating systems
(4) Fate of contaminants
(5) Pilot effort required to obtain design criteria
(6) Land requirements
(7) Other physical limitations
(8) Complexity of the system
(9) Manpower requirements
(10) Costs
(11) Utility and energy requirements
(12) Compatibility with existing plant and future requirements

Process reliability and operating flexibility are significant considerations in evaluating viable biological alternatives. Can the individual process variables be adjusted to respond to anticipated or emergency process conditions?

- Step 11: Tertiary treatment requirements

Discharges from biological treatment systems could require further treatment mandated by regulatory requirements reflecting the water quality standards governing the final receiving streams. The most predominant control required is nutrient control for both phosphorous and nitrogen reduction.

Effluent imposed limitations could include

(1) Total nutrient reductions, including both phosphorus and nitrogen removal
(2) Specific priority pollutant limits
(3) Specific metal concentrations

(4) Total DO level
(5) pH level
(6) Suspended solids level
(7) Dissolved solids level
(8) Total solids level
(9) Color or turbidity

REFERENCES

1. Baird, R. et al.: "Behavior of Benzidine and Other Aromatic Amines in Aerobic Wastewater Treatment," *Journal WPCF,* V 49, No 7, Pg 1609, July, 1977.

2. Barton, D.A.: "Intermedia Transport of Organic Compounds in Biological Wastewater Treatment Processes," *Environmental Progress,* V 6, No 4, Pg 246, November, 1987.

3. Barton, D.A., McKeown, H.J.: "Field Verification of Predictive Modeling of Organic Compound Removal by Biological Wastewater Treatment Processes," *Environmental Progress,* V 10, No 2, Pg 96, May, 1991.

4. Benefield, L.D., Randall, C.W.: *Biological Process Design for Wastewater Treatment,* Prentice-Hall, 1980.

5. Blackburn, J.W., Troxler, W.L.: "Prediction of the Fates of Organic Chemicals in a Biological Treatment Process—An Overview," *Environmental Progress,* V 3, No 3, Pg 163, August, 1984.

6. Blackburn, J.W.: "Prediction of Organic Fates in Biological Treatment Systems," *Environmental Progress,* V 6, No 4, Pg 217, November, 1987.

7. Chambers, C.W. et al.: "Degradation of Aromatic Compounds by Phenol-Adapted Bacteria," *Journal WPCF,* V 35, No 12, Pg 1517, December, 1963.

8. Cheremisinoff, P.N.: *Encyclopedia Environmental Control Technology,* Gulf Publishing Co., 1989; Volume 3, Chudoba, J. (Chapter 6): "Activated Sludge-Bulking Control."

9. Chudoba, J. et al.: "Control of Activated Sludge Filamentous Bulking. I. Effect of the Hydraulic Regime or Degree of Mixing in an Aeration Tank," *Water Research,* V 7, Pg 1163, 1973.

10. Chudoba, J. et al.: "Control of Activated Sludge Filamentous Bulking. II. Selection of Microorganisms By Means of a Selector," *Water Research,* V 7, Pg 1389 1973.

11. Chudoba, J. et al.: "Control of Activated Sludge Filamentous Bulking. III. Effect of Sludge Loading," *Water Research,* V 8, Pg 231, 1974.

12. Chudoba, J. et al.: "Control of Activated Sludge Filamentous Bulking—Experimentatal Verification of a Kinetic Selection Theory," *Water Research,* V 19, No 2, Pg 191, 1985.

13. Chudoba, J. et al.: "Control of Activated Sludge Filamentous Bulking-VI, Formulation of Basic Principles," *Water Research,* V 19, Pg 1017, 1985.

14. Eckenfelder, W.W. Jr.: *Principles of Water Quality Management,* CBI Publishing Company, 1980.

15. Eikelboom, D.H.: "Filamentous Organisms Observed in Activated Sludge," *Water Research,* V 9, Pg 365, 1975.

16. Gerhold, R.M., Malaney, G.W.: "Structural Determinants in the Oxidation of Aliphatic Compounds by Activated Sludge," *Journal WPCF,* V 38, No 4, Pg 562, April, 1966.

17. Grady, A.F., Grady, E.T.: *Microbiology for Scientists and Engineers,* McGraw-Hill Book Co., 1980 (1963).

18. Hartmann, L., Laubenberger, G.: "Influence of Turbulence on the Activity of Activated Sludge Flocs," *Journal WPCF,* V 40, No 4, Pg 670, April, 1968.

19. Hobson, T.: "Process Control Fundamentals, Microscopic Examination of Activated Sludge and Control of Aeration Rates," *Operations Forum,* Pg 22, November, 1986.

20. Kincannon, D.F., Gaudy, A.F. Jr.: "Some effects of High Salt Concentrations on Activated Sludge," Proceedings of 20th Industrial Waste Conference, May 1965, Purdue University.

21. Lawrence, A.W., McCarty, P.L.: "Unified Basis for Biological Treatment Design and Operation," *Journal Environmental Engineering, Proceedings ASCE,* V 96, Pg 757, 1970.

22. Lehninger, A.L.: *Principles of Biochemistry,* Worth Publishers, Inc., 1982.

23. Govind, R., Lei Lai, Dobbs, R.: "Integrated Model for Predicting the Fate of Organics in Wastewater Treatment Plants," *Environmental Progress,* V 10, No 1, Pg 13, February, 1991.

24. Ludzack, F.J., Ettinger, M.B.: "Chemical Structures Resistance to Aerobic Biological Stabilization," *Journal WPCF,* V 32 No 11, Page 1173, November, 1960.

25. Lyman, W.J. and Rosenblatt, D.H.: Handbook of Chemical Property Estimation Methods, McGraw Hill Book Company, 1982

26. Malaney, G.W., et al.: "Resistance of Carcinogenic Compounds to Oxidation by Activated Sludge," *Journal WPCF,* V 39, No 12, Pg 2020, December, 1967.

27. Mills, E.J. Jr., Stack, V.T. Jr.: "Biological Oxidation of Synthetic Organic Chemicals," Proceedings of Industrial Waste Conference, 1953, Purdue University.

28. Malaney, G.W., Gerhold, R.M.: "Structural Determinants in the Oxidation of Aliphatic Compounds by Activated Sludge," *Journal WPCF,* V 41, No 2, Pg R18, February, 1969.

29. Medcalf & Eddy, Inc.: *Wastewater Engineering-Treatment, Disposal, Reuse,* McGraw-Hill, 1991, Third Edition.

30. McKinney, R.E.: *Microbiology for Sanitary Engineers,* McGraw-Hill Book Co., 1962.

31. Gaudy, A.F. Jr., Kincannon, D.F.: "Comparing Design Models for Activated Sludge," *Water and Sewage Works,* Pg 64, February, 1977.

32. Namkung, E., Rittman, B.E.: "Estimating Volatile Organic Emissions from Publicly Owned Treatment Works," *Journal WPCF,* V 59, No 7, Pg 670, July, 1987.

33. Neufield, R.D., Hermann, E.R.: "Heavy Metal Removed by Acclimated Activated Sludge," *Journal WPCF,* V 47, No 2, Pg 310, February, 1975.

34. Neufield, R.D., et al.: "A Kinetic Model and Equilibrium Relationship for Heavy Metal Accumulation," *Journal WPCF,* V 49, No 3, Pg 489, March, 1977.

35. Patterson, J.W., Kodukala, P.S.: "Biodegradation of Hazardous Organic Pollutants," *Chemical Engineering Progress,* Pg 48, April, 1981.

36. Pitter, P.: "Determination of Biodegradability of Organic Substances, *Water Research,* V 10, Pg 231, 1976.

37. Porter, J.J., Snikder, E.H.: "Long-Term Biodegradability of Textile Chemicals," *Journal WPCF,* V 48, No 9, Pg 2198, September, 1976.

38. Rebhun, M., Galil, N.: "Inhibition by Hazardous Compounds in an Integrated Oil Refinery," *Journal WPCF,* V 60, No 11, Pg 1953, November 1988.

39. Richard, M.: "Activated Sludge Microbiology," *Water Pollution Control Federation,* 1989.

40. Rickard, M.D., Gaudy, A.F. Jr.: "Effect of Mixing Energy on Sludge Yield and Cell Composition," *Journal WPCF,* V 40, No 5, Pg R129, May, 1968.

41. Sawer, C.N., McCarty, P.L.: *Chemistry for Sanitary Engineers,* McGraw-Hill Book Company, 2nd Edition, 1969.

42. Stover, E.L., Kincannon, D.F.: "Biological Treatability of Specific Organic Chemicals Found in Chemical Industry Wastewaters," *Journal WPCF,* V 55, No 1, Pg 97, January 1983.

43. Tabak, H.H., et al.: "Biodegradability Studies with Organic Priority Pollutant Compounds," *Journal WPCF,* V 53, No 10, Pg 1503, October 1981.

44. Truong, K.N., Blackburn, J.W.: "The Stripping of Organic Chemicals in Biological Treatment Processes," *Environmental Progress,* V 3, No 3, Pg 143, August, 1984.

45. U.S. Environmental Protection Agency: *CERCLA Site Discharges to POTWs Treatability Manual,* EPA-540/2-90-007, August, 1990.

46. U.S. Environmental Protection Agency: *Design Guides for Biological Wastewater Treatment Processes,* 11010 ESQ 08/71, August 1971.

47. U.S. Environmental Protection Agency: *Process Design for Nitrogen Control,* PB-259-149/38A, October 1975.

48. U.S. Environmental Protection Agency, Gaudy, A.F. Jr., Gaudy, E.T.: *Biological Concepts for Design and Operation of the Activated Sludge Process,* Project 17090 FQJ, September 1971.

49. Vandevenne, L., Eckenfelder, W.W. Jr.: "A Comparison of Models for Completely Mixed Activated Sludge Treatment Design and Operation," *Water Research,* V 14, Pg 561, August, 1979.

50. Volskay, V.T. Jr., Grady, C.P.L. Jr., and Tabak, H.H.: "Effect of Selected RCRA Compounds on Activated Sludge Activity," *Research Journal WPCF,* V 62, No 5, Pg 654, July/August, 1990.

51. Volskay, V.T., Grady, C.P.L.: "Toxicity of Selected RCRA Compounds to Activated Sludge Microorganisms," *Journal WPCF,* V 60, No 10, Pg 1850, October, 1988.

52. Witmayer, G., et al.: "Effects of Salinity on a Rendering-Meat Packing-Hide Curing Wastewater Activated Sludge Process," *WPCF Penna. Magazine,* Pg 13, November/December, 1987.

53. Wood, D.K., Tchobanoglous, G.: "Trace Elements in Biological Waste Treatment," *Journal WPCF,* V 47, No 7, Pg 1933, July, 1975.

54. WEF Manual of Practice: *Design of Municipal Wastewater Treatment Plants,* Water Environment Federation, 1992.

55. WPCF Manual of Practice 11: *Operation of Wastewater Treatment Plants,* Water Pollution Control Federation, 2nd Printing, 1985.

56. WPCF Manual of Practice 8: *Wastewater Treatment Plant Design,* Water Pollution Control Federation, 2nd Printing, 1982.

57. Wu, Y.C., et al: "Control of Activated Sludge Bulking," *Journal Environmental Engineering, Proceedings ASCE,* 110, No 2, p 472, April, 1984.

58. Lesperance, T.W.: "A Generalized Approach to Activated Sludge," *Water Works and Waste Engineering,* Pg 52, May, 1965; Pg 34, July, 1965.

Activated Sludge System

Activated sludge treatment is primarily employed to destroy large quantities of biodegradable organics.

IGURE 3.1 illustrates the basic activated sludge treatment system in which influent enters a biological reactor containing mixed liquor suspended solids, remaining in contact with the solids for the average vessel retention time. Reactor flow containing 1500 to 4000 mg/L of solids is discharged to a clarifier. Treated effluent containing residual substrate and some suspended solids overflows from the clarifier. Thickened solids is recycled from the clarifier to the reactor. Because the process produces excess sludge, a portion of the clarifier bottoms is removed from the system.

BASIC CONCEPTS

Basic concepts for aerobic biological oxidation are discussed in Chapter II-2, which should be reviewed to understand the technical details presented in this chapter. Historically, municipal activated system solids retention time (SRT) and loading ranges are related to specific system configurations, as cited in Table 3.1 [12,25].

The relationship between food-to-microorganism (F/M) ratio or SRT and settling characteristics is illustrated in Figure 3.2, with three traditional operating regions (high rate, conventional, and extended aeration) producing acceptable settling identified. Improved loading effects resulting from increased temperature are indicated by a solid line.

Although loading rate can be used for process definition, SRT is a preferable criterion to define activated sludge system operating and design conditions. Because it is difficult to measure the quantity of microorganism required to define the F/M ratio, volatile suspended solids (VSS) concentration is usually assumed a direct measure of this quantity. The SRT minimizes the microorganism measure limitation, making the VSS mass a reliable parameter, as seen by the definition:

$$\text{SRT} = \frac{f \cdot \text{VSS stored}}{f \cdot \text{VSS wasted}} = \frac{\text{VSS stored}}{\text{VSS wasted}}$$

Defining f as the active microorganism fraction of the VSS, and reasonably constant in the system, the SRT is easily approximated as the ratio of the system and wasted cell concentrations. In addition, the SRT is easier to relate to the cell growth rate, the predominant consideration in the biological system.

REACTOR PROCESS KINETICS

The use of reactor process kinetic models for activated sludge design involves two critically important considerations, (1) identifying a reaction kinetic model to define the cell growth (and substrate utilization) rate, consistent with the industrial waste characteristics, and (2) selecting a reactor configuration to optimize the process. Biological reaction models were discussed in Chapter II-2. What is critical is that (1) the model limits and assumptions be understood and (2) that appropriate cell growth, half-life constants, and cell decay (or the equivalent kinetic) constants be selected. Once these basic process considerations have been applied to the specific industrial application, design details can be developed.

Significantly, many models were developed utilizing municipal waste or single substrate feeds so that reported growth constants cannot be assumed to be directly applicable to all industrial waste systems. Adjustment to the model constants must be made for the specific industrial waste characteristics. In most, if not all, cases, this means direct measurement of waste treatability parameters in pilot evaluations to establish the model constants. Next, the reactor configuration must be

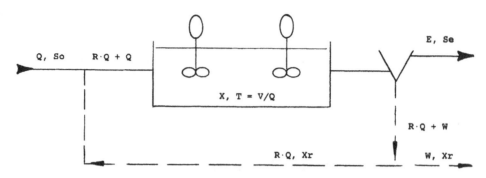

Figure 3.1 Activated sludge sketch.

established, based on two ideal limiting models, a completely mixed or a plug flow. The factors involved in this selection are complex and are discussed in the Process Design Variables section.

System Balances

Process performance relations can be developed using the following material balance around the biological reactor:

Mass accumulation or mass utilization =
reactor input − reactor output ± reaction formation   (3.1)

This can be expressed as follows:

Substrate utilization rate = $dS/dt \cdot V$

$$\text{or} \qquad (3.2)$$

Net cell production (R_{cn}) = cell growth − cell digestion

$$= dX/dt \cdot V - k_d \cdot V$$

The cell growth or substrate utilization kinetic models selected will define the reactor operating conditions. The

expressions discussed in this chapter adhere to those proposed by Lawrence and McCarty, as defined by Equation (2.4) in Chapter II-2.

Completely Mixed System with Recycle

A completely mixed system assumes (ideally) a uniform composition of substrate, oxygen, and suspended solids throughout the reactor, with all feed materials being immediately blended in the reactor. Under such conditions, the discharge stream has the same composition as the reactor contents. Lawrence and McCarty developed the following equations to define a process operating as a completely mixed system with sludge recycle [12]:

$$S = \frac{K_s \cdot [1 + k_d \cdot \text{SRT}]}{\text{SRT} \cdot [Y \cdot k - k_d] - 1} \qquad (3.3a)$$

$$X = \frac{Y \cdot [S_o - S]}{1 + k_d \cdot \text{SRT}} \cdot \frac{\text{SRT}}{\text{HRT}} \qquad (3.3b)$$

$$P_x = \frac{Y \cdot Q \cdot [S_o - S]}{1 + k_d \cdot \text{SRT}} \qquad (3.3c)$$

$$\frac{1}{\text{SRT}} = \frac{Y \cdot k \cdot S}{K_s + S} - k_d \qquad (3.3d)$$

TABLE 3.1. Activated Sludge Configurations Performance Data (adapted from References [12, 25]).

	Feed Loading kg BOD₅ fed/ kg MLVSS/day	Efficiency %	SRTd days	SRT_d/ SRT_m	Removed kg BOD₅/ kg MLVSS/day
Extended aeration (complete mix)	0.05–0.2	75–95	20–30	>70	0.04–0.19
Conventional (plug flow)	0.2–0.5	85–95	5–15	20–70	0.17–0.48
Tapered aeration (plug flow)	0.2–0.5	85–95	5–15	20–70	0.17–0.48
Step aeration (plug flow)	0.2–0.5	85–95	5–15	20–70	0.17–0.48
Contact stabilization (plug flow)	0.2–0.5	80–90	5–15	20–70	0.16–0.45
Pure oxygen (complete mix)	0.2–1.0	85–95	8–20		0.17–0.95

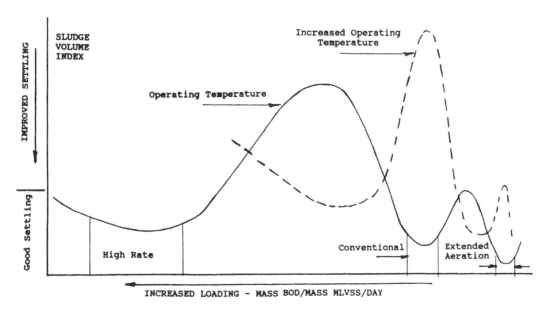

Figure 3.2 F/M performance curve (adapted from Reference [29]).

where HRT is the hydraulic retention time, time, Q is the influent volume/time, Y is the yield ratio, mass cells produced/mass substrate used, S_o is the feed substrate concentration, mass/volume, S is the reactor substrate concentration, mass/volume, k is the maximum substrate utilization rate, 1/time, K_s is the substrate concentration when dX/dt is 0.5 k, X is the reactor microbial mass concentration, mass/volume, k_d is the microorganism decay coefficient, 1/time, and P_x is the excess microorganism rate, mass/time. All units must be consistent.

Utilizing the general equation for SRT and assuming that the influent and effluent exit conditions are equal under washout conditions, the SRT for *washout conditions* (SRTw) can be defined as follows:

$$\frac{1}{\text{SRT}_w} = \frac{Y \cdot k \cdot S_o}{K_s + S_o} - k_d \qquad (3.4)$$

The *limiting SRT* (SRT$_l$) can be established for first-order reaction conditions, when $S_o \gg K_s$, so that

$$\text{SRT}_l = \frac{1}{[Y \cdot k - k_d]} \qquad (3.5)$$

The use of these relations should be undertaken with the understanding that the ideal conditions assumed are never achieved because the energy cost would be prohibitive. In full-scale systems some substrate leakage will occur; therefore, adjustment must be made to the estimated quantities.

Plug Flow Reactor

A plug flow system represents an idealized flow pattern devoid of mixing and in which an element (plug) is subjected to increasing residence time proportional to the distance traveled from the inlet to the outlet. This is typically symbolized as a long rectangular tank with the inlet at one end, a feed element traveling the length of the tank reacting with the biomass and oxygen, and exiting at the opposite end. As in the case of an idealized completely mixed reactor, a "true" plug flow process cannot occur because some turbulence will be encountered as a result of the aeration devices and the need to assure some intimate contact between the vessel reactants.

Lawrence and McCarty offered a series of design equations for a plug flow reactor, based on the microorganism concentration in the reactor not significantly changing, the SRT/HRT being greater than 5 and the recycle ratio being less than 1. Under such conditions the equation for reactor solids, excess solids production, and HRT are similar to those for a completely mixed system. Equations for effluent waste concentration are difficult to arrive at mathematically but can be approximated by assuming an infinite number of completely mixed systems in series.

The general equation for SRT, based on the recycle ratio being less than one, can be expressed as follows:

$$\frac{1}{\text{SRT}} = \frac{Y \cdot k \cdot (S_o - S)}{(S_o - S) + K_s \cdot \ln (S_o/S)} - k_d \qquad (3.6)$$

All variables are as previously defined. Equations describing the minimum or limiting SRT are mathematically difficult to define.

Selected Model

Most design texts present suspended growth design equations based on completely mixed operation, which whether

TABLE 3.2. Industrial Waste Kinetic Constants
(adapted from Reference [28]).

		K_d, 1/hr	Y, mg/mg	
(1)	Brewery		0.44	BOD$_5$
(2)	Poultry	0.03	1.32	BOD$_5$
(3)	Pulp and paper mill	0.0083	0.47	BOD$_5$
(4)	Pulp and paper kraft	0.0015		
(5)	Refinery	0.0104	0.53	BOD$_5$
(6)	Shrimp process	0.0667	0.5	BOD$_5$
(7)	Soybean	0.006	0.74	BOD$_5$
(8)	Textile, combined operation	0.030	0.38	COD
(9)	Textile, polyester dying	0.05	0.32	COD
(10)	Textile, wool dying	0.04	0.69	COD
(11)	Textile, nylon	0.0014	0.25	BOD$_5$
(12)	Thiosulfate	0.000417–0.000833	0.029–0.035	COD
(13)	Vegetable and food processing	0.00117–0.00792	0.32–0.88	COD
(14)	Whey, cottage cheese	0.00229	0.40	BOD$_5$

always accurate or not can be justified on the basis that (1) plug flow conditions cannot be achieved because of aeration basin mixing and (2) criteria developed on the basis of a completely mixed system will be more conservative than plug flow reactor estimates. Because some safety factor is recommended in developing the design basis from either complete mix or plug flow estimates, where the "judgment factor" is incorporated is academic. Therefore, the use of design models must be based on the realization that neither completely mixed or plug flow conditions can be achieved, some degree of axial mixing occurs in all reactors, and the calculated result is an approximation that must be factored for expected conditions.

Finally, considerable details have been presenting describing *loading factors, reaction kinetics,* and *design models* extensively utilized in municipal plant design. The indiscriminate application of any model is risky for industrial wastes whose characteristics are not defined and dependent on varying manufacturing conditions. The kinetic and design relations discussed are presented because the Process Engineer will be frequently confronted with them and should therefor understand their origin and limitations. They are not presented to suggest a substitute for specific waste related data. Significantly, the accuracy of reactor design estimates is highly dependent on the constants selected or determined from pilot studies. For the equations cited these include the Y, K_s, k_d, MCRT, and HRT values. As a basis for comparison,

relative values for municipal and industrial wastes are indicated in Table 3.2 through 3.5 [2,12,14,28].

Values for K_s vary over a wide range for different wastes and even for the same waste classification. For this reason, engineering estimates are best based on the SRT (reciprocal of the cell growth), defined by the equation:

$$1/\text{SRT} = Y \cdot U_a - k_d \qquad (3.7)$$

where U_a is the applied F/M ratio defined by Equation (2.14); all variables are as defined previously. Where first-order kinetics can be applied, engineering estimates can be conducted assuming the kinetic constant values indicated in Table 3.5 [2].

CELL GROWTH VIABILITY

Selected design criteria must assure cell growth in the quantity and quality required for effective substrate reduction and to develop a settleable floc. This requires an understanding of the cell growth and substrate utilization dynamics discussed in Chapter II-2, represented by the expression

$$\mu = \frac{Q \cdot Xe + W \cdot X_r}{X \cdot V} = \frac{1}{\text{SRT}} \qquad (3.8)$$

TABLE 3.3. Domestic Wastes Monod Constants
(adapted from References [2, 14]).

k, 1/day	2–10
K_s, mg/L BOD	25–100
K_s, mg/L COD	15–70
Y, mg VSS/mg BOD$_5$	0.4–0.8
Y, mg VSS/mg COD	0.35–0.45
k_d, 1/day	0.025–0.1

TABLE 3.4. Domestic Wastes Nitrification
Constants For Pure Cultures @20°C
(adapted from References [14, 21]).

k, 1/day	0.3–3
K_s, mg/L NH$_4^-$N	0.2–5
k_d, 1/day	0.03–0.06
Y, mg VSS/mg NH$_4^-$N	0.1–0.3

TABLE 3.5. Industrial Waste First Order Kinetic Constants
(adapted from Reference [28]).

Waste Type	k, 1/mg-hr	Component
(1) Ammonia base, semichemical	0.00046	BOD_5
(2) Brewery	0.00022	BOD_5
(3) Chemical industry	0.00014–0.00020	BOD_5
(4) Coke plant, ammonia liquid	0.00110	COD
(5) Organic chemical	0.00005–0.000073	BOD_5
(6) Petrochemical	0.00024–0.00028	BOD_5
(7) Pharmaceutical	0.00021–0.00057	BOD_5
(8) Phenolic	0.000092	BOD_5
(9) Paper and pulp	0.000417	BOD_5
(10) Refinery	0.00035	BOD_5
(11) Rendering plant	0.0015	BOD_5
(12) Tetraethyl lead	0.00071	BOD_5
(13) Textile, nylon	0.00015	BOD_5
(14) Thiosulfate	0.00011–0.00021	COD
(15) Vegetable oil	0.00031	BOD_5

What should be obvious from this equation is that when the Process Engineer arbitrarily or deliberately selects an SRT (or F/M ratio) as a design basis, a *cell growth rate* is selected. If the SRT chosen is not greater than the minimum required to develop a viable cell growth (carbonaceous or nitrifying) for the operating plant range, the reactor will "washout," and no substrate will be utilized.

Lawrence and McCarty suggested applying the minimum SRT as the basis of design [12]. This is defined as the point of initial washout out, where the cell removal first exceeds cell generation and the effluent concentration is equal to the influent concentration. Equations (3.4) and (3.5) represent appropriate equations for estimating minimum and limiting SRT. A minimum SRT for plug flow involves a mathematical relation too complex for practical design application. A conservative approximation can be made assuming a completely mixed system or a series of completely mixed reactors. In any estimate of applied SRT, the key word is approximation. The estimated value must be converted to a design SRT applying a safety factor, that is, to assure that cell growth conditions are met in full-scale facilities. Lawrence and McCarty suggested the concept of a safety factor, defined as:

$$\text{Safety factor} = \text{design SRT/minimum SRT} \quad (3.9)$$

Some suggested safety factors are given in Table 3.6.

ACTIVATED SLUDGE SYSTEMS

A variety of activated sludge systems, defined by the aeration basin configuration, has evolved from vast municipal waste treatment experience. These systems designed to economically treat large volumes of relatively consistent wastes are the principal treatment units of many industrial waste treatment complexes. Reactor selection requires matching the waste characteristics, process requirements, and the reactor configuration. Generally, in an activated sludge system the significant process factors include:

- Sludge quality control
- Process stability
- Energy requirements
- Reactor size
- Short circuiting
- Effects of influent variation

Activated sludge configurations will be described as employed for municipal systems and applied to industrial treatment [14,25]. However, it should be emphasized that in practice these reactors are not so distinctly defined, most operating in a "nonideal" regime. In reality, site restrictions affect selected reactor dimensions, reactor dimensions affect aeration and mixing, and the degree of turbulence affects the reactor kinetics, all of which are cost driven.

TABLE 3.6. SRT Design Safety Factors
(adapted from Reference [12]).

	SRT_d/SRT_m design/minimum
Carbonaceous Oxidation	
Extended aeration	>70
Other activated sludge configurations	20–70
Short term aeration	4–20
Nitrification	
Minimum safety factor of 2, based on SRT_d/SRT_{lim}; with SRT_d at least 10 days	

COMPLETELY MIXED REACTORS

A completely mixed activated sludge system (Figure 3.3 and Table 3.7) assumes immediate dispersion of feed substrate and return sludge into the reactor liquid contents. Under ideal conditions uniform aeration basin liquor contents are achieved, equal to the vessel discharge. One of the major considerations in developing completely mixed systems was the ability to maintain relatively uniform reactor contents, with a minimum variation, over a wide range of feeds.

The predominant characteristics of a completely mixed system are

(1) Immediate dispersion of feed streams

(2) Dilution of high content substrate

(3) Dilution of influent toxic substances to minimize their effects

(4) Good contact of reactor solids, substrate, and oxygen

(5) Effective oxygen utilization resulting from constant demand throughout the reactor

(6) Inherent equalization capabilities

(7) Potential bleed-through of unmixed and unreacted influent

(8) Low reactor substrate concentrations favoring pinpoint cell growth and the potential for poor sludge quality

Design Characteristics

A completely mixed system reactor size will normally be larger than other systems, although theoretically total volume could be reduced by using units in series. Practical design considerations limits the number of units employed because of (1) construction costs, (2) the removal rates in successive reactors could decrease in proportion to remaining substrate concentration, and (3) at some point the number of units in series will effectively result in plug flow kinetics with each unit representing a plug in the reactor. The physical design of a CMAS process depends on specific site limitations and the aeration devices selected. With mechanical aeration, the length-to-width ratio should be less than 3 to 1, no greater than 5 to 1. Long, narrow configurations, whether utilizing mechanical or diffused aeration, are commonly designed by

TABLE 3.7. CMAS Municipal Design Characteristics (adapted from Reference [14,25]).

F/M, kg BOD applied per kg MLVSS-day	0.2–1
SRT, days	1–15
HRT, hr	3–5
MLSS, mg/L	1500–6000
Return sludge, %	25–100
BOD removal, %	85–95

integrating a pattern of aeration "blocks," simulating a series of completely mixed units.

Sludge quality control is a common problem with industrial wastes because of low reactor substrate concentration and resulting loading or as the result of poor operating conditions. This promotes filamentous growth, final clarifier bulking sludge, and the inability to control sludge recycle or aeration basin solids level. A series of designs, referred to as *selector design,* has been suggested to minimize or eliminate this problem. This subject is covered in detail in the Process Design Variables section.

CONVENTIONAL CONTINUOUS OPERATION

The alternative to CMAS is a flow-through aeration system (plug reactor), with feed and return sludge introduced into one end and effluent discharged from the other (Figure 3.4 and Table 3.8). Plug flow system reaction kinetics are similar to a batch system, with substrate reduction a function of time, which is a function of the reactor length and the distance traveled by a plug. The oxygen demand is high at the inlet, decreasing with reactor length, being minimal at the exit. Ideally, a plug flow reactor operates with each segment independent of the upstream or downstream conditions.

These systems commonly employ diffused air aeration devices, which permit varying the diffuser density or output to better match oxygen demand. Practical design considerations involving total required volume, site limitation, and vessel construction costs often result in multiple basins (or compartments), operated in parallel and series. The resulting multistage smaller volume basins, particularly when in series, results in more turbulence than an ideal plug path.

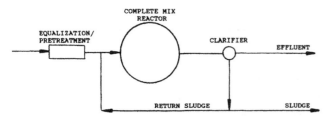

Figure 3.3 Complete mix configuration.

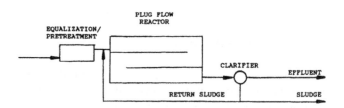

Figure 3.4 Conventional system configuration.

Conventional System Municipal Design Characteristics (adapted from Reference [14,25]).

F/M, kg BOD applied per kg MLVSS-day	0.2–0.4
SRT, days	3–15
HRT, hr	4–8 hours
MLSS, mg/L	1500–3000
Sludge return, %	25–75
BOD removal, %	85–95

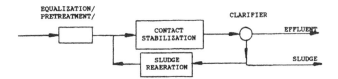

Figure 3.5 Contact stabilization configuration.

Tapered aeration and step feed are common modifications employed to improve aeration effectiveness.

- *Tapered aeration* incorporates an air supply system designed to better distribute oxygen in proportion to the waste load. More oxygen is supplied to the feed inlet where the load is highest and less at the discharge end.
- *Step feed* involves a different approach to air distribution, injecting the waste load at various points throughout the tank to more effectively utilize oxygen and reduce installed air capacity. Because the influent reaction time varies with the point of injection, this system has limited application in industrial wastes, unless the waste biodegradability is equal to, or better than, municipal wastes.

Many of the operating characteristics of a completely mixed system apply to a conventionally configured system, the exceptions being that the plug flow operation:

(1) Offers better protection against the bleed-through potential of unmixed influent
(2) Reduces the potential for filamentous growth because of the higher substrate concentration at the inlet section
(3) Results in poor surge protection against high influent concentration or toxic shock
(4) Results in poor oxygen utilization or distribution, unless a sophisticated configuration or dissolved oxygen (DO) control system is used

Sludge quality control is less of a problem than with completely mixed systems because a high concentration substrate is in contact with the recycle sludge at the inlet, discouraging filamentous bacterium growth. In fact, first-stage, plug flow configuration is one of the selector design alternatives to enhance completely mixed systems.

Physical Design

Theoretically, the plug flow reactor size is smaller than a completely mixed system. In practical process design, the reactor volumes for a completely mixed and plug flow system are generally not dramatically different. Plug flow reac-

tor has become an expression of operating philosophy, basin geometry, or both as opposed as an approach to minimizing basin size. However, a plug flow is an economically superior process to aerate a final stage or process segment to achieve a final dissolved oxygen effluent requirement (as opposed to aerating a completely mixed volume) or for continuation to nitrification treatment.

CONTACT STABILIZATION

Contact stabilization (Figure 3.5 and Table 3.9) was incorporated in municipal design to remove large biological oxygen demand (BOD) quantities in short contact periods. BOD is physically removed from the waste and adsorbed by the floc, but not stabilized, in a relatively small primary contact vessel. The primary contact vessel discharge flows to a clarifier where the sludge is settled, and the settled sludge is stabilized in a separate tank. The total vessel volumes are usually less than that of the conventional aeration basin. The process is effective if a significant quantity of BOD is in the colloidal state, with required contact time increasing with increased soluble substrate, until the advantages over a conventional system diminishes. Where applicable, the resulting basin volumes could be 30 to 40% less than that required in an extended aeration system. Because most industrial wastes involve high dissolved and little suspended BOD, and component biodegradability can vary, contact stabilization is generally not considered viable. Application is also limited because of the potential of unadsorbed material leaking to the effluent (in the sludge separation step) or inadequate time to stabilize poor biodegradable substrate components.

Contact Stabilization Municipal Design Characteristics (adapted from Reference [14,25]).

F/M	0.2–0.6
STR, days	5–15
Contact HRT, hr	0.5–1, forward flow
Stabilization HRT, hr	3–6, recycle flow
MLSS, mg/L	1000–3000
Contact, mg/L	4000–9000
Sludge return, %	50–150
BOD removal, %	80–90

SEQUENTIAL BATCH REACTORS

A sequential batch process involves performing all the activation sludge operations in a *single vessel*. The operation involves the following sequences:

- Influent is fed into the reactor, with the air supply and mixing off. The anoxic fill strategy allows a favorable microorganisms selection.

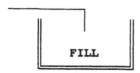

- The biological oxidation is initiated by starting the air flow, and mixing the reactor contents. The batch reaction continues until effluent quality is achieved.

- The vessel contents are allowed to settle at the completion of the reaction.

- After the solids settle, effluent is withdrawn from the vessel.

- The vessel awaits new waste. Sludge is wasted to maintain the required loading rate.

Sequential batch reactors (SBRs) are basically fill-and-draw systems operating on a batch basis. These systems have come into their own for industrial waste volumes of less than 3800 m³/day (1 MGD), based on some favorable inherent characteristics (Table 3.10):

TABLE 3.10. **SBR Municipal Design Characteristics** (adapted from Reference [14]).

F/M kg BOD applied per kg MLVSS/day	0.05–0.30
HRT, days	12–50
BOD removal, %	85–95

(1) SBR can be tailored to produce a maximum quality effluent by constantly monitoring the contents to the point that an acceptable effluent is achieved.

(2) The *optimum* microorganism population can be cultivated by adjusting the loading rates.

(3) Loading rates can be adjusted by controlling the reactor sludge inventory.

(4) Clarification equipment is eliminated.

(5) Influent short circuiting is eliminated.

The primary variable in a SBR system is time, as opposed to flow rate in a continuous system. By being able to continuously monitor each operation, a better effluent quality can be achieved, time permitting! Capital cost savings and controlled performance are the basis for selecting this system. The saving is in the clarifier and the sludge recycle system. This must be balanced against the required vessel(s) sizes because all operations must be completed in time to accommodate the waste generated and avoid excessive accumulation. The vessel can be decreased by reducing the batch sizes, operating multiple batches per day, but at increased operating costs. The potential savings in capital costs for batch systems may be offset by the fact that the same waste volume ranges may easily be treated in preengineered, package treatment units. These package units are completely integrated equalization, aeration, clarification, and sludge handling units. In some cases these units could be cheaper (overall) than batch systems.

Integrated Nitrification/Denitrification

A SBR system can include nitrification and denitrification. However, elaborate design or operating modifications may be required to accommodate the three reactor conditions necessary for carbonaceous oxidation and total nitrogen removal. As an alternative, a fixed-film aerobic system can be employed as a separate final nitrification stage, as illustrated in Figure 3.6.

As before, wastewater in the vessel is prepared for rapid uptake of the carbonaceous organics, with the reaction stage stopped short of completion. Waste is then pumped to a rotating biological disc or trickling filter for nitrification and returned to the vessel. At this point a denitrification reaction is initiated, with the remaining unmetabolized substrate acting as the carbon for the nitrification step under anaerobic conditions. This system, when combined with multiple batch reactors, greatly enhances operating flexibility.

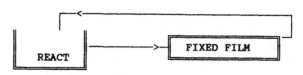

Figure 3.6 Complex SBR treatment system.

Engineering Considerations

Designing the system to accommodate the various operating steps in a reasonable time frame is critical. Typically, times for the various operations when treating municipal waste are indicated in Table 3.11. Industrial facilities, or those employed for remediation, would typically operate within the same time ranges as those cited for municipal systems. However, the reaction time would be sensitive to the waste biodegradability and in some cases extend the total cycle time. As an example, a reported groundwater remediation SBR system employed a 24-cycle time; with 6 to 10 hr of fill, 10 to 12 hr of reaction, 1 to 2 hr of settle, $\frac{1}{2}$ to 5 hr of decant, and $\frac{1}{2}$ to 1 idle hr [26].

Biological reaction time is the important consideration in establishing the vessel size. However, some of the cycle time must be allocated for the fill, discharge, settling, and idle (preparation) steps. The vessel fill and discharge times are related to the design influent and discharge rate, which are affected by the selected pump size and cost, or the available hydraulic head for gravity systems. The settle time depends on the sludge quality, idle time on required preparation time. The remaining cycle time is that available for biological treatment, the length of which establishes the batch size, required number of daily batches, or both. Vessel sizing is critical because it establishes treatment capacity. Operating flexibility can be increased by using two vessels, allowing time for recovering from any upset.

Batch sequences can be completely automated. However, whether the process is manually or automatically controlled, operating time will be required to monitor the intermediate stages. If quality control time is not included, with the ability to override any automated step, an entire treated batch could be out of compliance. This could result in costly reprocessing or disposal.

OXIDATION DITCHES

Oxidation ditches employ a rectangular basin, with a circular racetrack type configuration, into which the feed is introduced in one sector (Table 3.12). The liquid flows in a circular pathway under continuous aeration, and the mixed liquor discharged in the opposite end to the feed. Wastewa-

TABLE 3.12. Oxidation Ditch Municipal Design Characteristics (adapted from Reference [14]).

F/M, kg BOD applied per kg MLVSS/day	0.05–0.3
SRT, days	10–30
HRT, hr	8–36
MLSS, mg/L	1500–5000
Sludge removal, %	70–150
BOD removal, %	75–95

ters circulate at 0.2 to 0.3 m/s (0.8 to 1 fps). The structure is designed with a hydraulic gradient, which with the surface aeration pumping pattern results in a continuous movement of the waste from the inlet to the outlet. The unit operates as an extended aeration process, at an HRT of 24 hr and SRT ranging from 20 to 30 days, accommodating low organic loadings. The factors favoring oxidation ditches include:

(1) The potential for reduced capital and operating costs.
(2) The system is suitable for combined carbonaceous, nitrification, and denitrification.
(3) The system is applicable to wastes requiring long retention times.

PURE OXYGEN SYSTEMS

As the name implies, this system employs high-purity oxygen in place of air. Although the system is physically similar to conventional activated sludge systems, the manufacturers promoting these systems report that the system can operate at mixed liquor solids concentrations of 5000 mg/L or higher, with reduced retention times, more effective treatment efficiencies, less sludge, and improved sludge quality (Table 3.13). The units have been considered for upgrading existing municipal systems and could prove useful for high concentrated industrial wastes with a high oxygen demand. Reduced mixing requirements resulting from reported smaller aeration basin volumes can result in a more economical system. Potential cost reductions resulting from small aeration and clarification equipment must be balanced against the added cost of enclosing the aeration basin to minimize oxygen loss and the cost of the oxygen production and storage equipment. Not enough independent operating

TABLE 3.11. SBR Cycle Times (adapted from Reference [1,11]).

	Maximum % of			
	Volume	Cycle Time	Air	Municipal
Full	25–100	25	On/off	1–3 h or continuous
React	100	35	On/cycle	1–3 h
Settle	100	20	Off	1–3 h
Draw	100–35	15	Off	3/4–1 h
Idle	35–25	5	On/off	3/4–1 h
Total				typical 5–8 h

TABLE 3.13. Pure Oxygen Municipal Design Characteristics (adapted from Reference [14,25]).

F/M, kg BOD applied per kg MLVSS/day	0.25–1
SRT, days	3–10
HRT, hr	1–3
MLSS, mg/L	3000–8000
Sludge return, %	25–50
BOD removal, %	85–95

data are available to adequately predict when oxygen systems would be more economic or more effective than air systems. Evaluations conducted cite a variety of conclusions as to their applicability [14,25].

PACT™ SYSTEMS

The PACT™ system is a modification of the activated sludge system with provisions for carbon addition to the aeration basin, with the following potential process advantages:

(1) Improved removal of toxic or priority pollutants
(2) Improved operational stability for wastewaters with varying composition or concentration
(3) Improved chemical oxygen demand (COD) and BOD removal
(4) Effective color removal
(5) Improved solids settling
(6) Suppressed stripping of volatile organics

Carbon addition in the PACT™ system can be effective in reducing toxicity within a biological system. The carbon acts as an adsorbent to enhance removal of inhibitory or toxic components, reducing the effects of certain toxic compounds. When installed solely to improve effluent quality, the system is technically similar to adding a tertiary activated carbon system. The difference is for the most part an economical one, comparing the costs of separate tertiary activated carbon and regeneration with direct carbon addition and disposal.

One effect of carbon addition is to reduce the system's mixed liquor volatile suspended solids (MLVSS) capacity, a vital reactant in an activated sludge system. Activated carbon dosages as high as 2000 to 3000 mg/L have been utilized in PACT™ systems, which greatly reduces the active solids to 1000 to 2000 mg/L. As a result, the clarifier solids loading could be increased, and a larger excess waste sludge could be generated. Control of an activated sludge system has to be modified to adjust sludge recycle, provide for increased sludge wasting, and increased aeration solids loading to account for the inert (carbon) portion.

The system is best implemented in a completely mixed system where the reactor concentrations are low and the adsorbent is more effectively and economically applied.

If the carbon were subjected to large concentrations of organics, such as the head of a plug flow reactor, the carbon usage could be high because biological oxidation would not have been completed. The best addition in a plug flow type reactor would be at the outlet sections. Because this is a proprietary system, evaluation and design are best left with the suppliers.

PROCESS ENGINEERING DESIGN

The activated sludge system offers the operator a high degree of operating control and process flexibility, as illustrated in Table 3.14. However, its selection is greatly affected by the complex design involved to optimize each of the working elements and the higher installation and operating costs relative to some competitive biological processes. Although high treatment efficiencies are common, performance projections must be realistically based on achievable industrial waste effluent quality. A consistent high-quality effluent over a range of expected influent conditions requires that the influent variability be considered in the process design.

REPORTED PERFORMANCE DATA

Activated sludge systems are commonly designed for municipal plant removal rates, assuming that design efficiency is a "fixed value" resulting from good or poor operation. In reality, treatment performance is a product of all the influent and operating plant variations, some of which are imposed by design limitations or deficiencies. As a result, effluent quality is more likely a range specific to waste and treatment plant characteristics. As a frame of reference, results of approximately 100 industrial facilities reported in the U.S. Environmental Protection Agency (EPA) *Treatability Manual* are summarized in Table 3.15 [22].

The data reported are for industries such as paper mills, petrochemicals, iron and steel, hospitals, leather, textile, organic chemicals, pharmaceutical, rubber, and resin. For the most part the data reported are for full-scale facilities treating wastes from these manufacturing facilities and represent operating results. As a frame of reference, annual BOD and suspended solids municipal system performance data for specific configurations are cited in Table 2.3, in Chapter II-2. In addition, performances associated with specific configurations, commonly cited for municipal systems, are detailed in the activated sludge systems section.

These data are presented to demonstrate the range of performance possible with no assurance that the facilities reported were properly designed or operated, that these results are the best that can be achieved, or that these results can be achieved with other facilities where waste loading or characteristics are considerably different. Significantly, the effluent quality will be influenced by the suspended solids content, which will contribute to the BOD_5 and COD

TABLE 3.14. Activated Sludge Operating Characteristics.

Variable	Operator Controllable	Control Critical?
Waste Characteristics		
Quantity generated	No	Yes
Composition	No	Yes
Concentration	No	Yes
Biodegradability	No	Yes
Toxicity	No	Yes
Operating Characteristics		
Flow rate	Minimal	No
SRT	Yes	Yes
HRT	Depends on flow rate	No
MLVSS level	Yes	Yes
Operating Characteristics		
Sludge recirculation	Yes	Yes
Mixing	Within design	—*
Oxygen capacity	Within design	Yes
Wasting	Yes	—**
Nutrients	Yes	Yes
Alkalinity	Yes	Yes
Extreme temperatures	No	Yes

* Absolute level not critical, minimum required to assure reactants contact and provide oxygen.
** Must be balanced over a long run to assure that removal is equivalent to generation to avoid solids accumulation or depletion.

concentrations. This reinforces the notion that performance is a measure of the removal rate of the entire system and not only the dissolved organics in the aeration basin.

REQUIRED PROCESS DESIGN DATA

Industrial wastewater treatment design criteria should be developed from pilot studies conducted with specific manufacturing wastes to identify site related criteria, developing a range of acceptable operating conditions. Data generated from pilot studies are commonly collated on the basis of cell growth and substrate removal *kinetic models* combined with system configuration stoichiometric balances establishing a mathematical relation between growth rate and reaction time. Activated sludge design requires selecting parameters to assure a viable system, directly related to optimizing microorganism population and growth rates, thereby establishing effective *reactor* and *clarifier* conditions. The required design data are listed in Table 3.16.

WASTE CHARACTERISTICS

The most significant factors controlling treatment performance are the waste characteristics, their variation, and the presence of components detrimental to biological oxidation. Although a design is based on an assumed flow and composition, the influent range and limitations must be considered, with any large variances corrected by upstream equalization and pretreatment.

TABLE 3.15. Activated Sludge Industrial Performance Data
(adapted from Reference [22]).

	BOD₅	COD	TOC	TSS	O/G	TTL Phenol	TKN	TTL P
				Effluent, mg/L				
Minimum	5	45	35	6	<5	0.007	27	0.14
Maximum	4640	7420	1700	4050	303	500	322	47
Median	49	425	280	92	25	0.028	174	3.5
Mean	184	890	427	283	71	18.7	174	6.7
				Removal, %				
Minimum	17	0	8	0	6	0	26	0
Maximum	>99	96	95	96	>98	>99	63	97
Median	91	67	69	25	92	64	44	27
Mean	86	63	63	34	74	60	43	32

TABLE 3.16. **Required Design Data.**

Critical pilot plant treatability data specific to the waste
(1) Design temperature
(2) k_d coefficient, 1/days
(3) k_{max} dS/dt/wt, 1/time
(4) K_s concentration at 1/2 max, mg/L
(5) Y, cell yield, mg/mg

Waste solids characteristics that should be obtained from laboratory studies but can be estimated
(6) Fraction of effluent solids biodegradable
(7) Ultimate solids BOD per g solids
(8) Ratio of SBOD to ultimate BOD
(9) Ratio MLVSS/MLSS

Selected operating characteristics
(10) Required effluent quality
(11) Mixed liquor volatile suspended solids
(12) MRCT, calculated from treatability data
(13) Clarifier underflow MLSS, %

Operating characteristics that should be obtained from pilot studies but can be estimated from treatability data
(14) Summer temperature effects
(15) Winter temperature effects
(16) Minimum aeration basin DO, mg/L
(17) α factor, Kl_a waste/tap water
(18) β, C_s waste/C_s tap water
(19) Required oxygen loading
(20) Waste sludge generated
(21) Nitrogen required per pound SBOD removed
(22) Phosphorus required per pound SBOD removed

Waste Feed Flow Volume

Total waste volume is commonly estimated as the accumulative sum of identifiable plant waste sources and their estimated daily contribution; although individual waste volumes may be intermittent, infrequent, or spontaneous. The problem confronting the Process Engineer is establishing a design flow by identifying all contributing waste sources. This includes nonproduction waste sources such as utility blowdowns, dewatering streams, recycle streams, and laboratory discharges. Neglected process streams are an obvious problem. However, selecting too conservative a design basis is also a concern that the Process Engineer must consider because in biological treatment bigger is not always better. An incorrect design flow can carry cumulative errors throughout the system design because the flow value is used to calculate SRT, sludge wasting, sludge processing, aeration requirements, and recycle.

Waste Feed Concentration

Substrate concentration directly affects the reactor size, treatment efficiency, aeration requirements, and excess sludge produced. Industrial wastes are commonly low volume and high strength. BOD concentrations can range from less than 100 mg/L to greater than 1000 mg/L, much more variant and higher strength than municipal wastewaters.

However, waste volumes are controlled production discharges, usually much lower than those encountered in municipal systems. In such cases, high substrate concentration is compensated by increased reactor volume, diluting the reactor concentration to an acceptable loading. However, somewhere in the 3000-mg/L range the loading significantly affects reactor volume, construction, aeration equipment sizes, and sludge handling costs. Figure 3.7 illustrates the effect of influent concentration on HRT, and thereby reactor volume.

As illustrated, the required aeration volume increases proportionally to the influent substrate concentration. At an F/M ratio of 0.2 and an influent flow of 3800 m³/day (1 MGD), 9500 m³ (2.5 MG) of reactor volume is required for an influent substrate of 1000 mg/L, the volume further increasing with increasing influent flow. Significantly, the reactor volume can be decreased by increasing the loading (F/M from

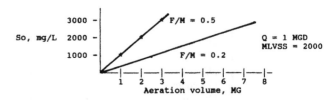

Figure 3.7 Influent concentration effects on aeration volume.

0.2 to 0.5) or reactor MLVSS (from 2000 to 5000). However, increased reactor loading decreases treatment efficiency, and increased mixed liquor solids level decreases the available process operating range.

Besides the reactor volume, high substrate concentration affects the system design in a variety of other ways:

(1) High influent concentration maximizes the aeration requirements at the reactor inlet, so that in many cases completely mixed systems are employed to dilute the effects. In conventional systems anaerobic conditions could occur if sufficient aeration is not applied at the inlet, frequently resulting in the entire system being over aerated.

(2) As the substrate loading increases aeration volume, aerator equipment sizing criteria shifts from oxygen to mixing requirements.

(3) Increased loading results in higher sludge generation, increasing the size, capital costs, and operating costs of sludge management equipment.

Because industrial biological treatment systems are commonly related to municipal performance, process differences should be emphasized, the major industrial differences being influent BOD or COD concentrations, BOD/COD ratios, the potential for high concentrations of inorganic or toxic components, and unpredictable substrate biodegradability. In fact, some waste characteristics or components may limit or preclude biological system consideration. These differences should emphasize that *all BODs are not created equal*, and using general waste treatment parameters without evaluation or modification could result in an erroneous design basis.

PROCESS DESIGN VARIABLES

Activated sludge process design involves applying the biological basic concepts discussed to the specific waste characteristics, in a series of sequential evaluations, to develop a complete system. The process steps include evaluating:

(1) System configuration
(2) Carbonaceous oxidation SRT
(3) Nitrification SRT
(4) Aeration basin mixed liquor solids level
(5) Aeration basin volume
(6) Aeration requirements
(7) Mixing
(8) Sludge wasting
(9) Final clarification
(10) Denitrification SRT
(11) Fate of contaminants

System Configuration

Developing an activated sludge treatment plant configuration involves

(1) Establishing the nitrification requirements
(2) Selecting a basic process configuration for cabonaceous oxidation
(3) Enhancing the basic configuration to optimize performance
(4) Establishing denitrification requirements

Nitrification

Where influent total nitrogen is considerably greater than the process nutrient requirement nitrification can occur, by design or as an inherent part of the reactor chemistry. Nitrification must be evaluated in the process design because it impacts oxygen requirements and sludge generation. When ammonia nitrification is part of the process design, the appropriate system configuration must take into account the corresponding organic and nitrogen levels. When the BOD_5 to TKN ratio is greater than 5, combined treatment is applicable; at less than 3, two-stage or separate treatment is employed.

Process Configuration for Carbonaceous Oxidation

A description of the available activated sludge systems is discussed in the section entitled Activated Sludge Processes and will not be repeated here. Process selection involves consideration of influent waste volume and concentration, compensating for the presence of toxic, inhibitory, or refractory constituents. These influent characteristics can be used as a *guide* in process selection.

- SBR is a viable process for treatment volumes less than 3800 m³/day (1 MGD) and for maximum treatment flexibility.

SBRs are a viable consideration for any industrial waste because of the operating flexibility inherent in the system to "tailor" a treatment cycle to specific waste characteristics, producing a quality effluent on a "batch-to-batch" basis. An SBR treatment cycle can be expanded to include nitrification and denitrification, as needed. Capital and operating costs are the primary limitation to this process. Some applicable considerations include

(1) Capital cost increases with reactor volume, with the size of single units limited by practical construction and mixing limitations. Multiple parallel reactors will greatly increase capital costs, eventually offsetting savings in sludge separation and handling equipment. Generally, the transition from batch to continuous treatment is at 3800 m³/day (1 MGD), with facilities reportedly op-

erating at 8 to 3100 m³/day (2000 to 829,000 GPD) [1,11].

(2) Although not offering the process flexibility, prefabricated activated sludge package units may be economically competitive to SBR units.

(3) The operating cost of manually operated batch systems will have to be compared with increased capital costs resulting from highly automated units.

(4) Unless treatment flexibility is the only consideration a critical economical evaluation between an SBR and continuous system must be conducted. Potential batch treatment capital savings could be quickly dissipated with increasing reactor volume, multiple units, automation, and increased piping complexity. Where a highly manual operating system is contemplated, operating costs must be carefully balanced with apparent (instrumentation) capital savings.

• Continuous treatment systems must be considered when daily waste volumes exceed 3800 m³/day.

Continuous treatment systems are the most commonly employed activated sludge systems, especially when waste volumes exceed 1 MGD, with the numerous configuration discussed in the Activated Sludge Processes section available.

• Complete mix systems are viable if the influent BOD₅ exceeds 1000 mg/L and to dilute the effects of toxic or inhibitory constituents. (See waste characteristics for limitations).

Completely mixed systems are commonly employed for industrial wastes, especially when the influent concentration is above 500 mg/L BOD_5, and exceed levels of 1000 mg/L. This configuration tends to alleviate many of the problems encountered with plug flow reactors by providing instant influent dilution and oxygen dispersion in reactor. The major process problem associated with these reactors involves low reactor concentration associated with influent dilution, potentially causing endogenous digestion conditions that could promote filamentous growth. Extensive studies have been conducted to establish causes for filamentous growth. One cause commonly mentioned is low reactor food-to-cell ratio. Based on this assumption, *selector design* should be studied as part of the inlet reactor system, as discussed subsequently.

In addition, the results of high mixing required for these systems must be carefully evaluated, some of which include

(1) Energy costs increases with increased mixing.

(2) Excessive mixing could shear the floc, causing poor settling.

(3) Increased mixing provides significant protection to varying, shock, and high influent loads. This is a major consideration in industrial facilities where influent concentrations are commonly high, and there could be a high degree of variability in (or spills to) the influent. In such cases, the benefits of increased reactor mixing have to be compared with upstream equalization, which provides both flow and loading protection.

(4) Increased mixing increases the system dilution capabilities, which could be beneficial in minimizing the effects of toxic or surges of inhibitory components.

• Plug flow configuration is viable if the influent BOD₅ is less than 500 mg/L and protection from influent bleed-through is desirable.

At influent BOD_5 values less than 500 mg/L a plug flow configuration permits a reasonable concentration gradient to be maintained in the reactor, allowing a moderate distribution of oxygen demand from the entrance to exit conditions. In addition, the configuration reduces the potential of substrate bleed-through because the long narrow configuration acts as a series of completely mixed units. However, employing a plug flow configuration is highly contingent on controlling the effects of influent surges, especially those containing inhibitory or toxic constituents that could radically reduce treatment effectiveness. Because mixing is minimal, the vessel must be designed with careful consideration to maximize biomass and substrate contact. In addition, reactor *enhancement* may be required to minimize process ineffectiveness resulting from oxygen deficiency in the early stages, overaeration in the latter stages, or both.

• PACT™ systems are viable if the influent contains large refractory concentrations.

PACT™ Systems are a proprietary process combining adsorption with biological treatment. This is a viable consideration when (1) toxic or priority pollutant effluent quality requirements are difficult to achieve, (2) influent toxic pollutants are detrimental to maintaining a stable biological process, (3) the system must be stabilized by "trimming" peak influent loads, and (4) volatile organic releases must be suppressed. The cost of makeup carbon and disposal of spent carbon is an important process and economical consideration in selecting this process.

• Oxidation ditches are an economic consideration in applying plug flow configuration and total nitrogen control in one facility.

Oxidation ditches are a viable alterative to achieve a plug flow, extended aeration configuration, accommodating nitrification and denitrification in the final sections. These advantages must be balanced with the lower loading and corresponding higher land requirements common with these systems. Oxidation systems may be applied as a final tertiary system to treat high concentration wastes to achieve reduced effluent BOD, COD, specific toxic organics, and total nitrogen concentrations.

• Pure oxygen systems are a viable alternative to optimize performance in special cases or to increase the capacity of an existing facility.

Pure oxygen systems have to be evaluated on a case-by-case basis, balancing increased capital and operating costs resulting from oxygen generation and upgraded aeration basin design against the *potential* for reduced savings resulting from increased performance or reduced reactor size.

Reactor Enhancements

Deficiencies cited for completely mixed or conventional configuration can be reduced by reactor design enhancement to "fine tune" the reactor performance.

• Complex mix systems prone to filamentous growth performance can be improved by eliminating detrimental feed and operating conditions or incorporating selector design alternatives.

Stabilizing a biological oxidation process requires aeration basin conditions producing a settleable floc capable of being recycled to control reactor solids level. This can be accomplished by minimizing filamentous bacterium growth, found in all activated sludge systems, a major cause of bulking sludge and a predominate problem in many completely mixed industrial treatment systems [9]. Bulking sludge, as the name implies, will not compact, diminishing recycle control. Before filamentous growth can be addressed in reactor design, their formation should be understood. Chudoba et al. proposed the following basic explanation of possible causes of filamentous bulking [3–8]:

(1) Filamentous microorganisms are present in all activated sludge, bulking problems result when their growth rate exceeds that of the floc formers.
(2) Filamentous microorganism growth is intensified by some wastewater components, low DO concentration, low reactor substrate concentration, low sludge loading (F:M) and high sludge age (SRT). Wastewater components favorable to filamentous growth include glucose-like saccharides such as glucose, saccharase, lactose, and maltose.
(3) Filamentous microorganisms are slow growers, whereas floc-forming cells are fast growers. Low substrate concentrations (less than 30 mg/L soluble degradable COD), resulting in low loading and high sludge age, have been identified as a major concern in intensifying filamentous growth. As illustrated in Figure 3.8, at low loadings the slow-growing filamentous microorganisms will remove substrate at a higher rate than the fast growing floc formers.
(4) Dominant filamentous growth rate will occur in completely mixed systems because of inherent low substrate concentrations in the reactor, whereas floc-forming mi-

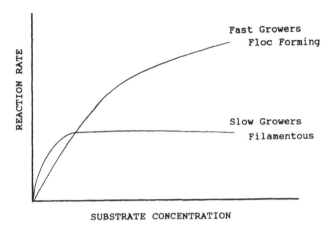

Figure 3.8 Filamentous-floc forming cell growth rate (adapted from Reference [7]).

croorganisms will dominate in plug flow configurations because of a substrate gradient allowing suppression of filaments. Settleable sludges with sludge volume index values of less than 100 were observed in hydraulic regimes approaching plug flow from multiple tests conducted with reactors operated at near plug, complete mix, and intermediate mix conditions [3].
(5) Finally, it should be noted that filamentous growth has also been attributed to low pH (less than 6), nutrient deficiency (nitrogen and phosphorus), and septic wastes (sulfide) [18].

Based on the potential problems with completely mixed configurations three broad reactor augmentations should be considered in designing such systems. First, waste characteristics should be screened for glucose-like saccharides components, high sulfide septic conditions, and toxic components. Individual streams should be screened and pretreated to remove (or neutralize the effects of) these components. Second, the design should assure an environment that suppresses filamentous growth by assuring an adequate oxygen level, a suitable pH range, and sufficient nutrients. Finally, where waste control is not feasible, provisions should be made to employ alternative configurations or include a first-stage process *selector.*

The work of Chudoba and other investigators has been summarized as guidelines for municipal treatment system alternatives to complete mix systems. These alternatives are cited in municipal design texts and include evaluating the following options [14,25]:

(1) A plug flow reactor to assure contact with the highest level substrate at the process inlet, encouraging the selective growth of flocculating sludge
(2) SBRs, where the F/M ratio can be chosen to ensure process stability
(3) Intermittent operation of a continuous system at oxygen deficient conditions, encouraging flocculating bacteria

growth that can exist under aerobic and anaerobic conditions, and destroying filamentous bacterium that tend to survive under aerobic conditions

Where these alternative configurations are not applicable, the complete mix basin design should be enhanced to minimize bulking sludge conditions. This involves adding a selector process, a separate initial reactor (*selector*) to cultivate flocculating and suppress filamentous, bacterium growth. The resulting cultivated population can then be fed to the main aerobic reactor to promote both high process efficiency and a settleable sludge. A *selector* can be a separate reactor or a segregated portion of the biological reactor, designed for high substrate to microorganism contact. The *selector* simulates a plug flow reactor, designed as either as a series of completely mixed reactors operating under controlled microorganism growth conditions or a segregated primary sections designed for near plug flow hydraulic conditions operating at a dispersion number below 0.2. The remaining reactor volume is designed to suit the desired configuration, as illustrated in Figure 3.9.

An example of reported selector design criteria can be summarized as follows [8]:

(1) The selector chambers should be capable of removing about 80% of the influent degradable COD.

(2) It should contain at least four compartments.

(3) It should accommodate a volumetric BOD_5 loading of 10 kg/m³/day.

(4) It should operate at a MLSS concentration of 3.3 kg/m³.

(5) It should accommodate a sludge loading of 3 kg/kg/day, based on BOD_5.

(6) It should have an oxygenation capacity of 4 kg/m³/day.

These parameters should assure a first compartment BOD_5 volumetric loading of 40 kg/m³/day and a sludge loading of

12 kg/kg/day. The whole aeration system design should fall within the following parameters:

(1) Volumetric BOD_5 loading of 1 kg/m³/day

(2) MLSS concentration of 3.3 kg/m³

(3) Sludge loading, based 0.3 kg/kg/day, based on BOD_5

(4) Oxygenation capacity of 2 kg/m³/day

Other investigators propose other selector criteria to suppress filamentous and promote floc-forming bacterial growth [8,25].

In fact, selector design is an art rather than a science and has not developed past suggested guidelines, which when applied to the investigator's study conditions proved effective. The reader should review the cited references for specific design details. It should be noted that a consensus has not been reached that selector design produces a viable biological system. This is because poor microorganism growth cannot always be attributed to substrate loading. Other causes include poor operating conditions and waste constituents inhibiting, retarding, or destroying microorganisms or "other possible causes." The Process Engineer should evaluate specific selector criteria in pilot studies when sludge bulking is evident and allow segregation of an initial stage of the aeration basin, allowing conversion to multiple chamber reactors, if necessary. As a final word, it is not astounding to those engaged in biochemistry, employed in industries such as the pharmaceutical or fermentation, that cultivation of specific bacterium could improve biological treatment; it is common practice in these industries to maximize yields and product quality.

- Plug flow configuration performance may be improved by step or tapered aeration to optimize oxygen utilization.

Oxygen distribution is a legitimate concern in plug flow configurations, requiring consideration of *tapered aeration* to balance oxygen requirements with the substrate gradient. *Step aeration* is another alternative, allowing feed at various points to balance oxygen availability. However, step aeration must be cautiously implemented for industrial wastes containing toxic or moderately biodegradable wastes because any introduction near the discharge point increases the potential for substrate "bleed-through," compounded because of reduced mixing and reaction time.

Denitrification Requirements

Where total effluent nitrogen level limits are imposed denitrification must be included as a final treatment step to reduce nitrates to elemental nitrogen. Design considerations are discussed in the denitrification section.

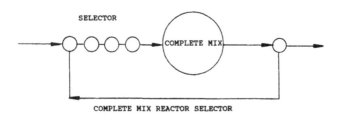

COMPLETE MIX REACTOR SELECTOR

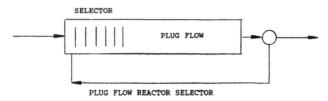

PLUG FLOW REACTOR SELECTOR

Figure 3.9 Selector configurations (adapted from Reference [4]).

Carbonaceous Oxidation SRT

Carbonaceous oxidation is discussed in the Basic Concepts section of this chapter, and in Chapter II-2. Applicable design relations can be summarized as follows:

(1) Cell growth rate relationship

$$\mu = 1/SRT = Y_b\, Q_b - k_d \qquad (3.10)$$

where Y_b is the kg of VSS growth per kg substrate removed, Q_b is substrate removal rate, kg BOD removed per kg VSS/day, k_d is the decay coefficient, 1/day, and μ is the net heterotrophic cell growth rate, kg VSS growth/day/kg VSS.

$$\mu = 1/SRT = Y_b\, E\, (F/M) - k_d \qquad (3.10a)$$

where F/M is the kg of substrate fed/day kg of VSS aerated and E is the substrate removal efficiency, kg of substrate removed per kg fed.

(2) A related kinetic equation for a completely mixed system is

$$\mu = 1/SRT = Y\,[(k\, S_i)/K_s + S_i)] - k_d \qquad (3.11)$$

Design conditions for an activated sludge system are based on evaluating washout and limiting conditions, washout conditions being defined as the point when the effluent concentration (S_e) is equal to the influent concentration (S_o), the limiting conditions, as applied to the Monod equation, being defined as the washout conditions in which S_o is much greater than K_s. Equations (3.4) and (3.5) define washout and limiting conditions for a completely mixed system.

Design conditions are established by applying a safety factor to the washout condition (SRT_m) or to the more conservative limiting condition (SRT_l). The relationship among the design SRT_d, the minimum SRT_m, and the limiting SRT_l is as follows:

$$SRT_d > SRT_m > SRT_l \qquad (3.12)$$

The safety factor (SF) is applied to the minimum SRT_m as indicated by Equation (3.12a).

$$SRT_d = SRT_m \cdot SF \qquad (3.12a)$$

A more conservative design is based on applying a safety factor to the limiting SRT_l, defined by Equation (3.12b).

$$SRT_d = SRT_l \cdot SF \qquad (3.12b)$$

TABLE 3.17. Municipal System SRT Values (adapted from Reference [14]).

Conventional plug system and modifications	3–15 days
Complete mix	1–15
Extended aeration	20–30
Oxidation ditch	10–30
High-purity oxygen	3–10

Safety factors proposed for municipal systems are indicated in Table 3.6. An excessive safety factor results in increased capital and operating costs, whereas too low a safety factor does not assure a stable system. SRTs employed for municipal design can vary according to the specific process configuration, generally in the range indicated in Table 3.17 [14].

Nitrification SRT

Nitrification basics are discussed in the Aerobic Biological Oxidation section of Chapter II-2; applicable design relations can be summarized as follows:

(1) Cell growth can be represented by the basic equation,

$$\mu_N = \acute{\mu}_N \cdot \frac{N}{K_n + N} \qquad (3.13)$$

(2) The corresponding nitrification rate can be represented by the relations:

$$dN/dt = \mu_N/Y_n \qquad (3.13a)$$

$$dN/dt = \acute{q}_N \cdot \frac{N}{K_n + N} \qquad (3.13b)$$

$$\acute{q}_N = \mu_N/Y_n \qquad (3.13c)$$

(3) The nitrifier cell growth rate, corrected for temperature, pH, nitrogen concentration, and DO concentration, can be represented by Equation (3.13e).

$$\mu_N = 0.047 \cdot [e^{0.098(T-15)}] \cdot [1 - 0.833\,(7.2 - pH)]$$
$$\cdot \frac{N}{N + 10^{(0.051T-1.158)}} \cdot \frac{DO}{DO + KDO} \qquad (3.13e)$$

where μ is the nitrifier growth rate, 1/day, T is the temperature in °C, N is the nitrogen concentration, mg/L, DO is the DO concentration, mg/L, and KDO is the half constant estimated at 1.3 for municipal systems, as reported by the EPA [21], 0.47 is the maximum nitrification rate at 15°C.

(4) The maximum possible nitrification rate at specific temperature, pH, and DO conditions, when N $\gg K_n$, can be represented by Equation (3.13f).

$$\mu_N = 0.047 \cdot [e^{(0.098(T-15))}]$$
$$\cdot [1 - 0.833 (7.2 - \text{pH})] \cdot \frac{\text{DO}}{\text{DO} + \text{KDO}} \quad (3.13\text{f})$$

The specified Cell growth equations and equivalent nitrification rate relations can be employed for nitrification reactor design.

Combined Nitrification System

The cell growth rate [Equation (3.13e)] can be related to the nitrification solids retention time (SRT_e), as with carbonaceous oxidation systems, by the expression:

$$SRT_e = 1/\mu_N \quad (3.13\text{g})$$

The design SRT can be obtained by applying a safety factor to the minimum or limiting SRT.

Separate Nitrification System

Separate nitrification system design can be based on the ammonia nitrogen oxidation rate (q_N), described in Chapter II-2, by Equations (2.24) through (2.27). In such cases the applied nitrification rate (R_N) is equal to the applied nitrification rate (q_N) adjusted by the fraction of available nitrifying microorganisms (f), so that

$$R_N = q_N \cdot f \quad (3.14)$$

where R_N is the nitrification rate, wt ammonia-nitrogen oxidized/wt MLVSS/day, q_N is equal to σ_N/Y_N, where σ is defined by Equation (3.13e) or (3.13f), and f is the fraction of nitrifying organisms in the MLVSS.

The peak nitrification rate is defined as

$$\dot{R}_N = \dot{q}_N \cdot f \quad (3.14\text{a})$$

where $\dot{R}_N$ is the peak nitrification rate and $\dot{q}_N$ is maximum ammonia nitrogen oxidation rate.

The nitrifier fraction (f) can be estimated by the proportional quantity of individual cells:

$$f = M_n/[M_n + M_c] \quad (3.14\text{b})$$

where M_n is the nitrifiers resulting from ammonia nitrification and estimated by $Y_n (N_o - N_e)$ and M_c is the microorganisms resulting from organic carbon oxidation and estimated by $Y_c (S_o - S_e)$, so that

$$f = 1/[(S_o/N_o) (Y_s/Y_n) + 1] \quad (3.14\text{c})$$

Based on this equation, calculated nitrifier fractions in MLVSS, as a fraction of the BOD_5 to TKN ratio are shown in Table 3.18.

General Design Considerations

Regardless of the design method employed, or whether a combined or separate system is utilized, the primary design consideration is accommodating the slower nitrification cell growth rate. Based on this criteria, nitrification is promoted by providing adequate nitrifying cells to assure nitrification process and assuring that the maximum nitrifiers growth rate is greater than that of the carbonaceous cell growth rate. This requires that the design solids retention time (SRT_d) be greater than the minimum design nitrification SRT (SRT_n). This condition can be accommodated by (1) separating the carbonaceous oxidation and nitrification steps (*separate nitrification*) or (2) by increasing the total MLVSS (biological solids) in the reactor, which can be accomplished in *combined nitrification* systems by increasing either the total volume or the reactor MLVSS concentration.

Based on municipal treatment experience, nitrification systems can be classified as separate or combined, based on the influent BOD_5 to TKN ratio. Separate systems operate at a ratio of less than three, with nitrification the dominant process, and the system operating with a predominance of nitrifiers. If the ratio is greater than 5, the system is considered a combined system. SRTs employed for municipal nitrification are generally in the range indicated in Table 3.19 [14].

TABLE 3.18. **Separate and Combined Nitrification Criteria** (adapted from Reference [21]).

	Two-Stage or Separate Treatment					Combined Treatment	
BOD_5/TKN	0.5	1.0	2.0	3.0	4.0	5.0	7.0
f	0.35	0.21	0.12	0.083	0.064	0.054	0.037

TABLE 3.19 **Municipal System Nitrification SRT Values (adapted from Reference [14]).**

Combined nitrification	8–20 days
Separate stage nitrification	15–100

Aeration Basin Mixed Liquor Solids Level

Reactor effectiveness is controlled by maintaining a suitable SRT, which is defined as the solids inventory divided by the wasting rate. Wasting rate is a function of the substrate concentration and the sludge conversion efficiency, both of which are not operating variables. This leaves the solids inventory, or the MLVSS concentration, as the primary operating control. Therefore, MLVSS establishes the system *sludge age, SRT,* and F/M ratio. MLVSS level is controlled by managing the clarifier return flow and concentration to the reactor. The recycle concentration is limited by the clarifier mechanical and process capabilities, which commonly can operate at influent levels up to 5000 mg/L, and achieve a maximum concentration of 10,000 mg/L. The recycle flow is limited by the recycle pump capacity.

Theoretically, the aeration basin solids level is fixed by simply selecting a number between 1500 to 5000 mg/L. The higher the loading, the smaller and less expensive the aeration basin. However, *system operability* must not be neglected. The operator must be able to manipulate the basins solids level to (1) compensate for an expected large increases in the influent loading, (2) increase the loading during periods of low biological activity such as at low temperatures, and (3) increase the inventory of fresh microorganisms during surges of toxic influent.

Basic principles applicable to aeration basin solids level can be summarized as follows:

(1) An increased MLVSS design concentration reduces the required reactor volume but limits the operating range, as illustrated in Figure 3.10.

(2) The reactor biomass capacity determines the quantity of substrate that can be stabilized. Based on the operating F/M loading:

Mass substrate stabilized per day
= F/M · reactor biomass capacity

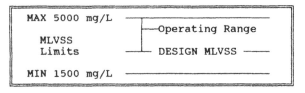

Figure 3.10 Reactor solids operating range.

(3) The aeration basin biomass can be expressed in terms of its capacity by the relationship:

Storage biomass = basin volume · basin concentration

or the equation

$$V X = \frac{Y \cdot Q \cdot \text{SRT} \cdot (S_o - S_e)}{1 + k_d \cdot \text{SRT}} \qquad (3.15)$$

All variables are as defined for Equations (3.3) through (3.5).

Finally, the microorganism level in a reactor is frequently described as either total, biomass, or MLVSS solids. However, it is possible to have a high total or MLVSS solids concentration, consisting of influent organic solids, and have a low active microorganism level. As part of an operating tool, frequent microscopic examinations should be employed to establish an adequate and active microorganism population, and not depend solely on analytical results or deteriorating clarifier performance. In the design stage, MLVSS solids is almost always used as a measure of the microorganism level.

Aeration Basin Volume

Utilizing the SRT as the design parameter, the reactor volume can be expressed as follows:

$$\text{Volume} = \frac{[\text{SRT}] \cdot [Q \cdot (S_o - S_e) \cdot Y]}{X \cdot (1 + k_d \cdot \text{SRT})} \qquad (3.16)$$

All variables are as defined for Equations (3.3) through (3.5).

Therefore, the reactor volume is related to the

- Influent characteristics, flow rate, and concentration
- Effluent concentration, a design requirement
- Conversion of the influent substrate to cells, Y
- Cell growth $(1/\mu)$ or the SRT_d
- Reactor solids concentration, X

Aeration

The basic criterion for sustaining aerobic conditions is maintaining an oxygen level of 1 to 2 mg/L. Generally, DO levels less than 0.5 mg/L are considered too low to maintain a viable aerobic biological system, producing a bulking sludge. The desired oxygen level can only be met by assuring that the carbonaceous and nitrification oxidation requirements are met, as detailed in Chapter II-2, Biological Aerobic Oxidation section. Applicable equations can be summarized as follows:

$$\text{kg O}_2 \text{ per day} = \frac{S_o - S_e}{\tau} - 1.42 \, W_x \qquad (3.17)$$

where, Q is the waste flow in m^3/day, So is the ultimate influent BOD, kg/m^3, Se is the ultimate effluent BOD, kg/m^3, W_x is the mass organisms wasted, VSS, kg/day, τ is a factor converting BOD_5 to BOD_{ult} (0.68).

Additional nitrification requirements can be estimated as follows:

$$NOD = dNH_4N \cdot 4.57 \qquad (3.17a)$$

$$NOD = TKNo \cdot 4.57 \qquad (3.17b)$$

where NOD is the nitrogen oxygen demand, mass per unit time, dNH_4N is the mass of ammonia nitrogen removed in a unit of time, and TKNo is the total Kjeldahl nitrogen in the influent, mass per unit of time.

Mixing

Activated sludge reactor mixing requirements are specific to the configuration employed; the minimum required energy is for a plug flow, and the maximum is for a complete mix system. Basically, the minimum energy that must be imparted is that required to prevent mixed liquor suspended solids settling and assure contact between the reactants and microorganisms. Generally, 20 to 30 m^3 of air per minute per 1000 m^3 aeration volume are adequate for spiral diffused air systems, 10 to 15 for grid air systems. Mechanically aerated systems require 15 to 40 kW per 1000 m^3 of aeration volume (0.6 to 1.5 HP/1000 cf) for completely mixed systems [14,25].

Sludge Wasting

Sludge generation is directly related to the net cell production rate as discussed in Chapter II-2, and defined by Equations (2.15) and (2.16). Expressed in terms of the Monod kinetic expression and the activated system material balance, the net cell production can be expressed as

$$dX/dt = \frac{Y \cdot Q \cdot (S_o - S_e)}{1 + k_d \cdot SRTc} \qquad (3.18)$$

All variables are as defined for Equations (3.3) through (3.5).

Final Clarification

Reactor mixed liquor processing involves a multitude of individual operations, which collectively define the activated sludge system operating stability. These operations include settling and separating the aeration basin discharge into a clarifier effluent overflow and a concentrated sludge suitable for recycle. The clarifier overflow can contain a suspended solids concentration ranging from 5 to 25 mg/L. The concen-

trated solids underflow will range from 6000 to 8000 mg/L, seldom greater than 10,000 mg/L. Part of the thickened sludge, representing the generated (excess) sludge, is wasted from the system; the rest is recycled back to the aeration basin to control MLVSS and SRT.

Final clarification design involves establishing the recycle capacity and determining the clarifier area. Recycle rate is frequently expressed in terms of the fraction of forward flow, denoted as the recycle ratio. The recycle ratio is related to the clarifier underflow (X_r) and required aeration basin (X) concentrations as follows:

$$\text{Recycle ratio } (R) = \frac{\text{recycle flow}}{\text{feed flow}} = \frac{X}{X_r - X} \qquad (3.19)$$

where X is the aeration basin VSS concentration, mg/L, and X_r is the clarifier underflow concentration, mg/L.

X_r/X is a measure of the clarifier performance, which depends on the sludge quality and the clarifier design. Assuming a maximum basin solids of 5000 mg/L and a clarifier underflow concentration of 10,000 mg/L, a recycle rate equal to 100% of the forward feed would be required. Based on a 2000 mg/L basin concentration and an underflow concentration of 10,000 mg/L, a recycle ratio of 0.25 or 25% of the feed flow would be required. However, as the underflow concentration decreases from the 10,000 mg/L level the required recycle flow increases. Poor clarifier performance is a result of poor sludge quality, attributable to poor aeration basin conditions, as discussed throughout this chapter. Significantly, as the biological system becomes unstable the sludge quality deteriorates, resulting in reduced underflow concentration, and an inability to maintain aeration basin solids concentration.

Because aeration basin concentrations generally range from 2000 to a maximum of 5000 mg/L, a minimum recycle ratio of 25% is required, approaching 100% as the basin concentration level reaches 5000 mg/L, exceeding 100% as sludge quality deteriorates. Generally, recycle pumping capabilities of 100% recycle are provided. *Clarification* basics are discussed in detail in Chapter II-8. Design recycle rates generally range from 50 to 100% of the forward feed, with clarifier underflow concentrations generally ranging from 6000 to 8000 mg/L. Applied clarifier overflow rates usually range from 340 to 545 $L/s/m^2$ (500 to 800 GPM/sf), and solids loading applied at 98 to 147 $kg/day/m^2$ (20 to 30 lb/d/sf).

Denitrification SRT

Effluent nitrogen control may be required for industrial discharges, necessitating reduction of the effluent total nitrogen content. Nitrification processes convert ammonia nitrogen to nitrate, but the total nitrogen concentration is not altered. Denitrification converts nitrate nitrogen to free nitrogen, thereby reducing the effluent nitrogen concentration. The general chemistry and kinetics of denitrification are discussed in Chapter II-2. Some important design considerations can be summarized as follows:

(1) The denitrification reaction can proceed in the presence of a broad range of nitrifiers, which can shift from aerobic to anaerobic conditions. However, the primary reaction occurs in an anaerobic environment, in the presence of biodegradable organic material acting as an electron donor.

(2) Denitrification is most effective at a pH range from 7 to 7.5, with the reaction suppressed to some degree below 6 and above 8.

(3) The reaction rate decreases with decreasing temperature, being most effective between 20 and 50°C.

(4) Electron donors evaluated for denitrification include municipal waste substrates, volatile acids, brewery wastes, molasses, and methanol. Methanol is the most effective and the most frequent selected.

(5) DO suppresses denitrification, competing with methanol as an electron donor, and must be removed prior to denitrification.

(6) Configurations for denitrification include (a) a separate final denitrification reactor following combined carbon oxidation/nitrification, (b) a separate final denitrification utilizing methanol, (c) an alternating aeration/anoxic process, and (d) proprietary processes.

Denitrification design criteria can be estimated using a combined kinetic expression:

$$\mu_d = \mu_d \frac{D}{K_d + D} \cdot \frac{M}{K_m + M} \qquad (3.20)$$

where μ_d is the rate of denitrifier growth corrected for nitrate and methanol, μ_d is the peak rate of denitrifier growth at temperature T, pH, and negligible DO, D is the concentration of the nitrate nitrogen, mg/L, K_d is the half-saturation constant, mg/L NO_3^-N, M is the methanol concentration, mg/L, and K_m is the half-saturation constant for methanol, mg/L.

The value of the half saturation constant, k_d, is very low; the EPA reports a 0.08 mg/L NO_3^-N value without solids recycle and 0.16 with solids recycle [21]. A low K_m value (K_m in order of 0.1) implies that only 1 mg/L of methanol need be in the effluent to achieve 90% of the maximum denitrification rate. Denitrification reactor design follows nitrification and carbonaceous oxidation suspended growth design procedures, based on an SRT value as defined by Equation (3.21).

$$1/SRT_d = Y_d \cdot q_d - b \qquad (3.21)$$

where SRT_d is the denitrification solids retention time, days, q_d is the nitrate removal rate, kg NO_3^-N removed/kg VSS/day, b is the decay coefficient, 1/day, and Y_d is the denitrifier gross yield, kg VSS growth per kg of nitrate nitrogen removed.

TABLE 3.20. EPA Reported Fate of Contaminants from Municipal Facilities (adapted from Reference [27]).

	Percent of the influent	
	Volatilized	Partitioned Sludge
Metals		
anion	0	50
cations		
antimony	0	0
barium	0	90
cadmium	0	27
chromium	0	70
lead	0	30
mercury	0	50
nickel	0	35
silver	0	90
Cyanide	1	90
PCB	9	8–34
Pesticides		
herbicides	0–57	4–35
organophosphorus	0	0–9
Semivolatiles		
acid	0	8–17
base	0	2–10
neutral	0–45	1–66
Volatiles	0–86	0–49

TABLE 3.21. Fate of Contaminants in Wastewater Treatment Plants (adapted from Reference [10]).

Compound	Total	Absorbed	Stripped	Biodegraded	
				Measured	Calculated
Acetone	91.7	0.6	1.8	94.1	89.3
Cyclohexanone	82.1			81.1	82.1
2-Butanone	96.7	0.6	0.8	94.3	95.3
4-Methyl-2-pentanone	97.5	1	2	95	94.5
Tetrahydrofuran	89.1	1.2	10.2	77.9	77.7
Carbon tetrachloride	98.8	0.6	103	5	?
Chlorobenzene	99.1	2.8	12.6	83.6	83.7
Chloroform	84.3	2	104	−21	?
1,2-Dichloroethane	53	3	67	−17	?
1,2-Dichloropropane	77.3	2	97	−25	?
Methylene chloride	73		24	1	49
Tetrachloroethylene	97.3	2.5	110.8	−16	?
Trichloroethylene	99.7	1.8	52.4	45.5	45.5
1,1,1-Trichloroethane	98.6	1.7	108.7	−11.9	?
1,1,2-Trichloroethane	70.8	2.1	37.0	31.7	31.7
Ethylbenzene	98.3	5.1	15.9	77.2	77.3
Toluene	99.0	1	25	72	73
Total xylenes	98.7	1	32	66	65.7
Bis-(2-ethylhexyl) phthalate	96.7	11		85	85.7
Butylbenzyl phthalate	91.0	11		81	80
1,4-Dichlorobenzene	94.3	2		93	92.3
Naphthalene	96.3	2		94	94.3
Nitrobenzene	95.7	1		93	94.7
Phenol	98	1		97	97

Experimental results obtained from pilot units operating at: a) 50,400 GPD, b) influent concentration of 0.25 mg/L of the compound tested, c) HRT of 7.5 hours, d) SRT of 4 days, e) 20°C, and f) simulated basin wind of 2 mph.
Discrepancies in data totals are a result of analytical difficulties and formation of by-products similar to the measured component resulting in negative removals, or individual measurements not totaling to the total removal percentage.
The theoretical removal by biological degradation is based on the total minus adsorption minus stripping.
Total removed is based on influent minus effluent concentration.
Question mark (?) indicates theoretical biodegraded removal cannot be estimated, or the result is questionable.

For engineering evaluations the EPA recommends Y values from 0.6 to 1.2 and b at 0.04 1/day.

The design SRT_d can be obtained by applying a safety factor to the minimum or limiting SRT.

Fate of Contaminants

The fate of contaminants is best evaluated in waste-specific pilot studies, analyzing the sludge for substrate adsorption. A separate parallel reactor without biological activity can be used to estimate stripping from the system. General fate model concepts for biological systems were discussed in Chapter II-2, Aerobic Biological Oxidation section. The EPA has developed a *fate model* for municipal systems using the relations discussed Chapter II-2 [27]. Industrial application of this model must be utilized with caution, requiring verification and calibration of the model constants. To illustrate the fate of individual components in an activated sludge facility, the fate of regulated pollutants estimated using the EPA fate model is listed in Table 3.20 [27]. Comparable results reported by Govind et al. are summarized in Table 3.21 [10].

GENERAL ENGINEERING CRITERIA

Figure 3.11 illustrates an activated sludge preliminary concept flowsheet, typically employed for industrial biological treatment. A complete activated sludge system could involve the following process components:

Equalization (optional)	Chapter I-4
Waste feed system	Chapter II-3
Aeration basin	
Aeration equipment	Chapter II-1
Sludge clarification	Chapter II-8
Sludge recycle	Chapter I-8
pH control (optional)	
Nutrient addition (optional)	Chapters II-8, III-7
Methanol feed for denitrification (optional)	

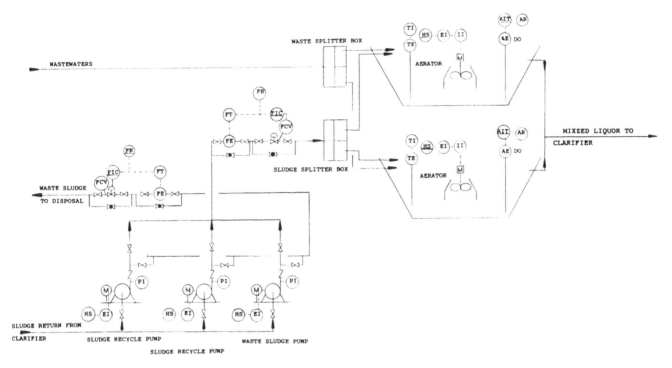

Figure 3.11 Activated sludge preliminary concept flowsheet.

Sludge treatment system

As indicated many of these components are discussed in separate chapters, with specific design criteria detailed.

AERATION BASIN DESIGN CONSIDERATIONS

Activated sludge system operating flexibility can be provided by dividing the aeration basin into multiple units, with provision for series or parallel operation, incorporating the following advantages into the system:

(1) A parallel configuration can be employed where production (and waste loads) variation is significant, with one parallel train run at optimum conditions and the other adjusted to accommodate varying surplus loads.

(2) A series configuration can be considered where toxic shock may be a problem, allowing recovery in the final basins.

(3) Two-stage operation could be more effective where high influent loadings and high effluent quality are a primary criteria. In such cases two stage operation could provide a better overall plant efficiency.

(4) Two-stage series operation provides a better system for nitrification, permitting the operator to set optimum conditions for first stage carbonaceous removal and second stage nitrification.

(5) Multiple units allows repairs to one basin without complete shutdown.

A word of caution about two stage series operation. The second stage tends to be unstable, producing a poorer sludge quality because of underloading. This condition can be remedied by including "bypass" provisions to bleed raw influent into the second stage. In addition, provision should be made for independent sludge return and wasting to maintain proper loading rates and avoid mixing of "biological populations" in multicultural systems.

There is no doubt that multiple units increase treatment plant capital costs and cannot always be justified. However, for industrial waste systems, with stringent effluent limits imposed and heavy noncompliance penalties possible, the operator should be given maximum process flexibility.

Aeration Basin Configuration

Aeration basin geometry is based on factors such as available land, reactor kinetics, optimizing mixing energy, and minimizing heat loss. The configuration can be circular or square, associated with a completely mixed system; long and narrow, associated with plug flow reactors; or an intermediate geometry representing nonideal and less definable kinetics. Besides affecting the reactor kinetics, a round or square configuration improves mixing and reduces the above-ground heat loss surface area (because of minimum area for equal volumes). Heat loss can be minimized by increasing the below ground basin depth and thereby reducing the above-ground vessel surface area. Although deep tanks can improve heat conservation they are more difficult to aerate and mix.

The physical aspects of a reactor, length-to-width dimensions, are frequently developed to adhere to specific kinetic requirements. Complete mix reactors should have a length to width of less than 3:1, no greater than 5:1, and ideally 1:1. Plug flow configurations can be achieved by maintaining a length-to-width ratio of at least 5:1, and in many cases over 10:1. Complete mix reactor effectiveness can be improved by increasing mixing energy to the practical economical limits and designing the vessel to optimize mixing and turbulence. Mixing theory is discussed in Chapter I-5.

Plug flow process efficiency can be assumed to be affected by the basin geometry, improving with increasing hydraulic efficiency (HE). The HE is defined as

$$HE = \frac{\text{hydraulic mean retention time } (t)}{\text{volumetric residence time } (T = V/Q)} \quad (3.22)$$

The geometry of a plug flow reactor has been defined in terms of a distinguishable concentration gradient throughout the aeration basin length and controlled mixing. Mixing is minimized to that required to assure oxygen distribution and the reactants contact, providing less than ideal hydraulic efficiency. Although the resulting reactor dimensions are difficult to relate to hydraulic efficiency, a correlation has been proposed relating hydraulic efficiency to flow width and length, defined by Equation (3.23) [19]:

$$HE = 0.84 \, [1 - \exp(-0.59 \, L_f/W_f)] \quad (3.23)$$

where L_f and W_f are the length and width of flow.

As illustrated in Figure 3.12, increases of L_f/W_f up to a value of approximately 10 produce appreciable HE improvement, although available land may limit this option.

An alternative to increasing the length-to-width ratio is to create a carousal-type path by adding internal dikes, as illustrated in Figure 3.13.

The flow length-to-width (L_f/W_f) ratio depends on the number of dikes (N), with a ratio frequently in the range of 0.8 to 1. This ratio can be expressed as

$$L_f/W_f = [L/W] \cdot r \cdot (N + 1)^2 \quad (3.24)$$

The optimum area involves selecting a basin length, width, and the number of internal dikes, based on costs and available land area. Shields et al. detailed the geometric and economic considerations involved in evaluating basin configuration, optimizing costs with geometric configuration [19]. For any required residence, flow rate, and depth, the approximate basin area required can be estimated using the suggested equation:

$$L \cdot W \equiv \frac{a}{\{1 - \exp\,[-0.59\,(L/W) \cdot r \cdot (N + 1)^2]\}} \quad (3.25)$$

where a is equal to $(t \cdot Q)/(0.84 \cdot D)$, t is the required residence time, time, Q is the influent flow rate, m^3 (ft^3)/time, and D is the basin depth, m (ft).

The parameters that must be evaluated to optimize the basin construction costs include the number of dikes (N), the L/W ratio, and the ratio of the dike length to the basin length (r).

Aeration Basin Construction

An aeration basin can be constructed of steel, concrete, or earthen material, with vessel stability and integrity of paramount importance. Current (or anticipated future) concerns about possible seepage of toxic materials into subsurface waters have greatly increased the cost and potential future liability of earthen basins. As a result they are no longer "simple and inexpensive" and therefore not the primary construction material of choice. Economics drive the choice between steel and concrete basins, with steel tanks common for small systems involving flow rates of 3800 m^3/d (1 MGD) or less. Generally, large basins are constructed of reinforced concrete, with maximum consideration for long-term physical stability under varying normal and upset conditions.

Diffused air systems are constructed 5 to 8 m (15 to 25 ft) deep, with 0.3 to 0.6 m (1 to 2 ft) freeboard. The width to depth for spiral flow mixing can vary from 1:1 to 2.2:1, with 1.5:1 common. The depth of mechanically aerated basins depends on the device mixing limits. For mechanical aerated systems a rectangular "footprint" should be subdivided into square subunits to accommodate complete mix dynamics and device characteristics.

Municipal systems are commonly designed with at least two tanks, whereas the size of industrial systems establishes the practicality of multiple units. Where multiple units are considered the facility should be designed so that

(1) Common walls can withstand full hydrostatic pressure, which would occur when one chamber is emptied.
(2) Flow distribution is equalized to the multiple compartments.
(3) Foundations are designed to prevent settlement or flotation resulting from subsurface water pressure.
(4) Provision is made to dewater each compartment.
(5) Walkways are provided to service, monitor, and maintain the units.

PROCESS CONTROL

The design of an activated sludge system is as good as it can be operated. The lag time in an activated sludge

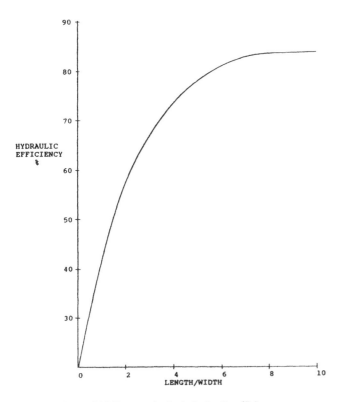

Figure 3.12 Rectangular basin hydraulic efficiency.

system, especially the aeration basin, makes automatic control impractical because (1) conditions do not change instantaneously and (2) automation is an invitation to reduce operator manpower and attention. Whether highly automated, continuously monitoring is warranted or not, the engineer can assist the operator in controlling the plant by providing capabilities for the following:

(1) Flow control (and recording) of influent and sludge recycle flows and monitoring and recording effluent flow.

(2) Flow totalizing of sludge wasting and recycle.

(3) Sample points to collect waste influent. At a minimum this should include the ability to obtain a "grab sample." Ideally automatic samplers should be included to col-

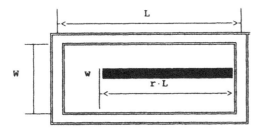

Where   L is the physical length, meters (feet)
        W is the physical width, meters (feet)
        w is the dike width, meters (feet)
        r is ratio of dike length to basin length

Figure 3.13 Rectangular basin configuration.

lect composite samples over an operator controlled time period.

(4) Influent pH control monitoring and alarming for any conditions outside the specified limits.

(5) Sampling and measuring aeration basin dissolved oxygen level. This could include automatic DO monitoring and controlling aeration device output. However, aeration basin probes are difficult to maintain. Where applied they should be located so that they can be readily accessed, monitored, and replaced. In completely mixed systems the effluent DO is representative of the "bulk" basin oxygen level. In plug flow systems DO measurements must be made at critical basin sections, and probes must be located to monitor representative areas influenced by the aeration device.

(6) Sampling the aeration basin discharge to measure the reactor mixed liquor solids content as well as sampling critical reactor areas and depths.

(7) Sampling and evaluating the aeration basin oxygen respiration rate.

(8) Level indication of the aeration basin volume and all critical storage tanks and process vessels and alarming for low and high level conditions.

(9) Sample points to collect waste effluent. At a minimum this should include the ability to obtain a grab sample. Ideally, automatic samplers should be included to collect composite samples over an operator controlled time period.

(10) Effluent pH control monitoring and alarming for any conditions outside the specified limits.

(11) Sampling of the clarifier sludge and monitoring the blanket level.

(12) Running lights for all major equipment.

(13) Monitoring clarifier torque and an early alert of potential excessive (damaging) torque.

Operator monitoring provisions require safe access to collect samples and observe the operation.

The question of how much should be continuously monitored, recorded and automatically controlled should be tempered as follows:

(1) Very few process variables need be instantaneously monitored, measured, or corrected. The process inherently reacts slowly to change so that constant adjusting is an illusion.

(2) What is the reliability of the measuring probe and the required manpower to maintain the system? Is high technical instrumentation manpower replacing operator time?

(3) What is the purpose of continuously measuring a variable? How will the instantaneous measurements be interrupted, and how will it assist in controlling the plant? Can the process limits of a variable be related to system performance?

Where the "drive" is persuasive for automatic controls, the stage is commonly set to utilize a data bank for continuously storing, analyzing, and preparing continuous report. The question to ask is what will the operators do with a mile and a half of daily strip charts and the resulting reports?

PIPING AND PUMPS

Industrial waste treatment plant piping systems, and especially waste sewer systems, generally follow the manufacturing facility hydraulic design, using materials of construction consistent with the plant piping. The activated sludge system can generally adhere to municipal system design standards, although there are some inherent differences governing their application; such as

(1) Industrial systems generally do not involve large collection and transfer systems common in municipal systems. In addition, modern plants include a dedicated industrial sewer collecting point source discharges, limiting runoff volumes to that collected in small diked areas.

(2) Plant capacities are commonly a fraction of municipal systems so that pipe sizes are relatively small (and short), making consistency with manufacturing facility standards a more economic consideration to avoid separate spare parts inventory.

Waste sewer piping material selection upstream of the treatment plant will be dictated by the waste corrosive properties and easily identified by the production piping selected for the processes generating the wastes. Industrial facilities commonly utilize carbon steel piping for waste facilities, a general material suitable for many inplant materials. Municipal treatment plant piping is commonly cast iron or ductile iron, with modern designs using carbon steel, coated, or Teflon piping [25].

General piping system design criteria, whether for industrial facilities or municipal treatment systems, are detailed in many design manuals and will not be repeated here [14,16,25]. However, some critical engineering considerations are listed as a checklist for the Process Engineer.

Pressure or Gravity Systems

Ideally, wastewater should flow by gravity through the wastewater treatment plant, i.e., the primary treatment, aeration basin, final clarification, and discharge point. This usually means gravity flow (or pumping) from point sources to an industrial sewer line to a waste collection tank, with adequate collection tank elevation to allow gravity flow control through the downstream treatment facilities. Site conditions may require some exceptions, such as

(1) The waste collection tank cannot be economically located at an elevation allowing controlled gravity flow to the facility. In such cases, the waste is pumped to the primary treatment system, with provisions for an "operating" spare pump.

(2) There is inadequate head in the final treatment vessel to flow to the discharge point. Final discharge pumps (operating and operating spares) must be provided.

(3) In some rare cases site irregularities require intermittent pumping stations.

Aeration basin discharge to the final clarifier should never be subjected to high shear forces and never pumped, and the head should be minimized to avoid excessive velocities.

General Piping Design

Generally, piping design should incorporate the following criteria:

(1) All critical piping should be installed in pipe galleries for easy access; buried piping should be avoided.

(2) Bypass lines should be included around all control valves, pumps, and critical equipment to allow replacement without a total system shut down.

(3) Sampling lines should be short, avoiding stagnant areas which could promote biological activities. Provisions should be made so that these lines can be readily flushed to obtain representative samples.

(4) Drains should be provided for all critical tanks and equipment.

(5) Operating valves must be operator accessible.

(6) Orifices require at least a 15 diameter straight run upstream and 5 diameters downstream.

(7) Pump piping should be designed for easy access and pump removal.

(8) Liquid lines should be freeze protected for operating and shutdown conditions, at a minimum allowing for free flow, gravity draining when not in operation.

(9) Low pipeline points may require drains.

(10) High pipeline points may require a means of venting.

(11) Reciprocating equipment piping should be independently supported to avoid vibration to other lines or equipment.

Sludge piping requires special consideration because (1) solids concentration changes the flow characteristics, technically from a single to a two-phase flow system, (2) sludge floc characteristics should not be altered to preserve settling characteristics, and (3) solids can settle in the lines and restrict flow, especially when the system is shutdown. Piping criteria applicable to sludge lines include

(1) Abrupt changes or restrictions should be avoided to minimize sludge collection and flow restriction. All elbows should be wide sweep for the same reason. Check valves should be avoided.

(2) Control valves, manual or automatic, should be full flow plug or ball.

(3) Provisions should be made for cleanouts and water flushing, with all flush water directed to a collection system.

Pipe Sizing

Pipe sizing follow the economic consideration generally applicable for process piping and will not be repeated here [16]. However sludge or high solids content waste transmission lines require special consideration. Generally, these lines are sized for flow velocities ranging from 0.6 to 2.4 m/s (2 to 8 fps); the optimum velocity depends on the sludge concentration and composition as well as the pump limits. A minimum 15 cm (6 in.) pipe size is recommended [24]. However, these criteria may present a problem for smaller industrial waste systems where 15-cm lines could result in less than recommended flow velocities. In many industrial systems this criteria has been relaxed to allow a 10-cm (4 in.) pipe and a minimum 0.6 m/s velocity.

The solids and two-phase characteristics of sludge lines result in head losses higher than those predicted for water lines, with values for activated sludge 10 to 25% higher, digested sludge two to three times higher, and concentrated sludge much higher [14]. Suggested Hazen-Williams sludge coefficients are as follows [17]:

(1) $C = 90$ for activated sludge with up to 1% solids
(2) $C = 55$ for primary sludge with up to 4.5% solids
(3) $C = 35$ for thickened or digested sludge with up to 8% solids

Detailed hydraulic calculating methods are available in municipal system design manuals [14,17,25].

Gravity Flow Criteria

Hydraulic flow considerations are site specific, but all incorporate some common general criteria:

(1) Provision must be made for treatment plant flood protection from the water way into which the effluent is discharged.

(2) Gravity piping must be designed to minimize losses.

(3) Piping systems should be engineered to provide uniform distribution to the individual plant elements.

(4) Industrial sewer systems should be designed to prevent backup into the manufacturing facility.

(5) Controls should be provided to maintain critical water levels for each of the treatment plant elements.

(6) Piping design must consider flow patterns, accounting for minimum and maximum waste generation and units out of service.

Site-specific conditions result in at least four critical levels to be considered in developing hydraulic evaluations. These include

(1) The water level of the (receiving) stream into which effluent is discharged
(2) Design flood level of the receiving stream
(3) Maximum inlet sewer level
(4) Minimum inlet sewer level

Once the available minimum sewer and outfall flood level are established, the available head must be apportioned between the treatment plant components. This must take into consideration worst-case scenarios such as at the maximum flow with one clarifier out of service, maximum plant recycle, and one unit of multiple treatment trains out of service. Hydraulic profile calculations are detailed in municipal design manuals [14,17,25].

Sludge Pumps

For most cases, wastewater selection follows criteria for water pumps, with the materials selected consistent with the waste characteristics. Positive displacement piston and rotary pumps are commonly used for concentrated sludge, and low energy, nonclog centrifugal pumps are used for sludge recycle. Piston pumps can transport high concentrated sludge long distances and are commonly constructed of coated structural steel frames, with special alloy cylinders, Buna-n pistons, and hardened steel valves [25]. Rotary pumps are used when pulsating flows cannot be tolerated downstream. The body can be cast iron or stainless steel for severe service and the wetted parts alloy or stainless steel [25]. Centrifugal waste and return pumps are commonly cast or ductile iron construction, with stainless steel shaft and shaft sleeve [25].

COMMON ACTIVATED SLUDGE DESIGN DEFICIENCIES

Some design deficiencies common for municipal systems have been investigated by the EPA and are identified as a checklist of process and mechanical design considerations [23]. Reference is made to the cited literature for a more detailed discussion. The most common deficiencies cited are summarized subsequently.

Aeration Basin

(1) No provisions were included for process alkalinity control, especially where nitrification is part of the treatment process.

(2) No provision was made for foam control.

(3) Poor operational control results because multiple units were not provided, or poor piping design limits the operation of multiple units. Provisions were not made

for (a) operating in different process modes, (b) adjusting waste loads, (c) bypassing the aeration basin, (d) isolating a single chamber for cleanout or repairs, and (e) emptying the aeration basin or draining critical process vessels.

(4) Dewatering of a single cell is not possible because of structural limitations of the multiple chamber basins.

(5) Plugging is encountered because coarse screens were not provided to protect waste feed or sludge piping or aerators.

(6) The treatment plant capacity is inadequate because in-plant waste streams from filters, screens, dewatering, digesters, test laboratories, and similar ancillary plant facilities were neglected.

(7) Poor aeration device performance results because of inadequate capacity or poor diffuser or aerator spacing. Deficiencies result in (a) inadequate mixing and low dissolved oxygen, (b) solids deposition, and (c) "dead spots," causing poor suspended solids or DO distribution.

(8) No provision was made to prevent scum accumulation in flow splitter boxes.

Diffused Air System Design

(1) Blower inlet air cleaners were not provided.

(2) No provision was made for removing aeration basin air diffuser drops.

(3) No method was provided to control flow distribution to individual diffusers.

(4) Air drops cannot be seen when the aeration basin is operating, resulting in difficulties in replacing units.

(5) Diffuser plugging is a constant problem.

(6) Diffuser air blowers generate excessive noise.

(7) Improper air piping material results in excessive scaling, adding to plugging difficulties. This is especially evident where poorly coated pipe is used.

Mechanical Aerator Design

(1) Improper placing of the gear box results in oil draining into the aeration basin.

(2) Monitoring of individual units is difficult because amperage meters or running lights were not provided.

(3) No time delay relays were provided to dampen the effects of the start-up load or changes in the load.

(4) Splash shields were not installed to prevent freezing of exposed aerator section or freezing of adjacent areas.

(5) Floating aerators were not properly located or secured, causing collisions with reactor walls or other aerators.

(6) Quick disconnect electrical connections become wet and short.

Process Control

(1) No provision was made for accurate flow control or measurements.

(2) No provision was made to vary sludge or waste flow to aeration basins.

(3) No provision was made to split, measure, or control flow to multiple secondary clarifier units.

(4) No provision was made to measure, maintain, or vary the aeration basin level.

(5) The aeration basin DO cannot be controlled because diffuser blower or mechanical aerator output cannot be changed.

(6) Inadequate provisions were made to observe, sample, and test all sludge lines.

(7) Total sludge wasting flow cannot be controlled.

(8) Return sludge capacity to multiple basins cannot be controlled.

(9) The return sludge pump range is inadequate.

(10) Continuous treatment plant influent flows cannot be measured or inventoried. Intermittent flows, such as sludge wasting, cannot be totalized.

(11) Separate waste sludge discharge capabilities were not provided for multiple clarifiers.

Secondary Clarification

(1) No provision was made for chemical addition.

(2) The sludge removal mechanism is not compatible with the clarifier.

(3) Side water depth is inadequate.

(4) Access to the weirs is not provided.

(5) Running lights or high torque control were not provided to alert the operator of a problem.

(6) Piping was not provided to isolate individual clarifiers, or for sludge control. As a result individual clarifiers cannot be isolated for maintenance, and recycle cannot be controlled from any clarifier to any individual aeration basin.

(7) Clarifier hydraulic capacity is not adequate because treatment plant side streams were neglected in the design.

(8) Clarifier was sized for too high an overflow rate.

(9) Clarifier was sized for too high a solids loading.

(10) Clarifier short-circuiting was not considered in the design.

(11) Clarifier hydraulics were incorrectly evaluated; high inlet and outlet losses were not considered.

(12) Weir was improperly placed.

(13) The physical dimensions of the rectangular clarifier are incorrect for effective hydraulic efficiency.

(14) Excessive plugging of sludge lines results because flushing provisions were not included in the piping design. In addition, sludge return lines was poorly located, limiting access. In some cases, sludge lines are buried and completely inaccessible.

See Chapter II-8, Clarifier and Thickening section, for other reported design deficiencies.

CASE STUDY NUMBER 10

Develop a completely mixed activated sludge process design system for an industrial waste with the following characteristics:

Flow, MGD:	1
BOD$_5$, mg/L:	500
COD, mg/L:	1250
N-ammonia, mg/L:	50
Phosphorus, mg/L:	0
Alkalinity, mg/L CaCO$_3$:	280

Applicable design criteria include

Effluent BOD$_5$, mg/L:	20
Effluent ammonia, mg/L:	2
Effluent suspended Solids, mg/L:	20
Fraction solids biodegradable:	0.65
Ultimate BOD solids:	1.42
Solids BOD$_5$/BODu:	0.68
MLVSS/MLSS ratio:	0.8
Nitrogen required per 100 BOD$_5$ removed:	5
Phosphorus required per 100 BOD$_5$ removed:	1
Design temperature, °F:	68
Summer temperature, °F:	80
Winter temperature, °F:	40
Basin depth, feet:	10
Basin surface length to width:	2

Laboratory data obtained indicate a soluble feed BOD$_5$ of 500 mg/L can be treated to achieve an effluent of less than 10 mg/L utilizing an SRT of 9 days at 20°C. The kinetic coefficients developed to define the treatability characteristics can be summarized as follows:

k_d:	0.05 (1/day)
k_{max}:	4 (1/day)
K_s:	85 mg/L
Y:	0.5
MCRT:	9 days

PROCESS CALCULATIONS

Carbonaceous Oxidation Calculations

(1) Determine the effluent soluble BOD
  a. Soluble BOD$_5$ = TBOD$_5$ − BOD solids
  b. Effluent solids at 20 mg/L
  c. Biodegradable portion solids at 65%

$$20 \cdot 0.65 = 13 \text{ mg/L of biodegradable solids}$$

  d. Inert effluent solids

$$20 - 13 = 7 \text{ mg/L of inert solids}$$

  e. Ultimate BOD of solids
  BOD$_u$/Biodegradable solids = 1.42

$$13 \cdot 1.42 = 18.46 \text{ ultimate BOD of solids}$$

  f. BOD$_5$ of solids
  BOD$_5$/BOD$_u$ = 0.68

$$18.46 \cdot 0.68 = 12.55 \text{ mg/L BOD}_5 \text{ of solids}$$

  g. Allowable soluble effluent BOD$_5$

$$20 - 12.55 = 7.45 \text{ mg/L soluble (design) BOD}_5$$

(2) Evaluate SRT requirements
  a. Laboratory results indicate SRT of 9 days at 68 °F.
  b. Washout SRT

$$1/\text{SRT} = \frac{Y \cdot k \cdot S_o}{K_s + S_o} - k_d = \frac{0.5 \cdot 4 \cdot 500}{85 + 500} - 0.05 = 1.66$$

SRT = 1/1.66 = 0.6 DAYS
Assume a safety factor of 20
SRT = 20 · 0.6 = 12 days
Use 12 days SRT at 68 °F, allowing a 33% safety factor from the laboratory data.

(3) Select design MLVSS
Select a design MLVSS of 2000 mg/L.

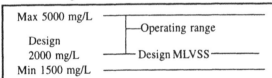

(4) Calculate aeration basin volume

$$\text{Volume} = \frac{[\text{SRT}] \cdot [Q \cdot (S_o - S_e) \cdot Y]}{X \cdot (1 + k_d \cdot \text{SRT})}$$

$$\text{Volume} = \frac{[12] \cdot [1 \cdot (500 - 7.45) \cdot 0.5]}{2000 \cdot (1 + 0.05 \cdot 12)} = 0.92 \text{ mg}$$

(5) Calculate hydraulic retention time

$$\text{HRT} = \frac{\text{volume}}{\text{flow}} = \frac{0.92 \text{ mg}}{1 \text{ MGD}} = 0.92 \text{ days}$$

(6) Calculate effluent concentration

$$S_e = \frac{K_s \cdot [1 + k_d \cdot \text{SRT}]}{\text{SRT} \cdot [Y \cdot k - k_d] - 1}$$

$$S_e = \frac{85 \cdot [1 + 0.05 \cdot 12]}{12 \cdot [0.5 \cdot 4 - 0.05] - 1} = 6.07 \text{ mg/L}$$

Based on the safety factor applied to the theoretical SRT, the effluent quality is any where from 6 to 7.45

(7) Calculate F/M ratio

$$F/M \text{ fed} = S_o / (X_v \cdot HRT) = 500 / (2000 \cdot 0.92) =$$

$$= 0.272 \text{ lb BOD}_5 \text{ fed/hr/lb MLVSS}$$

$$F/M \text{ used} = (S_o - S_e)/ (X_v \cdot HRT) = (500 - 6.07)/ (2000 \cdot 0.92)$$

$$= 0.268 \text{ lb BOD}_5 \text{ removed/day/lb MLVSS}$$

(8) Calculate effects of summer conditions
   a. F/M utilized (R) at 20°C is 0.268
   b. Summer condition at 80°F (26.7°C)
   c. $R_s = 0.268 \cdot 1.02 \char94(26.7 - 20) = 0.306$
   d. k_d at 80°F $= 0.05 \cdot 1.02 \char94(26.7 - 20) = 0.057$
   e. SRT at summer conditions

$$SRT = 1/(Y \cdot R_s - K_d) = 1/ (0.5 \cdot 0.306 - 0.057) = 10.42 \text{ d}$$

   f. Estimated required MLVSS at design S_e of 7.45 mg/L.

$$X = \frac{Y \cdot (S_o - S_e)}{1 + b \cdot SRT} \cdot \frac{SRT}{HRT}$$

   7.45 mg/L is assumed as an effluent concentration because an accurate adjustment of the coefficients, especially k, is not certain.

$$X = \frac{0.5 \cdot (500 - 7.45)}{1 + 0.057 \cdot 10.42} \cdot \frac{10.42}{0.92} = 1750 \text{ mg/L MLVSS}$$

(9) Calculate effects of winter conditions
   a. R_d at 20°C is 0.268
   b. Winter condition at 40°F (4.44°C)
   c. $R_s = 0.268 \cdot 1.02 \char94(4.44 - 20) = 0.197$
   d. k_d at 40°F $= 0.05 \cdot 1.02 \char94(4.44 - 20) = 0.037$
   e. SRT at summer conditions

$$SRT = 1/(Y \cdot R_s - K_d) = 1/ (0.5 \cdot 0.197 - 0.037) = 16.3 \text{ d}$$

   f. Estimated required MLVSS at design S_e of 7.45 mg/L. 7.45 mg/L is assumed as an effluent concentration because an accurate adjustment of the coefficients, especially k, is not certain.

$$X = \frac{0.5 \cdot (500 - 7.45)}{1 + 0.037 \cdot 16.3} \cdot \frac{16.3}{0.92} = 2722 \text{ mg/L MLVSS}$$

(10) Calculate sludge recirculation requirements

$$Recycle = \frac{recycle \ flow}{forward \ flow}$$

$$= \frac{basin \ VSS}{recycle \ SS \cdot (MLVSS/MLSS) - basin \ VSS}$$

MLVss/MLSS = 0.8

a. Design conditions

$$Recycle = \frac{2000}{10,000 \times 0.8 - 2000} \cdot 100 = 33\% \text{ MLVSS}$$

b. Summer conditions

$$Recycle = \frac{1748}{10,000 \times 0.8 - 1748} \cdot 100 = 28\%$$

c. Winter conditions

$$Recycle = \frac{2722}{10,000 \times 0.8 - 2722} \cdot 100 = 52\%$$

Design pumping equipment for 100% for operating flexibility. See clarifier calculation (Case Study No. 22) for a complete evaluation of recirculation effects.

(11) Nutrient Requirements
   a. Available nitrogen

$$1 \text{ MGD} \cdot 8.34 \cdot 50 \text{ mg/L} = 417 \text{ lb/day}$$

   b. Available phosphorus (0)
   c. BOD$_5$ removed

$$1 \text{ MGD} \cdot 8.34 \cdot (500 - 6.07) \text{ mg/L} = 4119 \text{ lb/day}$$

   d. Nitrogen Required

$$4119 \cdot 5/100 \text{ BOD}_5 \text{ removed} = 206 \text{ lb/day}$$

   e. Excess nitrogen

$$417 - 206 = 211 \text{ lb/day}$$

$$211 / (8.34 \cdot 1 \text{ MGD}) = 25.3 \text{ mg/L excess ammonia nitrogen.}$$
   f. Phosphorus required

$$4119 \cdot 1/100 \text{ BOD}_5 \text{ removed} = 41 \text{ lb/day}$$

(12) Estimate nitrification effectiveness
(13) Select safety factor

Safety factor of 2.5 selected.

(14) Select minimum aeration DO

Minimum DO of 2 mg/L.

(15) Estimate alkalinity destroyed

Alkalinity in influent adequate to maintain pH of 7.2.

(16) Estimate maximum nitrifier growth rate

$$\mu_N = 0.47 \cdot (e^{0.098 \, (T-15)}) \cdot [DO/(KO_2 + DO)]$$
$$\cdot [1 - 0.883 \cdot (7.2 - pH)]$$

at design temperature 20°C

$$\mu_N = 0.47 \cdot (e^{0.098(20-15)}) \cdot 2/(1.3 + 2)]$$
$$\cdot [1 - 0.883 \cdot (7.2 - 7.2)] = 0.465$$

(17) Calculate nitrification MCRT

$$MCRT = 1/\mu_N$$

Design MCRT = 1/0.465 = 2.15 days

(18) Select MCRT for both carbonaceous oxidation and nitrification

After 2.5 safety factor is applied carbonaceous design SRT of 12 days is greater than nitrification requirements and the design basis.

(19) Estimate effluent nitrogen concentration

$$\frac{1}{SRT_d} = \text{nitrogen cell growth rate} = \mu_N \cdot \frac{N}{K_n + N}$$

$$K_n = 10 \char`\^[(0.051 \cdot t) - 1.158]$$

At design conditions

$$1/SRT = 1/12$$

$$K_n = 10 \char`\^[(0.051 \cdot 20) - 1.158] = 0.728$$

$$\frac{1}{12} = 0.465 \cdot \frac{N}{0.728 + N}; \ N = 0.159 \text{ mg/L}$$

Effluent requirement of 2 mg/L of ammonia nitrogen met!

(20) Estimate sludge wasting

    a. VSS (P_x) generated per day in summer

$$P_x = \frac{[Q \cdot 8.34 \cdot (S_o - S_e) \cdot Y]}{(1 + k_d \cdot SRT)} = \text{lb VSS per day}$$

$$P_x = \frac{[1 \cdot 8.34 \cdot (500 - 7.45) \cdot 0.5]}{(1 + 0.057 \cdot 10.42)} = 1289 \text{ lb VSS/day}$$

MLVSS/MLSS ratio: 0.8

Total solids per day = 1289/0.8 (inerts) = 1611
Assume 1% solids = 1611 / 0.01 = 161,100 lb/day
At 8.34 lb/gal = 161,100/8.34 = 19,317 gal/day

    b. Sludge at design conditions is 1284 lb VSS/day.

(21) Estimate carbonaceous oxidation oxygen requirements
Based on a design S_e of 6.07 and sludge at 1284 lb VSS/day
At design conditions,

    a. BOD$_5$ removed = 1 MGD $\cdot$ 8.34 $\cdot$ (500 − 6.07) = 4119 lb/day

    b. Ultimate BOD removed

$$4119/0.68 = 6057 \text{ lb/day}$$

    c. Solids removed equals 1284 lb/day.

    d. Ultimate BOD of solids, lb/day

$$1284 \cdot 1.42 = 1823 \text{ lb/day}$$

    e. Total oxygen requirement

$$6057 - 1823 = 4234 \text{ lb/day}$$

(22) Estimate nitrification oxygen requirements
4.57 pounds of oxygen required per pound of ammonia nitrogen removed. Assume effluent zero.

$$1 \text{ MGD} \cdot 8.34 \cdot 50 \cdot 4.57 = 1906 \text{ lb/day oxygen.}$$

(23) Estimate total oxygen requirements

$$4234 + 1906 = 6140 \text{ total pounds of oxygen required.}$$

(24) Basin design (see Case Study 9 for details)

	Diffused air	Mechanical aeration
Volume, MG	0.92	0.92
Depth, ft	15	10
No basins	2	2
Operation	Parallel	Parallel
Unit area, sq ft	4100	6150
Unit width, ft	30	55
Unit length, ft	140	112
Over "footprint" width, ft	60	110
length, ft	140	112

(25) Aerator summary (see Case Study 9 for details)
6,140 pounds per day of oxygen
Diffuser system
192 diffusers per basin
384 diffusers per system
Diffuser capacity @ 3840 scfm
Two blowers, each at 75 hp
System oxygen efficiency @ 1.7 lb O$_2$/hr/hp
Turbine system
4 turbines each at 500 scfm and 25 hp per basin
8 total turbines @ 4000 scfm and 200 hp
Two 75 hp compressors
System oxygen efficiency
1.7 lb/hr/compressor hp
1.3 lb/hr/turbine hp
0.7 lb/hr/total hp
Surface aerators
2 @ 75 hp aerators per basin
4 @ 75 hp aerators total
Oxygen efficiency @ 0.85 lb/hr/hp

(26) Clarified equipment (see Case Study No 22)
Two units, each 68 ft in diameter, 3600 ft^2, and 11 ft deep.

CASE STUDY NUMBER 11

Based on the complete mixed design of Case Study 10 establish the need for a influent selector chamber.

ESTABLISH THE POTENTIAL FOR BULKING SLUDGE

Flow, MGD:	1
Feed concentration, mg/L:	500
MLVSS, mg/L:	2,000
HDT, days:	0.92
Feed (BOD$_5$) F/M:	0.27

The feed F/M may not adequately establish the potential for filamentous growth because the feed concentration is instantaneously diluted to the effluent concentration of 6.07 mg/L, resulting in a "working" F/M of 0.003. The relatively low reactor substrate concentration could result in a high potential for filamentous growth.

SELECTOR CHAMBER DESIGN

Some common guidelines will be evaluated to approximate a selector design. The need, and appropriate design, should be verified in pilot studies.

(1) Select a selector HDT: Assume 75 min, 0.052 days.
(2) Apply a loading of 10 kg/cu m/d (10,000 mg/L/d)

Actual: 500 mg/L / 0.052 = 9615 mg/L/day ≡ 10,000 mg/L/d.
(3) Apply a BOD$_5$ feed F/M of at least 3

Actual: 500 mg/L / (2000 mg/L MLVSS · 0.052 days) = 4.8

(4) Apply a MLSS loading of 3.3 kg/CuM (3300 mg/L)

Actual: 2000 mg/L MLVSS/0.68 = 2900 mg/L MLSS
3000 · 0.68 = 2250 mg/L MLVSS
2000 mg/L is too low to meet 3300 mg/L MLSS condition.

(5) Utilize at minimum of four compartments

1,000,000 GPD · 0.052 days = 52,000 gal = 6952 ft^3

6952/10 ft deep = 695 ft^2

Assuming a 2:1 length to width ratio,
  Width = 19 feet
  Length = 37 feet
Based on a 20 × 40 foot selector, four chambers can be included, each 10 × 20 feet.
(6) Summary
  Based on some general guidelines the selector size computed should be evaluated at the pilot scale level to (1) establish that the completely mixed basin conditions of Case Study 8 would produce a poor sludge quality and (2) verify (or establish) the selector design criteria.

CASE STUDY NUMBER 12

Evaluate an SBR for the influent in Case Study 10, assume that the treatment coefficients are valid, but that the excess sludge generated is stored and discharged once every seven days. During that period the feed flow remains constant at 1 MGD, but the influent soluble BOD$_5$ varies as follows:

Day 1:	500 mg/L
Day 2:	300 mg/L
Day 3:	200 mg/L
Day 4:	600 mg/L
Day 5:	700 mg/L
Day 6:	700 mg/L
Day 7:	500 mg/L
Avg:	500 mg/L

PROCESS CALCULATIONS

(1) Determine the effluent soluble BOD
  As indicated in Case Study 10 calculations, a total effluent BOD$_5$ requirement of 20 mg/L requires a soluble BOD$_5$ of 7.45 mg/L.
(2) Day one
(3) Calculate the daily applied SBOD loading

1 MGD · 500 mg/L BOD$_5$ · 8.34 lb/gal = 4170 lb BOD$_5$.

(4) Calculate the daily SBOD removal

1 MGD · (500 − 7.45) mg/L BOD$_5$ · 8.34 lb/gal
= 4108 lb BOD$_5$.

(5) Assume a suitable F/M or Fr/M ratio consistent with the expected effluent quality
  Assume an F/M of 0.270 pounds BOD fed/day/lb MLVSS as an initial loading, consistent with the design in Case Study 10.
(6) Calculate the quantity of MLVSS under aeration

BOD fed / F:M ratio = 4170/0.27 = 15,444 lb MLVSS

15,444 / 0.8 = 19,305 lb MLSS

(7) Approximate excess MLVSS generated

$P_r = Y ·$ BOD$_5$ removed $− k_d ·$ MLVSS under aeration

Use the initial MLVSS under aeration for each day evaluation as a conservative estimate of the excess sludge generated.

$P_r = 0.5 · 4108 − 0.05 · 15,444 = 1282$ VSS generated.

(8) Calculate total MLSS generated
Mixed liquor suspended solids, assume MLVSS/MLSS = 0.8
MLSS generated = 1,282 / 0.8 = 1603 lb/day.
(9) Total solids under aeration

	VSS	Total SS
Initial	15,444	19,305
Generated	1,282	1,603
Total	16,726	20,908

(10) Estimate of reactor MLSS concentration

$$\frac{20,908 \text{ lb solids/day}}{1 \text{ mg} · 8.34 + 20,908/1,000,000} ≡ 2,500 \text{ mg/L MLSS}$$

(11) Solids balance for seven days

Day	BOD removed	Solids produced MLVSS	Solids produced MLSS	Total Solids End MLVSS	Total Solids End MLSS
Base				15,444	19,305
1	4108	1282	1603	16,726	20,908
2	2440	384	480	17,110	21,388
3	1606	−53	−53	17,057	21,335
4	4942	1618	2023	18,675	23,358
5	5776	1954	2443	20,629	25,801
6	5776	1856	2321	22,485	28,122
7	4108	930	1163	23,415	29,284
	28,756	7971	9979		

Final MLSS concentration
= 29,284/(1 MGD · 8.34 + 29,284/1,000,000) ≅ 3,500 mg/L

Procedures for Daily Estimates
a. BOD Removed = 1 MGD flow · 8.34 · (Co − 7.45)
b. Sludge produced (MLVSS) = BOD Removed · 0.5 − VLSS · 0.05
 where VLSS is the previous hour day VLSS under aeration
c. Sludge produced MLSS = MLVSS/0.8
d. Total sludge inventory at the end of the day

$$+ \frac{\text{Previous day MLVSS or MLSS}}{\text{Current day MLVSS or MLSS}}$$

$$\overline{\text{End of current day MLVSS or MLSS}}$$

(12) Assume a reasonable MLSS concentration after decant
 Assume 10,000 mg/L
(13) Calculate the total MLSS volume after decantation
 a. Total gallons of MLSS

$$29{,}284 \text{ lb solids/day} \cdot \frac{1{,}000{,}000}{10{,}000} \cdot \frac{1}{8.34} = 351{,}000 \text{ gal}$$

 b. Volume new sludge generated

$$351{,}000 \cdot [9{,}979/29{,}284] = 120{,}000 \text{ gal}$$

 c. Volume base sludge generated

$$351{,}000 \cdot [19{,}305/29{,}284] = 231{,}000 \text{ gal}$$

(14) Working volume after settling (final day before decant)

 1,000,000 + 351,000 = 1,351,000 gal/day (see discussion)

(15) Estimate the reactor volume
 Number of tanks to be used: assume 3
 Cycles per reactor per day: 1
 Determine the reactor volume at maximum sludge storage:
 1,000,000/3 + 351,000/3 = 450,000 gal each reactor.
 Each reactor ≅ 333,000 decant + 117,000 @ 1% sludge
(16) Determine reactor dimensions
 a. Assume reactor depth; assume 20 ft
 b. Determine reactor area

$$450{,}000 \cdot 0.1337 \text{ ft}^3/\text{gal}/20 = 3008 \text{ ft}^2$$

c. Establish diameter or width and length

 Three 40 × 80 rectangular basins, with a *L/W* 2:1.

d. Determine liquid decant level

$$20 \text{ feet} \cdot [333{,}000/450{,}000] = 14.8 \text{ ft}$$

e. Total sludge level

$$20 − 14.8 = 5.2 \text{ ft sludge storage}$$

f. Determine excess sludge level

$$5.2 \cdot [9{,}979/29{,}284] = 1.8 \text{ ft or 22 in.}$$

g. Determine residual sludge level

$$5.2 − 1.8 = 3.4 \text{ ft}$$

(17) Operations

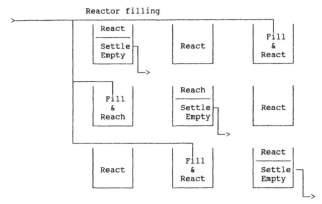

(18) Operating Parameters

Reaction time:	12 to 16 hr per reactor
Full time:	8 hr per reactor
Settling time:	2 hr per reactor
Decant time:	2 hr per reactor

(19) Decant pumpout:

 Daily Decant: 333,000/(120 mins) = 2775 GPM.

(20) Accumulated sludge pumpout:
 Once per week, 120,000/3 = 40,000 gal/reactor
 40,000/15 min ≅ 2700 GPM, similar to the decant pump rate.

(21) Aeration requirements
 Maximum BOD$_5$ removed per day: 5776
 Maximum waste solids per day: 1954
 BOD$_5$ to ultimate BOD ratio: 0.68
 Ultimate BOD per waste solids: 1.42
 Daily carbonaceous oxygen demand:

$$5776/0.68 − 1954 \cdot 1.42 = 5719 \text{ lb/day}$$

(22) Daily nitrification oxygen demand:

$$1 \text{ MGD} \cdot 8.34 \text{ lb/gal} \cdot 50 \text{ mg/L nitrogen} \cdot 4.57 \text{ mg O}_2/\text{mg } N$$
$$= 1906 \text{ lb/day}$$

(23) Maximum oxygen demand:

$$5719 + 1906 = 7625 \; total \text{ pounds oxygen per day.}$$

$$7625 / 3 \text{ Tanks} = 2542 \text{ pounds per day per tank.}$$

Reaction for 12 to 16 hr, assume 12 hr: so that, 2542/12 hr $\equiv$ 212 pounds per hour *demand* each reactor.

(24) Mixing requirements
(25) Reactor working volume: $\equiv$ 450,000 gal $\equiv$ 60,000 ft^3.
(26) Mixing intensity: 0.7 to 1.5 hp/1,000 ft^3.
(27) Mixing requirements: 60,000 $\cdot$ 1.5/1,000 $\equiv$ 90 hp
45 to 90 hp required for mixing.

(28) Nutrient requirements
(29) Maximum SBOD5 removed: 5776
(30) Nitrogen required:

Available nitrogen: 1 MGD $\cdot$ 8.34 $\cdot$ 50 mg/L = 417 lb/day

Required nitrogen: 5776 $\cdot$ 5 lb/100 BOD = 289 lb/day

Excess nitrogen: = 128 lb/day

(31) Phosphorus required

5776 BOD $\cdot$ 1 lb *P*/100 BOD = 58 lb *P* per day.

DISCUSSION

In estimating the total reactor volume an adjustment was not made for the water included in the 1% generated sludge, which is also in the 1,000,000 gallon decant volume. If the water is deducted from the excess sludge the reactor volume is reduced to 411,000 gallons, which within the accuracy of this calculation is close to the 450,000 gallons selected, allowing 10% additional working volume. In fact, the operating volume will be controlled by the selected cycle, especially the frequency of decant and excess sludge discharge. Finally, at a million gallons a day three chambers of 450,000 gallons was estimated, a total volume of 1,350,000 gallons, which is larger than the 920,000 gallon aeration basin for the activated sludge system. An cost analysis must be made to determine the difference in capital and operating costs for the two systems.

CASE STUDY NUMBER 13

Develop a suspended growth denitrification system for the effluent discharged from Case Study 10, with the following characteristics:

Flow, MGD: 1
BOD$_5$, mg/L: 500
COD, mg/L: 1250

Assume all residual ammonia nitrogen is converted to nitrate nitrogen: 25 mg/L nitrate nitrogen

Applicable design criteria include

Effluent N-nitrate: 1.25 mg/L or less; 95% destruction
Design temperature, °F: 68
Basin depth, feet: 10
Basin surface length to width: 2
Base the design on the following kinetic constants at 20°C:

k_d: 0.04 (1/day)
Peak nitrate removal rate, q_{max}: 0.20 (1/day)
K_s: 0.15 mg/L
Y: 0.9

PROCESS CALCULATIONS

(1) SRT Evaluation
(2) Calculate cell growth rate:

$$\mu = Y \cdot q_{max} - k_d$$

$$\mu = 0.9 \cdot 0.2 - 0.04 = 0.14 \text{ lbs/day-lb}$$

(3) Calculate minimum SRT:

$$SRT_e = 1/\mu = 1/0.14 = 7.1 \text{ days}$$

Using a safety factor of 2, $SRT_e = 14$ days

(4) Calculate design nitrate "q" removal rate

$$1/SRT = Y \cdot q - k_d$$

$$1/14 = 0.9 \cdot q - 0.04$$

q = F/M applied = 0.12, nitrate removed/lb MLVSS/day.

(5) Calculate effluent nitrogen/nitrate concentration

$$q_d = 0.12$$

$$q_{max} = 0.20$$

$$K_s = 0.15$$

$$q_d = q_d \, q_{max} \frac{D}{K_s + D}$$

$$0.12 = 0.20 \cdot D/(0.15 + D)$$

Solving for D, $D = 0.23$ mg/L nitrate nitrogen; *meeting* the required 1.25 mg/L design criteria.

(6) Select a MLVSS concentration design MLVSS at 2500 mg/L.
(7) Calculate HRT

$$F/M = q_d = (D_o - D_e)/(X \cdot HRT)$$

$$0.12 = (25 - 0.23)/(2500 \cdot HRT)$$

$$HRT = 0.08 \text{ days} = 2 \text{ hr}$$

(8) Calculate reactor volume

$$1 \text{ MGD} \cdot 0.08 \text{ days} = 0.08 \text{ MG volume}$$

(9) Calculator reactor sludge inventory

Aeration basin sludge inventory = MLVSS $\cdot$ volume (MG) $\cdot$ 8.34

$$= 2500 \cdot 0.08 \cdot 8.34$$

$$= 1668 \text{ lb MLVSS}$$

(10) Wasting rate

Sludge inventory/SRT = 1668/14 = 119 lb MLVSS/day wasted

Assume 0.8 mg MLVSS/ mg MLSS

119/0.8 = 149 lb MLSS/day

Assume 10,000 mg/L (1%) from clarifier underflow

149 / 0.01 = 14,900 lb slurry per day

14,900/8.34 lb/gal = 1787 gal/day.

(11) Methanol requirements per day

Pounds nitrate nitrogen removed = 1 MGD $\cdot$ 8.34 $\cdot$ (25 − 0.23)

$$= 207 \text{ lb/day}$$

Assume 3 mg methanol/mg nitrate nitrogen removed.

3 $\cdot$ 207 = 621 lb methanol per day

(12) Alkalinity produced per day

Assume 3 mg $CaCO_3$ produced/mg nitrate nitrogen removed.

3 $\cdot$ 207 = 621 lb $CaCO_3$ produced per day

Equivalent to 621/(1 MGD $\cdot$ 8.34) = 74 mg/L in influent. Influent will have to be adjusted accordingly if equivalent alkalinity is not depleted in nitrification process.

(13) Basin Design

Required working volume: 0.08 MG = 10,700 ft^3

Number of basins: assume two compartments for flexibility

Basin working depth: 10 ft

Projected *total* surface area: 1070 ft^2

Projected surface area per basin: 535

Assume 2:1 length to width

Surface width each basin: 20 ft

Surface length each basin: 40 ft

Freeboard: 2 ft

Total basin depth: 12 ft

REFERENCES

1. Arora, M.L., Barth, E.F., Umphres, M.B.: "Technology Evaluation of Sequencing Batch Reactors," *Journal WPCF*, V 57, No 8, Pg 867, August, 1985.

2. Cheremisinoff, P.N.: *Encyclopedia Environmental Control Technology*, Gulf Publishing Co., 1989; Volume 3, Chapter 7: Biokinetic Constants in Activated Sludge.

3. Chudoba, J., et al.: "Control of Activated Sludge Filamentous Bulking. I. Effect of the Hydraulic Regime or Degree of Mixing in an Aeration Tank," *Water Research*, V 7, Pg 1163, 1973.

4. Chudoba, J., et al.: "Control of Activated Sludge Filamentous Bulking. II. Selection of Microorganisms By Means of a Selector," *Water Research*, V 7, Pg 1389, 1973.

5. Chudoba, J., et al.: "Control of Activated Sludge Filamentous Bulking. III. Effect of Sludge Loading," *Water Research*, V 8, Pg 231, 1974.

6. Chudoba, J., et al.: "Control of Activated Sludge Filamentous Bulking—Experimental Verification of a Kinetic Selection Theory," *Water Research*, V 19, No 2, Pg 191, 1985.

7. Chudoba, J., et al.: "Control of Activated Sludge Filamentous Bulking. VI. Formulation of Basic Principles," *Water Research*, V 19, Pg 1017, 1985.

8. Cheremisinoff, P.N.: *Encyclopedia Environmental Control Technology*, Gulf Publishing Co., 1989; Volume 3, Chudoba, J. (Chapter 6): "Activated Sludge-Bulking Control."

9. Eikelboom, D.H.: "Filamentous Organisms Observed in Activated Sludge," *Water Research*, V 9, Pg 365, 1975.

10. Govind, R., et al.: Integrated Model for Predicting the Fate of Organics in Wastewater Treatment Plants," *Environmental Progress*, V 10, No 1, Page 13, February, 1991.

11. Irving, R.L., et al.: "Analysis of Full Scale SBR operation at Grundy Center, Iowa," *Journal WPCF*, V 59, No 3, Pg 1132, March, 1987.

12. Lawrence, A.W. and McCarty, P.L.: "Unified Basis for Biological Treatment Design and Operation," *Journal Environmental Engineering, Proceedings ASCE*, V 96, No 6, Pg 757, 1970.

13. Levenspiel, O.: *Chemical Reaction Engineering*, John Wiley and Sons, 1972, 2nd Edition.

14. Medcalf & Eddy, Inc.: *Wastewater Engineering-Treatment, Disposal, Reuse*, McGraw-Hill, 1991, Third Edition.

15. O'Connor, J.T. (Ed.): *"Environmental Engineering Unit Operations and Unit Processes Laboratory Manual,"* Association of Environmental Engineering Professors, 1972.

16. Perry, J.H.: *Perry's Chemical Engineers' Handbook*, McGraw-Hill Book Company, Inc., 1950, Third Edition.

17. Qasim, S.R.: *Wastewater Treatment Plants; Planning, Design, and Operation*, Technomic Publishing Co., Inc., 1994.

18. Richard, M.: *Activated Sludge Microbiology*, Water Pollution Control Federation, 1989.

19. Shields, F.D. Jr., Thackston, E.L.: "Designing Treatment Basin Dimensions to Reduce Cost," *Journal San Eng Div., Proceedings ASCE*, V 117, No 3, p 381, May/June, 1991.

20. U.S. Environmental Protection Agency: *Design Guides for Biological Wastewater Treatment Processes*, 11010 ESQ 08/71, August 1971.

21. U.S. Environmental Protection Agency: *Process Design for Nitrogen Control*, PB-259-149/38A, October 1975.

22. U.S. Environmental Protection Agency: *Treatability Manual*, EPA-600/8-80-042a, 1980.

23. U.S. Environmental Protection Agency: *Handbook for Identification and Correction of Typical Design Deficiencies at Municipal Wastewater Treatment Facilities*, EPA-625/6-82-007, 1982.

24. WPCF Manual of Practice No 8: *Wastewater Treatment Plant Design*, Water Pollution Control Federation, 1982, 2nd Edition.

25. WEF Manual of Practice: *Design of Municipal Wastewater Treatment Plants*, Water Environment Federation, 1992.

26. Ying et al.: "Biological Treatment of Landfill Leachate in Sequencing Batch Reactors," *Environmental Progress,* V 5, No 1, Pg 41, February, 1986.

27. U.S. Environmental Protection Agency: *CERCLA Site Discharges to POTWS Treatability Manual,* EPA 540/2-90-007, August 1990.

28. Mynhier, M.D. and Grady, C.P.L.: "Design Graphs for Activated Sludge Process," *Journal Environmental Engineering, Proceedings ASCE,* V 111, p 829, 1975.

29. Lesperance, T.W.: "A Generalized Approach to Activated Sludge," *Water Works and Wastes Engineering,* Pg 52, May, 1965.

Biological Oxidation: Lagoons

Lagoons are primarily employed to destroy limited quantities of biodegradable organics, based on a reasonable sized system.

BASIC CONCEPTS

GENERAL DESCRIPTIONS AND DEFINITIONS

LAGOONS are economical methods of biologically treating organic wastes but have definite limitations as sole systems because the pond activity cannot be adequately controlled. Effluent quality from industrial systems treating highly variable industrial wastes can be erratic because process control is limited to manipulating flows to individual basins and remedial action when apparent pond activity is reduced. As effluent requirements become more stringent, lagoons become less viable as primary treatment units and are better applied as contributing elements of more complex systems.

Figure 4.1 illustrates the basic aerated lagoon and stabilization pond. Feed (Q, S_o) enters the aeration basin having a capacity of V gallons and a corresponding retention time of t days. Stabilization ponds are similar to aerated lagoon, except that mechanical aeration is limited or not employed at all and oxygen supply is often governed by natural diffusion. Lagoons operate in a once-through aerobic, anaerobic, or combined mode, employing uncontrollable food-to-microorganism (F/M) contact mechanisms. The principal design variable is pond volume, usually divided into subunits to improve system effectiveness.

In essence, lagoons substitute *time* for the many control variables available in an activated sludge system. They are an early attempt by engineers to substitute a single unit operation to accommodate the multistage municipal treatment processes, such as primary treatment, biological reaction, final sludge settling, sludge recycling, sludge wasting, and sludge digestion. They employ a natural oxygen supply, relatively low energy mechanical aeration, or a combination. The price for eliminating sludge management and high-capacity aeration equipment is greater land area and less process control.

Lagoon (or pond) classifications are confusing and not always consistent in the literature. For the purpose of industrial design, they can be classified into four categories:

(1) (Mechanically) aerated lagoon
(2) (Aerobic/anaerobic) facultative pond
(3) Aerobic (photosynthesis) pond
(4) Retention pond

In addition, enhanced lagoon designs include dual-power aerobic units, where the system performance and economics are optimized by using multiple units combining completely and partially (or no) mixing units operating in series.

There are *three* factors that characterize the differences among an aerated lagoon, a facultative pond, or an aerobic pond:

(1) *Solids storage* capacity
(2) *Aerobic* or *anaerobic conditions*
(3) *Oxygen* supply

The specific operating mode is established by controlling the degree of external aeration (and mixing) and the basin depth. The suspended solids generated depends on the influent suspended solids concentration and biological conversion of degradable organics to waste sludge. Generally, ponds or lagoons do not employ direct solids separation. Generated solids are either discharged with the effluent or deposited at the basin bottom for anaerobic decomposition and eventual dredging.

 AERATED LAGOON

A partial or completely mixed lagoon is one in which aerobic conditions are maintained throughout the basin.

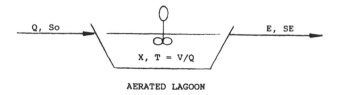

AERATED LAGOON

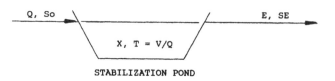

STABILIZATION POND

Figure 4.1 Lagoon or pond sketch.

When the energy level is adequate to maintain the solids in complete suspension, the system is referred to as an *aerated lagoon* (or an *aerobic lagoon*). As the power level is reduced solids settle and the system is referred to as a *facultative lagoon*. In an aerated lagoon all generated solids are dispersed, some destroyed aerobically as part of the endogenous reaction, and excess solids discharged with the effluent. This type of lagoon generally requires the least detention time because the system utilizes the biomass (suspended solids) to the fullest. However, of the various types of lagoons, the aerated lagoon power requirements are maximum because mixing and aeration are supplied by applied external mechanical energy.

FACULATIVE LAGOON

An *absence* of mixing in a *deep* basin results in a *facultative stabilization pond* where solids settle and two basin zones are evident, an aerobic upper layer for substrate removal and an anaerobic bottom layer for solids digestion. Solids are not continuously removed from the system.

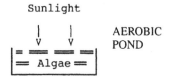

AEROBIC POND

The *absence* of mixing in a *shallow* pond results in a *high rate stabilization* or an *aerobic pond*. It is maintained aerobic solely as a result of photosynthesis reactions producing an abundance of algae growth. The pond depth establishes whether it functions as a facultative or an aerobic system. As long as oxygen can be supplied to the entire system depth it will be aerobic, with facultative conditions developing

with either heavy dissolved and suspended organic loads or depths greater than that influenced by the photosynthesis reactions.

RETENTION POND

Under suitable climatic conditions, and where land is available, systems are sometimes designed on the basis of facultative criteria and operated either as controlled discharge ponds or complete retention ponds. Retention ponds serve as solids settling treatment units, improving effluent quality and eliminating or controlling final discharge volumes. When these ponds operate under aerobic conditions for final effluent treatment, they are referred to as polishing ponds.

Anaerobic ponds are not discussed because the many regulations governing volatile emissions and odor make anaerobic treatment best conducted in closed vessels where the operation can be better controlled. The Process Engineer should refer to Chapter II-8, Anaerobic Processes section for applicable design procedures.

The previous discussion is intended to theoretically describe commonly accepted definitions but probably do not completely apply to most operating pond systems. In evaluating or classifying facilities, the Process Engineer is cautioned to review the operating parameters of any system, and not the name used, because *pond, lagoon, aerated, aerobic, anaerobic,* and *facultative* are often used as a matter of convenience. The reasons being that the primary and overwhelming "desktop" consideration for pond design is costs, and the most common criterion is loading rate. In actual operation, these systems are routinely upgraded to include recirculation, minimal mechanical aeration, *odor control,* sludge dredging, etc., in an attempt to meet the proposed "desktop" criteria and retain the treatment system in the available land area.

One overwhelming consideration *restricting* the initial use of these systems for industrial facilities is the manufacturing production value of land for future earnings. The primary reason for *abandoning* existing facilities is the overwhelming number of neighboring complaints and the legal costs of defending permit noncompliances.

REACTION KINETICS

Reaction kinetics for ponds and lagoons are related to a once-through reactor configuration. They reflect the degree of mixing applied as defined by the two limiting ideal models—a completely mixed or plug flow system or nonideal models interpreting the effects of partial mixing. In the absence of mechanical aeration, the efficiency of converting energy exerted onto the pond into oxygen must be reflected in the kinetics.

In one form or another, investigators have proposed completely mixed kinetic relations as a method to estimate the performance of just about any lagoon or pond, utilizing *modified* first-order constants to account for mixing, substrate, and oxygen level. These models are commonly discussed in municipal treatment texts [2,8,20,21]. The suggested procedures, for one reason or another, have been refuted as not realistically describing operating conditions [14,15]. Indeed, kinetic evaluations of operating municipal systems suggest that none of the models proposed, or generally applied design criteria, are universally acceptable in predicting the complexity of stabilization basin variables and corresponding performance [1,4]. However, suitable alternatives are either not available or too qualitative in nature. Needless to say, none of the kinetic models cited should be used for industrial application without extensive site-specific pilot evaluation.

Industrial pond design is extremely difficult because interpretation of available operating data is a formidable task, requiring that the Process Engineer utilize reported, and sometimes conflicting, loading and retention time data, without any assurance of effluent quality. The preferred alternative is to develop site-specific data, in which case the model used becomes academic because test results can be collated to one of the common kinetic expressions and used to estimate pond requirements in the range evaluated. In any case, the Process Engineer must consider the limits of proposed kinetic model or design criteria, compared with the specific application. The following considerations are critical in designing a lagoon or pond:

(1) Comparison of industrial waste characteristics relative to the design model, or proposed design criteria limitations

(2) Whether the pond is to operate in a completely aerobic, anaerobic, or combined condition

(3) The availability of oxygen to the system, whether external (controllable) or from natural sources (uncontrollable)

(4) The degree of mixing in the system

(5) The depth of the reactor

LAGOON PROCESS EVALUATIONS

Table 4.1 summarizes some general aerated and stabilization ponds characteristics, as cited by the U.S. Environmental Protection Agency (EPA) [20].

AERATED LAGOON

An aerated lagoon can be used to treat industrial wastes containing dissolved organics and relatively low nondegradable solids content. Low nondegradable influent solids are essential because they will add to the effluent suspended solids concentration, impacting effluent quality. The area dimensions of a lagoon are not generally critical, configured to appropriate dimensions to fit available land area. Depths can range from 2 to 6 m (6 to 20 ft), governed by required retention times, selected surface area, and the aeration device limits.

Low reactor microorganism concentration dramatically affects lagoon performance, especially in winter conditions. Treating high influent biological oxygen demand (BOD) concentration, whether dissolved organics or degradable solids, is difficult because of the low microorganism concentration. Sizing the basin for effective treatment of extremely high organic loading is impractical because at some influent condition the cost of reactor volume, and corresponding large mixing requirements, exceeds the anticipated savings in sludge management and recycle costs. In most cases, the principals governing aerated lagoon design follow those discussed for activated sludge, except that the hydraulic retention time governs microorganism growth. Aerated lagoons are theoretically described as *completely mixed systems without recycle*. The design of an aerated lagoon involves most of the activated sludge design considerations governing (1) substrate removal and effluent criteria, (2) minimum mixing requirements, (3) oxygen capacity and the selection of suitable aeration devices, and (4) evaluation of stability factors such as temperature and washout conditions.

Influent solids control is an important consideration in aerated lagoon design. Because biodegradable substrate de-

TABLE 4.1. General Pond and Lagoon Criteria (adapted from Reference [20]).

	Facultative	Aerated	Aerobic	Anaerobic
LOADING				
kg BOD$_5$/ha/day	22–67		85–170	
lb/acre/day	20–60		75–152	
kg BOD$_5$/1000 CuM/day		8–320		160–800
lb BOD$_5$/1000 CF/day		0.5–20		10–50
DETENTION				
days	25–180	7–20	10–40	20–50
DEPTH				
m	1.2–2.5	2–6	0.3–0.45	2.5–5
ft	4–8	6–20	1–1.5	8–16

struction requires aeration and basin volume, primary settling is a necessary consideration, sometimes included as a first stage quiescent chamber. Because some degree of mixing is inherent in an aerated lagoon, most excess solids generated are transported with the effluent. Because aerated lagoons do not include a final clarifier, effluent solids control may be necessary to achieve stringent effluent quality for direct discharges. Effluent solids include the feed nonbiodegradable solids and the net waste sludge generated under endogenous conditions.

Aeration and mixing is supplied by external aeration devices, although some minor photosynthesis and surface reaeration could occur in the upper 0.30 to 0.45 m (1 to 1.5 ft) levels. Because of the basin sizes the applied aeration level is cost driven; systems must be economically evaluated on the basis of a completely mixed, partially mixed, or dual-power systems. The input energy is governed primarily by the oxygen requirements, supplemented by the desired mixing to establish the required level of solids suspension. Generally, 3 to 6 kW/1000 m³ (15 to 30 hp/Mgal) of reactor volume are reported as adequate for solids suspension or 1 to 3 kW/1000 m³ (5 to 15 hp/Mgal) for oxygen dispersion and partial solids suspension [6,20].

Completely Mixed Systems

Completely mixed aerated lagoons are designed as once-through biological reactors, with effluent volume the same as the forward flow and its concentration equal to the reactor contents. BOD removal and the effluent characteristics are commonly estimated assuming Monod or first-order reaction kinetics.

Monod Kinetics

With the understanding that in an aerated lagoon the hydraulic retention time (HRT) is equivalent to the solids retention time (SRT), the Lawrence and McCarty Monod operating equations for effluent concentration, sludge generation rate, cell growth and SRT, washout SRT, and limiting SRT are identical to those suggested for Activated Sludge in Chapter II-3 [7]. The equation for reactor microorganism concentration is as follows:

$$X = \frac{Y \cdot (S_o - S_e)}{1 + k_d \cdot \text{SRT}} \qquad (4.1)$$

with variables as defined for the equations cited.

First-Order Kinetics

First-order kinetics are commonly assumed in completely mixed aeration basin design, directly derived from the Monod relations, based on the substrate constant being negligi-

ble relative to the half-velocity constant. The equation for completely mixed reactors operating in series is as follows [20]:

$$C_n/C_o = \left[\frac{1}{1 + \dfrac{K_c t}{N}} \right]^N \qquad (4.2)$$

where C_n = effluent BOD$_5$ concentration in cell n, mg/L, C_o = influent BOD$_5$ concentration, mg/L, K_c = complete mix first reaction rate constant, day^{-1}, t = total hydraulic residence time in pond system, days, and N = number of ponds in series.

If unequal volume ponds are considered, the general equation can be expressed as follows:

$$\frac{C_n}{C_o} = \frac{1}{1 + k_{ci}\, t_i} \times \frac{1}{1 + k_{cii}\, t_{ii}} \times \cdots \frac{1}{1 + k_{cn}\, t_n} \qquad (4.3)$$

where K_{ci}, K_{cii}, K_{cn} = complete mix first-order reaction rate constant in each pond. Because of the lack of better information, all are generally assumed to be equal. t_i, t_{ii}, t_n = HRT in each pond, days.

Theoretically, the total pond area is dramatically affected by the number of (series) basins utilized. Operationally it is ineffective to have a single lagoon and seldom economical or practical to have more than two or three basins in the system.

Significantly, as the number of units in series increases the process configuration approaches that of a plug flow system.

The simplicity of these equations invites trouble, implying that a "guesstimate" of k results in an effective design, further simplified because k values are available in a multitude of publications and a supposedly "conservative" value can be selected by the inexperienced. The correct selection of a kinetic rate constant, consistent with the actual operating variables, is a critical and controversial variable in lagoon design, as discussed below.

Kinetic Rate Constants

Most completely mixed aerated lagoon procedures assume a Monod or a first-order kinetic relation established for municipal design, with the assumption that these relations apply for industrial wastewaters. Where these models are questioned, alternatives offered are vague and usually summarized with a need for more investigative efforts. Assuming that first-order kinetics are applicable, the Process Engineer should be aware of the following reported limitations:

(1) Because of low microorganism concentration, the basic first-order $k \cdot x$ term has been replaced with the product term "K." This could lead to the interpretation that microorganism are not vital in an aerated lagoon, which

in fact is not the case. The reality is that solids concentration is not a controllable variable, difficult to accurately sample and measure, and often neglected.

(2) Results at low temperatures may not be appropriately predicted by a first-order mechanism [1].

(3) The constant may not be an independent variable but highly affected by substrate concentration, temperature, retention time, pond loading, basin configuration, and mixing level [1,6,15–17].

(4) When the basin configuration deviates from complete mixing a more complex model may be required [15].

(5) In multiple-stage operation the coefficient may not remain constant for all completely mixed stages or for combined mix or partial mix systems [6].

(6) Finally, as with all industrial systems the validity of selected model constants must be verified.

The selection of the model constants is a significant design consideration, whether Monod or first-order kinetic relations are utilized. As a frame of reference, typical Monod constants are cited in Tables 3.2 through 3.5 in Chapter II-3. For a first-order relationship, a municipal system K_c value of 2.5 days^{-1} is considered suitable for an initial estimate [20]. Typical first-order constants for completely mixed first stage (domestic waste) systems have been reported as ranging from 3.0 to 4.8 days^{-1} at 20°C. Second-stage completely mixed lagoon constants have been reported as ranging from 0.39 to 0.49 days^{-1} at 20°C, and second stage partially mixed lagoon constants have been reported as ranging from 0.19 to 0.45 days^{-1} at 20°C [6]. These values are applicable to the conditions studied and should be subjected to the limitations previously cited for general application.

Studies involving completely mixed and partially mixed multiple ponds in series have indicated a variation of first-stage constants with loading, although secondary basin coefficients were somewhat constant and considerably lower [6].

Apparently, significant removal of biodegradable substrate components occurred in the initial stage resulting in a relatively constant second stage reaction rate and a corresponding substrate-independent constant. The variation of the first-order constant with biodegradability is significant for industrial wastes, as often reported in biological treatment design literature [2,6]. Study results for first- and second-stage completely mixed basins, suggested a variation of K_{20} with influent loading correlated as [6]:

$$K_{20} = K_s \cdot S_o^a \tag{4.4a}$$

$$\log K_{20} = \log K_s + a \cdot \log S_o \tag{4.4b}$$

where a is a constant reported at 0.8, K_s is reported at 0.0556 1/day at 20°C, and S_o is the influent BOD$_5$, mg/L.

The indication being that for highly biodegradable constituents the coefficient can be assumed to be first order, provided it is corrected for initial loading and temperature.

Temperature

It is generally assumed that the kinetic constant varies with temperature according to the relation:

$$K_t = K_{20} \cdot \S^{(T-20)} \tag{4.5}$$

where K_t is the removal rate at the lagoon temperature, K_{20} is the removal rate at 20°C, 1/day T is the temperature in °C, and $\S$ is the temperature correction coefficient cited as 1.085, ranging from 1.015 to 1.085 [6,20].

The temperature within an aerated lagoon can be estimated by conducting an energy balance around the basin, resulting in the equation [8,20]:

$$T_w = \frac{A \cdot \int \cdot T_a + Q \cdot T_i}{A \cdot \int + Q} \tag{4.6}$$

where T_w is the basin temperature, °F (°C), T_a is the ambient air temperature, °F (°C), T_i is the inlet wastewater temperature, °F (°C), A is the pond surface area, ft^2 (m^2), Q is the wastewater flow, Mgal/d (m^3/day), A is the surface area, ft^2 (m^2), $\int$ is a proportionality factor 0.000012 (0.5) for eastern United States, t is lagoon retention time, days, and d is basin depth, ft (m).

Mixed Liquor Solids

Unlike an activated sludge system, the mixed liquor volatile suspended solids (MLVSS) concentration is not a controllable variable but an equilibrium value dependent on the retention time, which Lawrence and McCarty expressed as Equation (4.1) [7].

Oxygen Requirements

The basic criterion for sustaining aerobic conditions is maintaining an oxygen level of greater than 0.5 mg/L, generally between 1 and 2 mg/L. The desired oxygen level can only be met by assuring that the carbonaceous and nitrification oxidation requirements are met, as detailed in Chapter II-2 and summarized by these equations.

$$\text{kg O}_2 \text{ per day} = \frac{Q(S_o - S_e)}{\tau} - 1.42 W_x \tag{4.7}$$

where Q is the waste flow in m^3/day, S_o is the influent BOD$_5$, kg/m^3, S_e is the effluent BOD, kg/m^3, W_x is the mass organisms wasted, VSS, kg/day, τ is a factor converting BOD$_5$ to BOD$_{ult}$ (commonly 0.68).

Where nitrification is significant, the added oxygen requirements can be estimated using the expression:

$$NOD = d\text{NH}_4^-\text{N} \cdot 4.57 \qquad (4.7a)$$

or a conservative estimate calculated using the expression:

$$NOD = TKN_o \cdot 4.57 \qquad (4.7b)$$

where NOD is the nitrogen oxygen demand, mass per unit time, $d\text{NH}_4^-\text{N}$ is the mass of ammonia nitrogen removed in a unit of time, and TKN_o is the total Kjeldahl nitrogen in the influent, mass per unit of time.

Sludge Wasting

The amount of excess sludge generated can be estimated using the Monod related expression:

$$P_x = \frac{Y \cdot Q \cdot (S_o - S_e)}{1 + b \cdot HRT} \qquad (4.8)$$

where HRT is the SRT, time, Q is the influent mass/time, Y is the yield ratio, mass cells produced/mass substrate used, S_o is the feed substrate concentration, mass/volume, S_e is the reactor substrate concentration, mass/volume, b is the microorganism decay coefficient, 1/time, and P_x is the excess microorganism rate, mass/time. All units must be consistent.

The reactor microorganism concentration can be estimated using Equation (4.1). However, because in a completely mixed system the reactor and effluent concentration are theoretically equal, the sludge wasted is approximately equal to

$$dX_v/dt = Q \cdot X \qquad (4.8a)$$

where X is the reactor microbial mass concentration, mass/volume.

Effluent Concentration

Effluent soluble BOD$_5$ concentration can be estimated using the appropriate kinetic model, as discussed previously. The *total* effluent concentration from an aerobic lagoon can be estimated using the Eckenfelder equation [2]:

$$BOD_5 \text{ effluent} = S_e + 0.3\, TSS_{eff} \qquad (4.9a)$$

or the Balasha and Sperber relationship [2]:

$$BOD_5 \text{ effluent} = S_e + 0.54\, VSS_{eff} \qquad (4.9b)$$

where TSS or VSS are in mg/L.

The effluent from aerobic lagoons are high in BOD$_5$ and suspended solids. The BOD$_5$ being high as a result of re-

maining dissolved organics and suspended microbial solids, the total solids being high as a result of no final clarification. Effluent clarification is difficult because of the dispersed nature of the mixed liquor solids, a result of reactor extended aeration conditions. One possible solution is to provide for an aerobic final settling basin to enhance solids separation. The design of such a final basin is critical because conditions favoring photosynthesis reactions, and algae production, must be prevented to avoid additional effluent contaminants.

Partially Mixed Systems

Mixing level in partially mixed systems is based on maintaining aerobic reactor conditions, with no attempt to completely disperse the suspended solids. The design is commonly assumed to follow first-order, completely mixed kinetics, utilizing an adjusted reaction rate coefficient [20]. Because of settled solids degradation these systems are frequently referred to as facultative aerated systems, and the reader should review that section for further design information.

Dual Power Aerated Lagoons

Dual-power aerated lagoons (see Figure 4.2) combine the criteria of series, completely mixed, and partially mixed lagoons, the initial stage being a completely mixed lagoon and the final stages being partially mixed. These units have received considerable attention, based on cost effectiveness and reduced energy considerations [5,6].

Kinetic expressions proposed for dual-power lagoons follow the first-order relationship generally proposed for ponds:

$$S_e = S_o/[(1 + K_i \cdot t_i) \cdot (1 + K_{ii} \cdot t_{ii})^n] \qquad (4.10)$$

where t_i is the detention time in the completely mixed tank, days, t_{ii} is the detention time in the each of the partially mixed tanks, days, K_i is the reaction rate coefficient for the completely mixed tank, 1/days, K_{ii} is the reaction rate coefficient for the partially mixed tanks, 1/days, S_o is the soluble BOD$_5$ influent concentration, mg/L, S_e is the soluble BOD$_5$ effluent concentration, mg/L, and n is the number of equally sized partially mixed reactors following one completely mixed reactor.

As with any pond system, the selection of a reaction coefficient for each stage is critical, as discussed previously.

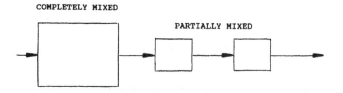

Figure 4.2 Dual power aerated lagoons (adapted from Reference [12]).

HIGH-RATE (AEROBIC) STABILIZATION PONDS

High-rate ponds, also called *aerobic ponds,* maintain aerobic conditions through photosynthesis reactions initiated by using energy exerted on the pond. As a result, ponds are limited to locations where sunny climates are present for significant periods of the year. Characteristically, these ponds are not deep, usually 30 to 45 cm (12 to 18 in.), to allow full-depth light penetration. The chief advantage of these systems are low operating costs. Their disadvantages include (1) intermittent operation because of their dependency on environmental conditions (available energy), (2) additional pond capacity is required for inactive or low activity periods, (3) large land requirements, (4) large quantities of algae must be harvested and disposed, and (5) poor operating control. Aerobic ponds will be effective only if considerable land area is available and located where climatic conditions favor significant energy capture to sustain the process. Aerobic pond performance is difficult to manage or predict because of the inability to control one of the process reactants, *oxygen.* As a result these systems are generally not considered as sole treatment but as either polishing ponds or a definition of the upper level of one variation of a facultative pond.

Aerobic ponds have limited use in industrial application as principal industrial treatment plants where strict effluent quality must be achieved or where large waste storage for nonproductive days is not feasible. In addition, applied wastes should be low in BOD, the bulk of which should be soluble. High suspended solids or dissolved organics could prohibit their use. Substantial required land area could eliminate large potential space for manufacturing expansion, or they could be located in remote areas where only neighboring complaints result in any attention.

Although aerobic ponds stabilize organics to carbon dioxide and water, similar to other aerobic biological processes, the reaction path critical to sustaining the process is significantly different from other carbonaceous oxidation processes. This is evident in the photosynthesis reactions describing the process:

(1) Organic carbon oxidation in stabilization ponds are similar to any other biological system, producing carbon dioxide and water. The significant difference is that ponds use these final products for the oxygen production. In the case of sewage this relationship can be represented as [10]:

$$C_{11}H_{29}O_7N + 14\ O_2 + H^+ = 11\ CO_2 = 13\ H_2O + NH_4^+$$

(2) Carbon dioxide is used to form algae, which in turn releases oxygen according to the relation:

$$NH_4^+ + CO_2 = algae + O_2 + H^+ - 886\ Kcal$$

(3) Theoretically, 2.3 g of carbon dioxide are required and 1.7 g of oxygen produced for every gram of algal cell synthesized.

(4) The sulfide reaction can be completed in two environments, aerobic or anaerobic. In an aerobic environment the sulfide is converted to sulfate. In anaerobic conditions, sulfides are converted to hydrogen sulfide.

Process design is at best an approximation involving estimates of (1) the effective energy used by the pond, (2) required pond area to capture available energy, (3) algae concentration, (4) hydraulic loading, and (5) critical pond depth. The difficulty in estimating pond sizes is establishing the energy level exerted to the pond, which is dependent on the plant location and which will change from daylight to night time, daily, monthly, and seasonally. In addition, an energy estimate must be made of the photosynthetic efficiency in utilizing the available energy. Because total yearly solar energy is maximum in sites located at 0 to 30 degrees latitude, the use of aerobic lagoons is questionable outside this latitude because of reduced energy levels from October to March. The design of facultative ponds are detailed in a study presented by Oswald [10].

FACULTATIVE PONDS

The facultative pond is the most common system employed because of costs, ease of operation, and its high loading capacity. They can be used for dissolved BOD industrial wastes with high solids content. Minimum oxygen is required as a result of anaerobic digestion of settled solids and adsorbed organics settling to the pond bottom. Settling is enhanced by long retention time and quiescence conditions. Facultative lagoons consist of an upper aerobic layer, a lower anaerobic layer, and a middle or transitional layer buffering the two environments. These systems are generally 1.2 to 2.5 m (4 to 8 ft) deep. Solids management is a general concern, with settled nondegradable solids requiring dredging and the upper aerobic layer prone to algae generation.

As commonly employed these systems are oxygen limited, dependent on photosynthesis processes to maintain adequate dissolved oxygen in the upper level. However, whether by design or upgrading, some facultative lagoons are provided with external aeration to supplement the highly variable photosynthesis processes. Where a supplementary oxygen source is not provided, total system performance will vary with diurnal and seasonal conditions, and excess storage capacity will be required to allow for low or negligible oxygen periods. Mechanically aerated systems are commonly referred to as facultative *lagoons,* and have the advantage of not generating algae. However, excessive mixing must be avoided to prevent "feedback" organic loads to the aerobic section from the anaerobic bottom.

The major design criterion for facultative ponds is substrate soluble BOD removal, although the system design can

TABLE 4.2. 1969 Facultative Stabilization Pond Study (adapted from Reference [3]).*

	North	Central	South
No. of States	18	17	15
Loading, kg/ha/day (lb BOD$_5$/acre/day)			
Mean	29 (26)	37 (33)	49 (44)
Range	18.7–45 (16.7–40)	19.5–90 (17.4–80)	34–56 (30–50)
Median	24 (21)	37 (33)	56 (50)
Detention time, days			
Mean	117	82	31
Range	30–180	25–180	20–40
Median	125	65	31

*Survey data reported in British units.

be significantly affected by organic feedback resulting from renewed and extensive sludge digestion during the summer months. For that reason, the system design has to be adjusted for this expected organic overload. The performance is also affected by thermal gradients in the spring and autumn, which can result in resuspension of disturbed bottom layer solids. These effects are difficult to include in the design basis and are either ignored or factored into the final basin size based on experience or as an "educated guess."

In general, proposed design methods fall into three categories (1) state or local requirements, as reported in Table 4.2, (2) areal loading rates, as reported in Table 4.3, or (3) empirical relations. State or local requirements are commonly developed for municipal design, stated as allowable winter and summer areal loading rates, which regulatory agencies may attempt to directly apply to industrial waste treatment ponds. The Process Engineer should be aware that because of varying industrial wastes biodegradability such loadings may not produce reported effluent quality, resulting in an ineffective design. General application of "commonly" reported surface loadings rates cited by the EPA, such as listed in Table 4.3, have the same limitations as mandatory state or local requirements, their questionable validity to all industrial wastes [20].

TABLE 4.3. EPA Cited Loading Rates (adapted from Reference [20]).

Temperature below 0°C (32°F)	BOD$_5$:	11–22 kg/ha/day 10–20 lb/ac/day
First cell limit	BOD$_5$:	40 kg/ha/day
First cell retention	Days:	120–180
Temperature 0–15°C (32–60°F)	BOD$_5$:	22–45 kg/ha/day 20–40 lb/ac/day
Temperature above 15°C (60°F)	BOD$_5$:	45–90 kg/ha/day 40–80 lb/ac/day
First cell limit	BOD$_5$:	100 kg/ha/day

The effect of climatic conditions on facultative stabilization ponds is evident from an extensive survey of reported regulatory requirements conducted by Canter and England [3] of the U.S. municipal facilities in 1969, summarized in Table 4.2. These criteria can serve as a guide for industrial system design.

Commonly cited design models include the complete mix, ideal plug flow, and nonideal models. These are summarized by the EPA in an extensive pond design manual [20] and discussed subsequently. The equations are based on the assumption that they can adequately describe the basin flow pattern and substrate utilization, assuming conditions that will maintain the integrity of both the aerobic and anaerobic sections.

Design Methods

Marais-Shaw Equation

This equation assumes a completely mixed model adhering to first-order kinetics.

$$C_n/C_o = \left(\frac{1}{1 + K_c T_n}\right)^n \tag{4.11}$$

where C_n = effluent BOD$_5$ concentration in cell n, mg/L, C_o = influent BOD$_5$ concentration, mg/L, K_c = complete mix first reaction rate constant, day^{-1}, T_n = hydraulic residence time in each pond, days and n = number of ponds in series.

As a practical matter, a maximum pond concentration of 55 mg/L is recommended for the initial stages to avoid complete anaerobic and odorous conditions. This limiting pond BOD$_5$ concentration (Ce_{max}^-) is related to the permissible depth by the equation:

$$(Ce_{max}^-) = 700 / (0.6 \cdot d + 8) \tag{4.11a}$$

where d is the depth in feet. The equation was developed for British units and should be used accordingly. Although the use of a completely mixed concept has been questioned, its use can be justified in that a complete mix configuration usually results in a conservative reactor volume estimate. However, a major limitation in the use of this equation is assigning a rate constant, which is difficult to establish even for domestic wastes. Benefield and Randall cite the following expression proposed by Mara for domestic wastes [2]:

$$K_t = 0.30 \cdot (1.05)^{(T-20)} \tag{4.11b}$$

Plug Flow Kinetics

This configuration assumes that a facultative lagoon performance is best defined by a plug flow configuration:

TABLE 4.4. Plug Flow Equation Coefficients (adapted from Reference [20]).

BOD₅ loading rate		
kg/ha/day	lb/acre/day	K_{20}, 1/day
22	20	0.045
45	40	0.071
67	60	0.083
90	80	0.096
112	100	0.129

$$S_e/S_o = \exp - K_{20} \cdot t \qquad (4.12)$$

The K_{20} coefficient is related to the influent loading as indicated in Table 4.4.

The coefficient can be corrected for temperature using the relation:

$$K_t = K_{20} \cdot 1.09^{(T-20)} \qquad (4.12a)$$

The plug flow relation will produce results similar to that projected for an infinite number of completely mixed reactors in series.

Nonideal Design Method

The use of an ideal complete mix model has been questioned by Thirumirthi and others, suggesting that facultative ponds operate neither as a complete mix or plug flow, rather in some intermediate flow pattern that can be related to the dispersion number [14,15], a completely mixed system having a dispersion number equal to infinity and plug flow system a value of zero. Two relations are proposed to express nonideal conditions.

(1) Accurate formula

$$\frac{C_e}{C_i} = \frac{4 \cdot a \cdot e^{1/(2d)}}{(1 + a)^2 \cdot e^{a/(2d)} - (1 - a)^2 \cdot e^{-a/(2d)}} \qquad (4.13)$$

(2) Modified Formula

$$\frac{C_e}{C_i} = \frac{4 \cdot a \cdot e^{1-a/2d}}{(1 + a)^2} \qquad (4.13b)$$

where C_e is the effluent BOD, mg/L, C_i is the influent BOD, mg/L, K is the first-order BOD removal coefficient, day^{-1}, t is the mean detention time, days, a is $(1 + 4 \cdot K \cdot t \cdot d)^{\frac{1}{2}}$, d is $D/U \cdot L = D \cdot t/L^2$, which is dimensionless dispersion number, D is the axial dispersion coefficient, m²/hr (ft²/hr), U is the fluid velocity, m/hr (ft/hr), and L is the length of fluid travel path from influent to effluent, m (ft).

A standard BOD removal coefficient (K_s) of 0.056/day at 20°C (ranging from 0.042 to 0.071/day) was determined

from municipal pond evaluations. The units were assumed to be operating under plug flow conditions and without any toxic components inhibiting algal growth. In fact, the standard coefficient varies with the pond hydraulics, and was measured as 0.5/day, based on linear BOD removal, at 20°C (ranging from 0.43 to 0.54) in subsequent complete mix aeration basin studies [14,22]. In the Thirumirthi method standard coefficients are based on the following conditions:

(1) The operating temperature is 20°C.
(2) The pond organic load is 672 kg per day per ha (60 pounds per day per acre).
(3) Industrial toxic chemicals are not present.
(4) A minimum (visible) solar energy of 100 langley per day is available.
(5) There is an absence of benthal load.

An applied corrected coefficient K is defined as

$$K = K_s C_o C_{te} C_{tox} \qquad (4.13b)$$

where K_s is the standard coefficient at 20°C for the appropriate pond hydraulics, 1/day, C_{te} is equivalent to $\sigma^{(T-20)}$, where σ is 1.036, and C_o is a correction factor for influent organic loading and is equivalent to:

$$C_o = 1 - [0.083 / K_s][\log (L'/L)] \qquad (4.13c)$$

where L' is 672 kg/day/ha (60 lb/day/acre), L is the organic loading in kg/day/ha (lb/day/acre), and C_{tox} is the toxicity correction factor imposed on municipal wastes resulting from industrial loads, as indicated in Table 4.5.

Theoretically, this method should better define the stabilization process if a suitable reaction coefficient and dispersion factor can be established. As with other theoretical methods

TABLE 4.5. Thirumirthi Toxicity Correction (adapted from Reference [14]).

Component	F mg/L	C_{tox}
Methanoic acid	180	2
	360	16
Ethanoic acid	270	1.6
Propanoic acid	180	2.65
Hexanoic acid	200	1.3
	300	3
1-Butanol	4000	2
Octanol	150	2
	200	4

Solutions of insecticides at 25% (or higher), at influent concentrations greater than 100 mg/L, resulted in complete algal destruction.
The compounds listed illustrate their toxic effects on pond activity. The reference should be reviewed for further details.
The toxic effects of any waste should be *experimentally determined*.
Stabilization lagoons are not an effective treatment method for toxic wastes, since the pond microorganism concentration is low, and the pond activity is easily terminated.

discussed, insufficient data are available to arbitrarily assign a rate coefficient for industrial facilities.

Nitrogen Removal

The EPA [20] proposed method for nitrogen removal is based on a first order ammonia removal model as defined by Equations (4.14a) or (4.14b). The relations approximates the ammonia loss in a pond by stripping and the combined biological processes.

For 1° to 20°C range

$$N_e/N_o$$
$$= \frac{1}{1 + A/Q \cdot (0.0038 + 0.000134 \times T) \cdot e^{(1.041 + 0.044T)(pH - 6.6)}}$$
$$(4.14a)$$

For 21° to 25°C range

$$N_e/N_o = \frac{1}{1 + A/Q \cdot (0.005035 \cdot e^{(1.540)(pH - 6.6)})} \quad (4.14b)$$

where A is the pond surface area, m^2, Q is the flow rate, m^3/day, T is the pond temperature, °C, N_o is the influent ammonia nitrogen, mg/L, and N_e is the effluent ammonia nitrogen, mg/L.

General Design Considerations

Limitations of the expressions cited for stabilization pond design are that (1) they were initially developed from municipal plant operating or laboratory data and (2) selection of the appropriate rate coefficient (or dispersion number) for an industrial waste is an educated guess. Selecting the appropriate constants may not be any more accurate than selecting a loading rate, retention time, and an appropriate depth and using as much available land as possible as a safety factor. Specific data can be obtained from pilot testing from which the data can be scaled up directly or "fitted" into any of the relations cited. The Process Engineer must be aware of the limitations of arbitrarily using the design methods discussed, proposing projected effluent quality as legal permit limits, without waste-specific evaluation.

Oxygen Requirements

Oxygen requirements and sludge production can be estimated utilizing the methods suggested for aerated lagoons. However, compensation must be made in facultative lagoons for soluble BOD feedback. Eckenfelder et al. [2] proposed the following expression:

Oxygen, kg (lb)/hr
$$= W \cdot F \cdot \text{ultimate BOD removed}/24 \quad (4.14)$$

where W is kg (Mlb) per day of feed, F is the feedback factor, 0.9 in winter and 1.4 during the summer, and ultimate BOD removed, ppm.

McKinney et al. proposed that the oxygen value can be estimated as 1.5 kg of oxygen per kg of ultimate BOD applied to the system [2].

COMPLETE RETENTION PONDS

Controlled discharge ponds are designed on the basis of the receiving waterway's capacity to assimilate the treated effluent. This results in restricted discharge periods, controlled storage volumes, well-defined treated effluent characteristics, and relatively short discharge periods. The general design criteria for these systems include

(1) Large basin volumes to allow adequate storage capacity.
(2) Allowance for restricted discharge on a seasonal basis; which could be a prolonged period covering months, twice a year, once a year, or site-specific imposed conditions. In many cases discharge periods are specifically detailed in the discharge permit, requiring prior discharge notification and approval, stringent effluent testing to meet specified quality limits, and monitoring of the receiving stream.

Process design criteria are not available for controlled discharge systems because the storage time is excessively greater than that of any kinetic considerations. Commonly applied criteria, cited by the EPA include [20]:

(1) Multiple cell design, a minimum of 3, with provision for flexible operation to include cell maintenance, parallel or series operation, and high storage capacity.
(2) Hydraulic detention times are site specific reflecting climatic conditions and expected discharge periods, with 6 months or more storage volume common.
(3) Initial stage liquid depths 2 m (6 ft) or less, with subsequent cell depths 2.5 m (8 ft) or less.
(4) Application of reaction kinetic considerations are not practical, but organic loading generally ranges from 22 to 28 kg BOD_5/ha/d (20 to 25 lb BOD_5/AC/d).

A special design consideration is a *total retention pond*, where *no* effluent is discharged. In such cases the wastewaters generated must be less than the annual natural evaporation, with adequate storage periods to balance waste generation with climatic related evaporation criteria and restricted seepage allowances. General system criteria include [20]:

(1) Annual moisture deficit must exceed 75 cm (30 in.).
(2) A large surface area should be provided to maximize the evaporation process.
(3) Design should be based on historical recorded maximum rainfall and minimum evaporation conditions.
(4) Provision should be made for alternate disposal methods.

Total retention ponds are especially useful if excess effluent, rich in nutrients, can be employed for irrigation purposes, with allowances for evaporation of excess volumes.

PROCESS ENGINEERING DESIGN

General process limitations and defining operating characteristics of lagoons and ponds are summarized in Table 4.6.
Aerated lagoons and ponds allow the operator minimum
flexibility in controlling the biological process parameters,
with the system effectiveness dependent on long retention
time and multistage configuration. Because of long retention
time, the resulting pond temperature is extremely sensitive to
climatic conditions and the heat loss resulting from aeration
equipment spraying. However long retention time does allow
acclimation and recovery to adverse influent conditions. A
secondary control is influent volume fed into the system, if
waste generation can be controlled. Pond effluent quality
is unpredictable, uncontrollable, and highly susceptible to
leakage, suspended solids carryover, and system upsets.
Overloading and anaerobic conditions are common in many
operating plants. The process can be cyclical with any number of conditions resulting in minimal biological activity
and poor treatment efficiency, with the only available remedy
being pond acclimation and regeneration.

REPORTED PERFORMANCE DATA

Based on results reported in the EPA Treatability Manual,
the effluent quality and treatment removal efficiencies of
operating aerated lagoons and ponds are summarized in Table 4.7 to 4.9 [18]. The data reported are for wastes from a
variety of industries and for the most part from full-scale
facilities. These data are cited to demonstrate the range of
performance possible, with no assurance that the facilities
reported were properly designed or operated or that these
are the best that can be achieved. In addition, performance
can be radically different for industrial facilities with higher

TABLE 4.6. Operating Characteristics.

Variable	Operator Controllable	Critical
Waste Characteristics		
Waste generated	No	Yes
Composition	No	Yes
Concentration	No	Yes
Biodegradability	No	Yes
Toxicity	No	Yes
Operating Characteristics		
Flow rate	Minimal	Yes
HRT		
Aerated lagoon	Depends flow	No
VSS level		
aerated lagoon	No	Yes
Mixing		
Aerated lagoon	Yes	Yes
Ponds	Recirculation	No
Oxygen capacity		
Aerated lagoon	Yes	Yes
Ponds	No	Yes
Wasting	No	Yes
Nutrients	Yes	Yes
Alkalinity	Yes	Yes
Temperature	No	Yes
Backfeed conc.		
Aerated lagoon	No	No
High rate pond	No	No
Facultative pond	No	Yes

loadings or wastes containing less biodegradable components.

REQUIRED PROCESS DESIGN DATA

Required process requirements are similar to those indicated for the activated sludge process, discussed in Chapter
II-3. Pilot plant procedures are detailed in municipal design
handbooks and in environmental laboratory manuals [8,9].
Resulting data can be tabulated or integrated into the Monod,
first order, or similar applicable lagoon models to obtain

TABLE 4.7. Aerated Lagoon Performance Data (adapted from Reference [18]).

	BOD₅	COD	TOC	TSS	O/G	TTL Phenol	TKN
Effluent, mg/L							
Minimum	6	92	47	3		0.003	22
Maximum	869	1610	573	1790		0.018	105
Median	90	591	126	155			
Mean	150	679	220	410	17	0.011	64
Removal, %							
Minimum	0	3	11	0		31	75
Maximum	>99	>99	99	99		>99	79
Median	78	63	46	24			
Mean	71	50	113	33	98	>65	77

TABLE 4.8. Polishing Lagoon Performance Data (adapted from Reference [18]).

	COD	TSS	TTL Phenol
Effluent, mg/L			
Minimum	142	22	0.028
Maximum	263	28	0.051
Mean	202	25	0.04
Removal, %			
Minimum	0	24	0
Maximum	52	76	46
Mean	26	50	23

design coefficients, establishing the required design data listed in Table 4.10.

WASTE EVALUATION

Although accounting for all applicable waste streams and establishing a appropriate design conditions is as important as in an activated sludge system, a high degree of ingenuity is seldom applied to designing these systems and where applied often ignored in their operation. Like other biological processes waste characteristics significantly affect lagoon operations. Lagoon and pond waste evaluation considerations are generally similar to those discussed in Chapter II-3 for activated sludge and will not be repeated.

Pond conditions will normally not be radically affected by minor waste flow and concentration variations or minimal toxic constituents because of the pond size. However, large flow surges could physically disturb the pond, causing increased influent leakage and bottom sludge disturbances because of excessive turbulence. Surges of biodegradable, nonbiodegradable, or toxic components are undesirable but not necessarily detrimental because of the inherent dilution capabilities of ponds if short-circuiting can be minimized. Constant organic overloading is a major pond problem because of a limited and uncontrollable oxygen supply. Anaerobic conditions in part or throughout the system is common

TABLE 4.9. Facultative Lagoon Performance Data (adapted from Reference [18]).

	BOD$_5$	COD	TSS	TKN
Effluent, mg/L				
Minimum	53	717	48	35
Maximum	274	2110	105	100
Median	152			
Mean	160	1410	76	68
Removal, %				
Minimum	77	55	74	33
Maximum	92	68	86	67
Median	87			
Mean	85	62	80	50

TABLE 4.10. Required Design Data.

> *Critical pilot plant treatability data specific to the waste*
> (1) Design temperature
>
> *Waste solids characteristics that should be obtained from laboratory studies but can be estimated*
> (2) Fraction of effluent solids biodegradable
> (3) Ultimate solids BOD per g solids
> (4) Ratio of SBOD to ultimate BOD
> (5) Ratio MLVSS/MLSS
>
> *Selected operating characteristics*
> (6) Required effluent quality
> (7) Pond depth
> (8) Appropriate surface loading
> (9) Appropriate retention time
>
> *Operating characteristics that should be obtained from pilot studies but can be estimated from treatability data*
> (10) Summer temperature effects
> (11) Winter temperature effects
> (12) Required oxygen loading
> (13) Available natural oxygen
> (14) Optional aeration requirements
> (15) Waste sludge generated
> (16) mg nitrogen required per mg SBOD removed
> (17) mg phosphorus required per mg SBOD removed

as a result of unmanageable and excessive feed conditions and poorly defined operating limits. Operator response is generally limited to remedial action after the pond condition has reached noticeable poor conditions, evident from a persistent odor or a visibly poor effluent.

PROCESS DESIGN VARIABLES

In developing a lagoon or pond design, the Process Engineer must consider criteria generally similar to an activated sludge system discussed in Chapter II-3. Those specific to lagoon or pond design include

(1) System configuration
(2) Uncontrolled variables
(3) Basin depth
(4) Mixing level
(5) Basin volume

System Configuration

Establish Whether a Lagoon or Pond is Applicable

Using loading rates indicated in Table 4.1 approximate the area required, and establish whether sufficient land is available.

Process Definition

Establish whether an aerated, facultative, or combined configuration is to be considered.

(1) If effluent quality is critical, then a multistage configuration is essential. An initial completely mixed stage followed by two or three stages of partially mixed or facultative processes should be considered. A final stage should be designed for solids settling.

(2) If low solids effluent quality is required, a final filter (or equivalent) should be considered.

Multistage Configuration

Develop a multistage configuration to optimize performance and operating flexibility.

Depth

Select the pond depths consistent with the aeration mode and storage (waste and sludge) requirements.

Mixing

Establish the mixing requirements consistent with the selected process.

Stability Considerations

Consideration should be given to influent deficiencies, including pretreatment for nutrient addition and alkalinity control as well as controlled feeding based on

(1) Seasonal or intermittent waste generation
(2) Cyclical influent waste characteristics
(3) Production of toxic waste influent on a cyclical or intermittent basis

Operational Enhancements

Operational flexibility should be evaluated, with emphasis on

(1) Effluent recycle to improve basin mixing and dilute influent conditions
(2) Supplementary aeration and mixing equipment

Solids Management

Solids management provisions should be included to harvest algae generated, intermittent bottoms dredging, and removal of the excess biological sludge generated. Final dewatering, treatment, utilization, and disposal is a critical consideration.

Controlled Discharge

Design attention should be given to a controlled discharge where effluent quality is stringent, discharge waterway quality highly regulated, or both.

Cost Evaluation

Establish the costs of the probable configuration, and viable alternatives. Do the economies warrant a pond instead of other biological alternatives?

Uncontrolled Variables

Selecting a pond as a sole treatment should be based on an understanding of the variables that significantly affect performance but are not controllable. These include oxygen supply, backfeed organics, pond temperature, and alkalinity.

Oxygen Supply

Under the best of conditions the maximum dissolved oxygen (DO) concentration available at ambient conditions is 8 mg/L. The ability to achieve this or any DO level is dependent on the applied method. Oxygenation rate is controllable when an external source is applied but uncontrollable and limited where photosynthesis or surface reaeration is the oxygen source. Unless wind activity is high, surface reaeration is insignificant and commonly neglected in the basin design. Photosynthetic activity depends on pond depth and available light. These environmental influences are difficult for the Process Engineer to estimate and impossible for the operator to control. Therefore, if natural aeration is a basis of design, the engineer must compensate by providing for adequate storage capacity and supplementary aeration.

Backfeed Organics

The unpredictability of an oxygen supply is further complicated by backfeed BOD resulting from overloaded, lower-level, anaerobic digestion. Pond overloading commonly occurs after reduced or dormant winter activity, followed by renewed high summer activity, and further intensified by high thermal gradient mixing. This process could result in an excessive backfeed of oxygen-demanding components to the upper aerobic layers and a potential overloading of the entire system. The engineer should provide for (1) excess treatment capacity, (2) alternate treatment methods, and/or (3) supplementary aeration.

Temperature

Long retention times results in the basin contents being driven toward ambient conditions, reduced temperatures during winter conditions, and elevated summer temperatures.

Process temperature affects (1) the selective biological species influencing the reaction, (2) the reaction rate, (3) the physical condition of the basin, and (4) basin backmixing resulting from density gradients. Seasonal changes result in reduced biological activities during cold conditions, accelerated activities during warm temperatures, and considerable pond turbulence during transitional periods. Adjusting the design criteria for these conditions is difficult, if not impossible, and further complicates process controllability.

Acidity/Alkalinity (pH)

Pond alkalinity is impacted by the system carbonate chemistry, represented by the primary equations [20]:

$$\leftarrow CO_2 + H_2O = H_2CO_3 = HCO_3^- + H^+$$

$$HCO_3^- = CO_3^= + H^+$$

$$CO_3^= + H_2O = HCO_3^- + OH^-$$

The removal of carbon dioxide through photosynthesis reactions triggers a chain of equilibrium reactions causing a decrease in the hydrogen and bicarbonate concentration, resulting in increased pH. This chemistry is affected by the photosynthesis process, the anaerobic digestion process, and the influent alkalinity. Therefore, factors that influence these variables will influence the system pH, which could vary considerably as a result of daily or seasonal changes. Provisions should be made to monitor pond alkalinity, and to adjust severe acidity or alkalinity conditions that could radically affect pond chemistry.

Depth

Basin depth is related to the selected process, as discussed under Process Configuration, and limited when natural aeration is the primary process mechanism. Generally, basin depths employed are in the range indicated in Table 4.11.

Factors influencing basin depth selection include

(1) Maximum depth is desired for winterization of the unit.
(2) Where external aeration is applied, the device operating characteristics could limit the depth.

(3) Stabilization ponds less than 0.9 m (3 ft) are difficult to maintain because high rate aerobic conditions occur, encouraging excessive vegetation growth.
(4) At depths greater than 1.5 m (5 ft) highly anaerobic conditions are difficult to avoid unless mechanical aeration devices are employed.
(5) In facultative lagoons, where aerobic and anaerobic conditions are encountered, a buffer zone is required to prevent oxygen transfer detrimental to the anaerobic layers. This results in depths of 1.5 m (5 ft) or higher.

All the design parameters are interrelated. Retention time establishes basin volume and loading establishes area requirements, both of which sets the required depth, which in turn defines the basin geometry. Any of these physical conditions may have to be altered because of dominant operating or design limitations. Except for mechanically aerated (completely mixed) systems, selected depth acutely affects the system oxygen capacity. When oxygen supply depends on photosynthesis reactions, operating depth affects pond stability and performance because light intensity decreases with increasing depth. As a practical consideration, a variable weir should be included to adjust the basin level, allowing adjustment of operating volume and pond depth. Depth control is essential in controlling pond freezing and excess algae growth, and excess volume allows waste storage during dormant conditions.

Mixing Level

There are two mixing levels that can be applied to a pond or lagoon treatment system, the minimum required for solids suspension and that required to maintain a uniform basin oxygen level. Rich proposed Equation (4.15) (Associated Water & Air Resources) to estimate low-speed mechanical power requirements to maintain lagoon settleable solids in suspension [12].

$$P = 0.004 \times MLSS + 5 \qquad (4.15)$$

where P is the power level, W/m^3 and MLSS is the mixed liquor suspended solids less than 2000, mg/L.

Based on Equation (4.15), the power required to maintain a 100 to 200 mg/L suspended solids system suspended is esti-

TABLE 4.11. Pond Depths.

| Type System | Depth Range | | Governed by |
	Meters	(Feet)	
High rate stabilization	0.3–0.5	(1–1.5)	Photosynthesis
Facultative ponds	1.2–2.5	(4–8)	Process capacity
Aerated lagoons	2–6	(6–20)	Aeration device
Dual power	Per configuration		Individual cells
Controlled discharge	2–2.5	(6–8)	Capacity
Total discharge	to suit		Capacity

mated at 6 W/m³ (30 hp/MG). Rich reports that a 1 W/m³ (5 hp/MG) is sufficient to disperse oxygen in a partially mixed basin [12]. The EPA manual cites 3 W/m³ (15 hp/mg) as the minimum energy level to maintain solids suspended in a completely mixed system [20]. These values are within the range generally reported as the practical applied energy level for adequate dispersion [1,6,20]. Levels ranging from 10 to 20 W/m³ (50 to 100 hp/MG) have been reported as the defining energy level for complete mixing, but this would be considered excessive for a pond [1,6]. Applied mechanical energy to a lagoon is frequently a function of the required DO level, with mixing a secondary consideration. The DO concentration in a pond will be governed by (1) the influent organic loading, (2) aerobic, anaerobic, or mixed system performance criteria, (3) the capacity and efficiency of any applied mechanical aeration, and (4) photosynthesis effectiveness if mechanical aeration is not used.

Where aeration devices are employed in complete mix configurations, they serve as a mixing and aeration device but can seldom be sized to optimize both requirements. Because the energy required for completely mixing is usually considerably more than that required for oxygen supply, a basin seldom operates at completely mix conditions, the energy being somewhere above the minimum required to supply adequate oxygen. Even in stabilization basins where mixing or external aeration is generally not provided, some operating flexibility should be included. This could consist of some recirculation capacity to supplement mixing and standby aerators to increase system oxygen level during dormant conditions.

Aeration Basin Volume

Installed basin volume is primarily dependent on the lagoon type, influenced by the available land area, required surface area, allowable liquid depth, required storage capacity, or a combination of some or all of these factors. Theoretically, the controlling design variable in a biological system is the minimum SRT (HRT in a once-through system) allowing the growth of the required biological population. Indeed, this is the intent in a completely mixed aerated lagoon. In the other lagoon systems the criteria for the required *total* retention volume is more complex, involving some or all of the following:

(1) In all once-through lagoons additional volume must be allowed for expected solids storage, which will increase with increasing influent solids and decreasing mixing energy.

(2) In systems dependent on natural aeration the volume may be established by area requirements to assure capturing adequate exerted energy and depth requirements to assure penetration to the aerobic levels.

(3) In partial or total retention systems the volume will be controlled by storage requirements.

(4) In facultative systems the complex processes of aerobic stabilization and anaerobic digestion and the storage

required because of seasonal variations make precise basin volume estimation extremely difficult. Generally, adequate *area* must be provided to assure capturing of sufficient energy exerted and a minimum depth to assure energy penetration to the aerobic depths. This defines the aerobic volume, which must be supplemented with adequate volume for anaerobic digestion, solids storage, and some waste storage. In addition, a protective interface depth must be provided to separate the aerobic and anaerobic sections.

Much of the general data available for lagoon design is generated from performance of existing facilities, where the design parameter is expressed either as loading rate or retention time. As a guide, data cited in Table 4.1 can serve as a frame of reference.

Solids Management

There is little operational control of solids accumulated in a pond or lagoon. Accumulation is a product of the influent nondegradable solids and the net solids resulting from the biological and digestion processes. By design, excess solids produced in these systems are either directly discharged with the effluent (for further treatment if necessary) or settle to the pond bottom to undergo anaerobic digestion. Any residual solids is periodically removed by dredging. In all cases, basin effluent solids concentration must be evaluated to establish if discharge limits can be met. Because of relatively low solids loading, clarification is usually not effective, necessitating either a final settling pond with high detention time or final filtration.

Aerobic stabilization systems generate large quantities of algal that must be harvested and removed from the system, or the pond will be pushed to anaerobic conditions. Net solids generation in a facultative system is the accumulative result of the influent nondegradable solids concentration and the systems solids anaerobic digestion capabilities. Theoretically, the facultative "anaerobic digester" can destroy a large quantity of accumulated solid. In reality, such systems require dredging capabilities and alternate means of sludge disposal.

GENERAL ENGINEERING CRITERIA

Figure 4.3 illustrates an aerated lagoon preliminary concept flowsheet. The process components specific to an aerated lagoon or pond system include the following:

Waste storage (optional)	Chapter II-1
Lagoon or pond	
Aeration equipment (optional)	
Recycle	Chapter I-8
pH control (optional)	
Nutrient addition (optional)	Chapter I-5

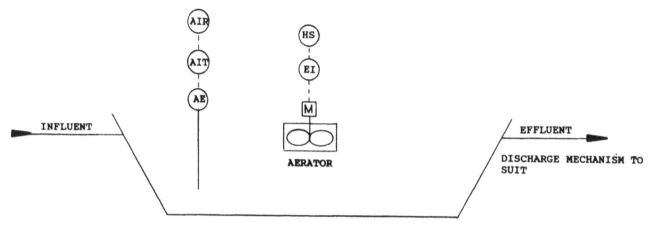

Figure 4.3 Aerated lagoon preliminary concept flowsheet.

Each of these elements, except the pond design, are discussed in the chapters indicated. The General Engineering Design Criteria section in Chapter II-3 describes applicable basin design, process control, piping and pumps, and layout considerations, which will not be repeated in this section.

POND AND LAGOON DESIGN CONSIDERATIONS

Pond Staging

Pond staging is a critical design variable allowing operator flexibility by providing (1) multiple-staged treatment efficiency, (2) protection against influent leakage through short-circuiting, (3) the ability to maintain a portion of the system operational in all kinds of process or maintenance circumstances, and (4) series or parallel operation.

Although achieving improved effluent quality in a multistage operation has been disputed, the fact is that single-flow through ponds offer little operator control, being solely dependent on unpredictable ambient conditions. Where ponds are installed to meet strict effluent quality, excess treatment volume and multiple units are the only "safety factor" available to the engineer. The obvious disadvantage of multistage units are increased construction costs. This is offset by the potential for improved process stability.

Process considerations in evaluating the number and characteristics of a pond include

(1) Single pond construction represents minimum capital expense and least process stability.
(2) Flow pattern control is difficult in single ponds, and plug flow characteristics are achieved with increased staging, minimizing influent leakage to the effluent. This is important in industrial waste application where varying waste loads and characteristics are to be expected.
(3) Single ponds do not allow for continued waste processing when maintenance is required.

(4) Larger ponds are more susceptible to wind effects such as increased (and possibly excessive) basin mixing, severe levee damage, and water drifting from the basin. In such cases, larger freeboard provisions must be included to control wave action.

Figure 4.4 illustrates some common configurations illustrating (1) a single pond, (2) multiple ponds in series, (3) multiple ponds in parallel, and (4) multiple ponds in combined parallel-series operation [20]. These configurations are enhanced by recyle capabilities. Figure 4.5 illustrates how four ponds can be piped to offer flexible operating capabilities to operate as single-, two-, three-, or four-stage series ponds, with provision for parallel distribution in the initial stages [10].

Pond Hydraulics

Hydraulic considerations are critical in pond design, especially the location of the inlet and outlet piping to provide a uniform velocity profile across the pond. The EPA Stabilization pond design manual [20] cites specific design details critical to achieving proper pond hydraulics, emphasizing head loss, multiple feed ports, prevention of short-circuiting, and recirculation. Inlet and outlets should be designed to limit head loss to 8 to 10 cm (3 to 4 in.) at 2/3 to 3/4 of full capacity, with special emphasis at peak forward and recycle flows. Multiple-inlet arrangements should be included in all ponds regardless of size, assuring adequate flow distribution to maximize retention time. The outlets should be located as far from the inlet as possible, with multiple depth and long longitudinal diffuser outlets covering a large portion of the basin cross-sectional area. The flow distribution should be designed to prevent channelled currents leading to short-circuiting. Internal baffling should be considered, and strategically located, to control (short-circuiting) directional flow within a pond. Short-circuiting due to wind action can be minimized by locating the inlet/

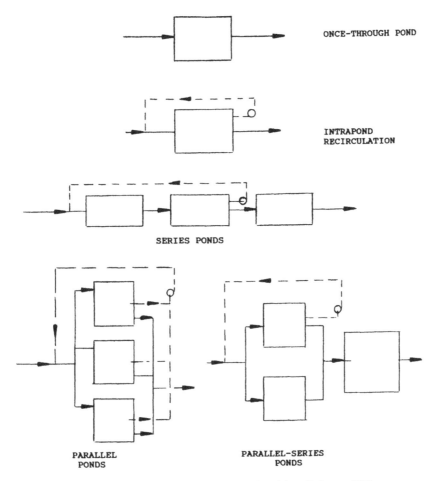

Figure 4.4 Common pond configurations (adapted from Reference [20]).

outlet axis perpendicular to the prevailing wind direction. All withdrawals should be a minimum of 0.3 m (1 ft) below the water surface. Provision should be made for effluent recirculation within a cell, or other cells, to enhance mixing, return additional (dissolved) oxygen, return active algal cells, and for influent dilution.

BASIN CONSTRUCTION CONSIDERATIONS

Pond construction normally implies inexpensive construction. The following points are worth considering in the structural design:

(1) Earthen constructions are discouraged unless adequate protection of subsurface waters can be assured. Forthcoming, RCRA requirements will probably impose stringent impervious construction or groundwater monitoring requirements.

(2) Safety must be a primary consideration in the design. Because these basins tend to be huge, provisions must be made to secure and restrict access to the area.

(3) The basin depth must be a balance between a deep basin, which will assist in heat conservation and mini-

mize algae growth, and a shallow pond, which will promote algae growth and aerobic conditions.

(4) The area required for basin capacity requires huge land usage. These basins should be isolated and away from populated areas.

(5) Dike stability depends on the internal and external slope, selected on the basis of basin construction and the lining employed. The pond bottom should be level.

(6) Provision should be made for at least 1 m (3 ft) freeboard, depending on the aeration device selected and pond configuration.

(7) Depths are normally not more than 1.8 to 2.4 m (6 to 8 ft), increasing when mechanical aeration is employed.

(8) Provision must be made for an access road to the ponds and access to the dikes.

(9) Ponds should be fenced, with access roads inside the fence and constructed to meet all local regulations.

(10) Erosion protection for pumped influents should include erosion pads and directing the inlet piping system upward.

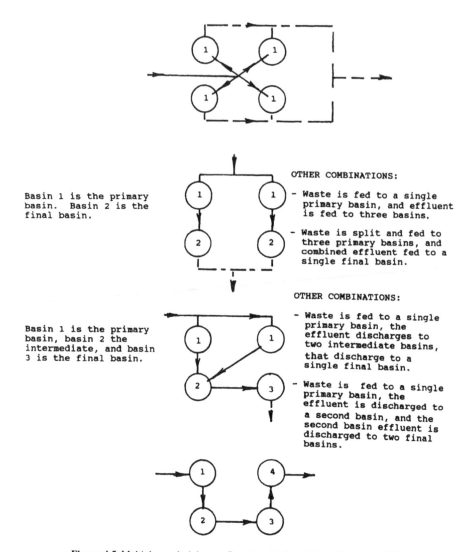

Basin 1 is the primary
basin.   Basin 2 is the
final basin.

OTHER COMBINATIONS:

- Waste is fed to a single
  primary basin, and effluent
  is fed to three basins.

- Waste is split and fed to
  three primary basins, and
  combined effluent fed to a
  single final basin.

OTHER COMBINATIONS:

- Waste is fed to a single
  primary basin, the
  effluent discharges to
  two intermediate basins,
  that discharge to a
  single final basin.

- Waste is fed to a single
  primary basin, the
  effluent is discharged to
  a second basin, and the
  second basin effluent is
  discharged to two final
  basins.

Basin 1 is the primary
basin, basin 2 the
intermediate, and basin
3 is the final basin.

Figure 4.5 Multiple-pond piping configurations (adapted from Reference [10]).

(11) Inlet and outlet piping systems should be at opposite ends and the outlet arranged so that withdrawal is from a middle depth.

(12) Baffles should be included to prevent solids and floating material from discharging with the effluent.

(13) The pond should be watertight, comply with applicable RCRA requirements, and provide for maximum protection against any transport to underground water systems.

POND OPERATING AND MAINTENANCE CONSIDERATIONS

The following operational and maintenance considerations should be part of the design:

(1) Heat conservation and minimum heat loss

(2) Maintaining aerobic conditions consistent with the pond design but at a minimum at the lagoon upper levels to minimize odor problems

(3) Adequate mixing in aerated lagoons to minimize solids deposition and dredging requirements

(4) Variable outlet weir to control basin depth

(5) Provisions for skimming and removing oil and grease

(6) Provision for liner maintenance

(7) Ability to drain the pond for maintenance and dredging

(8) Ability to remove debris from the pond

(9) Ability to store and treat incoming wastewaters under varying ambient and pond conditions

(10) Ability to add odor control chemicals

(11) Ability to repair mechanical equipment in position, or to remove and repair in a shop

(12) Control of dike weed growths

(13) Ability to harvest vegetation growth

PROCESS CONTROL

Pond system process controls are inherently minimal, relying mainly on large detention times (and process lag) to stabilize any system upsets. However, the pond systems should be configured to allow some basic operator flexibility to control process conditions, such as provisions for

(1) Controlling flow into the lagoon(s)

(2) Pretreatment

(3) Bypassing the biological basin and storing runoff, spills, or process wastes during emergency conditions

Some level of monitoring must be included in pond design to alert of "extreme" conditions.

(1) Provision for sampling waste influent, at a minimum as a "grab sample," preferably by automatically collecting composite samples over an operator controlled time period

(2) Provision for automatically monitoring influent pH and alarming for any conditions outside the specified limits

(3) Provision for collecting samples and measuring aeration basin dissolved oxygen level

(4) Provision to monitor the aeration basin level, and the levels of all critical vessels, and alarming low and high level conditions

(5) Provision for sampling waste effluent, at a minimum as a grab sample, preferably by automatically collecting composite samples over an operator controlled time period

(6) Provision for automatically monitoring effluent pH and alarming for any conditions outside the specified limits

(7) Running lights should be provided for any major operating equipment

COMMON POND DESIGN DEFICIENCIES

Some design deficiencies common for municipal system ponds and lagoons have been investigated by the EPA [19]. They are identified as a checklist of process and mechanical design considerations. Reference is made to the cited literature for discussions and proposed corrections of these potential problems.

(1) Inability to meet effluent quality at reduced pond temperatures

(2) Inadequate provision for groundwater pollution prevention as a result of (a) a poor or no liner, (b) no monitoring wells, (c) the pond being located too close or below seasonal high groundwater table, and (d) no

groundwater interception and collection system for lined lagoons

(3) No provision for varying the pond liquid depth for process control (or mosquito control)

(4) Single point feed entry results in overloading of pond sections

(5) Anaerobic conditions prevalent due to organic overloading, inadequate aeration, inability to control aeration, or inability to measure the pond DO profile

(6) Mixing for aerated lagoon systems was inadequate

(7) Inadequate provisions to prevent short-circuiting from large unmixed basins, or no provision for plug flow because baffles or multiple-basin construction was not employed

(8) Inadequate erosion control measures were employed

(9) No recirculation pumps were provided to allow for (a) process flexibility, (b) improved DO control, (c) improved feed (organic loading) distribution, and (d) improved mixing

(10) Inadequate freeze protection resulting from (a) the depth being too low, less than 0.9 m (3 ft), (b) inadequate freeboard for wind protection, at least 0.9 meters (3 ft), and (c) no provision for aerator ice shields

(11) Poor pond design results in operating and site problems because (a) single rather than multiple units were installed; (b) no depth control provisions were included; (c) recirculation control was not provided; (d) anaerobic conditions prevalent throughout the basin because of inadequate surface area, or no recirculation capabilities; (e) the sludge storage provided is inadequate; (f) provision for odor control is inadequate, and inactive pond "dead spots" exist; (g) the dike design is inadequate, with the width being too small to provide operator access or for transporting machinery; (h) erosion control is inadequate; (i) improper dike seeding causes structural problems; (j) potential rodent problems not considered; and (k) safety provisions to limit access to the pond area was not considered.

(12) Pond is located too close to residential areas.

CASE STUDY NUMBER 14

Based on the conditions and design criteria cited in Case Study 10 the use of an aerated lagoon will be evaluated as an alternative treatment to an activated sludge system. Applicable data can be summarized as follows:

Flow, MGD:	1
BOD_5, mg/L:	500
COD, mg/L:	1250
Effluent BOD_5, mg/L:	20
Effluent suspended Solids, mg/L:	20
Design temperature, °F:	68
Summer temperature, °F:	80
Winter temperature, °F:	40

The activated sludge design is based on an SRT of 12 days producing an soluble BOD of 7.45 or better. *Assume that an HRT equal to the SRT will produce equal results.* The following kinetic coefficients at 20°C can be assumed:

k_d: 0.05 (1/day)
Y: 0.5

PROCESS CALCULATIONS

(1) Determine the effluent soluble BOD
Based on the data cited, a 20 mg/L total BOD_5 effluent, with 20 mg/L solids, is equivalent to 7.45 mg/L soluble BOD_5; the design basis. See item 1, Case Study 10 for details.

(2) First-order kinetic constants

(3) Design conditions:

Temperature at 68°F (20°C)

$$K_1 = [(S_o/S_e) - 1] / HRT$$

$$[(500/7.45) - 1] / 12 = 5.5$$

$$k_d = 0.05 \text{ (1/day)}$$

$$Y = 0.5$$

MLVSS concentration

$$X = \frac{Y \cdot (S_o - S_e)}{1 + k_d \cdot HRT} = \frac{0.5 (500 - 7.45)}{1 + 0.05 \cdot 12} = 154 \text{ mg/L}$$

(4) Summer conditions: A first-order constant for summer conditions can be estimated as follows:

Temperature at 80°F (26.67°C)

$$k_{ls} = 5.5 \cdot 1.085^{26.67-20} = 9.50$$

$$k_d = 0.05 \cdot 1.085^{(26.67-20)} = 0.086 \text{ and } Y = 0.50.$$

a. Effluent concentration, mg/L

$$S_e = S_o \cdot [1/(1 + k \cdot t)]$$

$$S_e = 500 \cdot [1/(1 + 9.5 \cdot 12)] = 4.3 \text{ mg/L}$$

b. MLVSS concentration, mg/l

$$X = \frac{0.5 \cdot (500 - 4.3)}{1 + 0.086 \cdot 12} = 122 \text{ mg/L}$$

(5) Winter conditions:

Temperature at 40°F (4.44°C)

$$k_{lw} = 5.5 \cdot 1.085^{(4.44-20)} = 1.55$$

$$k_d = 0.05 \cdot 1.085^{(4.44-20)} = 0.014$$

$$Y = 0.50$$

a. Effluent concentration, mg/L

$$S_e = 500 \cdot [1/(1 + 1.55 \cdot 12)] = 26 \text{ mg/L}$$

b. MLVSS concentration, mg/l

$$X = \frac{0.5 \cdot (500 - 26)}{1 + 0.014 \cdot 12} = 203 \text{ mg/L}$$

POND EVALUATION

Pond sizes will be evaluated on the basis of achieving a required 7.45 mg/L soluble BOD effluent.

(6) Assume equal volume and constant K

$$500/7.45 = C_n/C_o = \left[\frac{1}{1 + \dfrac{K_c t}{N}} \right]^N$$

No. ponds, N	Design Conditions $K = 5.5$		Winter Conditions $K = 4.0$	
	t, total days	days/ pond	t, total days	Days pond
1	12	12	43	43
2	2.6	1.3	9.3	4.6
3	1.7	0.6	5.9	3
4	1.4	0.4	4.8	1.2

The effects of multiple units are illustrated by the theoretical calculations tabulated above. However, the assumption that the kinetic constant remains constant is probably not valid, with its value reduced considerably after large first stage organic reductions.

(7) Assume equal volumes but varying K

$$\frac{C_n}{C_o} = \frac{1}{1 + k_{ci} t_i} \times \frac{1}{1 + k_{cii} t_{ii}} \times \cdots \frac{1}{1 + k_{cn} t_n}$$

Because equal volume ponds are considered t = total time/N

The assumption will be made that the first stage design constant remains at 5.5, reducing dramatically to 0.5 in the remaining second, third, and fourth ponds. This corresponds to a winter first stage constant of 4.0, reducing to 0.37.

No. Ponds, N	Design conditions K = 5.5 First chamber 0.50		Winter Conditions K = 1.55 Other Chambers 0.14	
	t, total days	Days/ pond	t, total days	Days/ pond
1	12	12	43	43
2	8	4	28	14
3	7	2.3	25.5	8.5
4	7	2.3	24	6

The assumption of a reduced first-order constant dramatically reduces the effects of multichambered ponds. Where equal performance reduces the required theoretical retention time from 12 to approximately 1.4 days, reduced kinetic constants result in a reduction from 12 to 7 in winter conditions.

(8) Select total retention time

Taking into account winter conditions, and uncertainties in pond performance, four ponds totally 24 million gallons will be used.

(9) Basin Design

The effects of activated sludge recycle are dramatically demonstrated in the aeration basin sizes. Based on both systems operating at an SRT of 12, the activated sludge basin volume is 0.92 million gallons and that of the aerated lagoon 12 million or more gallons. The aeration system size can be reduced by using multiple chambers, although the extent of that reduction depends on the subsequent reaction rates, which are believed to be reduced in proportion to the substrate concentration. The selected 24 million gallon system should be adequate for winter conditions by providing four chambers. The resulting system will have the following dimensions. It should be noted that the winter results will be affected by the selected temperature correction coefficient, selected at 1.085 and ranging from 1.015 to 1.085. A reduced coefficient significantly decreases the required winter basin values.

a. Volume, million gallons = 9
b. Number of basin selected = 4
c. Each basin:
Working depth = 10 ft
Projected surface area = 80,200 ft^2
Surface width = 200 ft
Surface length = 400 ft
Freeboard = 2 ft
Total depth = 12 ft

Assuming 15 hp per 1,000,000 million gallons working volume, 90 hp are required per lagoon stage, or a total of 360 hp for the entire system.

(10) Other design criteria

See Case Study 10 for estimates of sludge wasting, nitrification, and oxygen requirements.

CASE STUDY NUMBER 15

Evaluate *facultative lagoon* sizes for the waste discussed in Case Study Number 15. Because laboratory data obtained are not directly applicable in establishing loadings and kinetic constants, assume theoretical values cited for comparison purposes.

AERATION BASIN EVALUATION

(1) Establish HRT, basin volume, depth, and retention time using loading rate criteria.
Assume a loading rate of 80 lb/ac/d at 68°F (20°C), and 40 lb/ac/d at 40°F (4.4°C).

(2) Determine the organic feed rate to the lagoon

$$1 \text{ MGD} \cdot 8.34 \cdot 500 \text{ mg/L} = 4170 \text{ lb/day}$$

(3) Determine the required lagoon acres
Design: 4170/80 = 52 acres
Based on a 5 ft height,
52 · 5 ft · 0.3259 MG/ac-ft = 85 MG
85 mg/ 1 MGD = 85 days
Winter: 4170/40 = 104 acres
Based on a 5 ft height,
104 · 5 ft · 0.3259 MG/ac-ft = 170 MG
170 mg/ 1 MGD = 170 days

(4) Establish HRT and retention time using the Marris and Shaw complete mix equation

$$C_n/C_o = \frac{1}{[1 + K_c\, T_n]^n}$$

where C_n = effluent BOD5 concentration in cell n, mg/L, C_o = influent BOD5 concentration, mg/L, K_c = complete mix first reaction rate constant, day^{-1}, T_n = hydraulic residence time in *each* pond, days, and n = number of ponds in series Assume that K_c is 1.2/days at 35°C; so that with K_t = $K_{35} \cdot 1.085^{(T-35)}$, K at design conditions of 20°C equals 0.35, and at winter conditions of 4.4°C equals 0.1. Assume that the coefficients remain constant and equal volume (multi) ponds are considered:

(5) Assume equal volume and constant K

No. ponds	Design total		Winter total	
	Days	MG	Days	MG
1	189	189	661	661
2	41	41	144	145
3	26	26	92	93
4	21	21	74	75

(6) Assume equal volume and varying constant K
Assume that the design coefficient varies as follows:

Pond	1	2	3	4
Design	0.35	0.2	0.1	0.1
Winter	0.1	0.06	0.03	0.03

No. ponds	Design total		Winter total	
	Days	MG	Days	MG
1	189	189	661	661
2	54	54	185	185
3	46	46	158	158
4	44	44	148	148

(7) Establish basin volume and retention time using the plug flow model

$$S_e/S_o = \exp^{-K20 \cdot t}$$

With the K coefficient at 0.129 at design conditions of 20°C and a loading of 100 lb/d/acre or more and 0.034 at 4.4°C.

(8) Design conditions

$$t = 33 \text{ days, requiring a 33 MG basin.}$$

(9) Winter conditions

$$t = 124 \text{ days, requiring a 124 MG basin.}$$

(10) Establish HRT and retention time using the Wehner-Wilhelm equation and Thirumurthi application

$$\frac{C_e}{C_i} = \frac{4 \cdot a \cdot e^{1/(2d)}}{(1 + a)^2 \cdot e^{a/(2d)} - (1 - a)^2 \cdot e^{-a/(2d)}}$$

$$a = (1 + 4 \cdot K \cdot t \cdot D)^{1/2}$$

Estimate the design and winter condition requirements, assuming a diffusion coefficient (D) equal to 0.05, 0.10, 0.50, or 1.

	D	Design $k_d = 0.15$		Winter $K_w = 0.087$	
		Days	mg	Days	mg
Approach plug	0.05	34	34	58	58
	0.10	39	39	67	67
	0.50	69	69	120	120
Complete mix	1.00	95	95	166	166

(11) Summary

The range of results obtained a single lagoon, *5-ft deep,* are summarized below.

Facultative Lagoon Results

	Design		Winter		
	Days	Acres	Days	Acres	Model
Loading method	85	52	170	104	None
Morris-Shaw	189	116	661	406	Complete mix
Plug flow model	33	20	124	76	Plug
Wehner-Wilhelm equation	34	21	58	36	Near plug flow
	95	58	166	102	Complete mix

Facultative Lagoon Results

Morris-Shaw	Design		Winter		
	Days	Acres	Days	Acres	Model
2 ponds	41–54	25–33	144–185	88–114	Complex mix
3 ponds	26–46	16–28	92–158	56–97	Complete mix
4 ponds	21–44	13–27	74–148	45–91	Complete mix

DISCUSSION

Based on the estimates made from the theoretical models cited, 13 to 114 acres would be required to treat the industrial waste described in Case Study 10 in a multiple-unit facultative pond. However, the wide disparity in estimated basin volume illustrates the danger in utilizing municipal lagoon models in establishing industrial treatment requirements. An accurate estimate can only be obtained by generating site-specific pilot data.

REFERENCES

1. Bartsch, E.H., Randall, C.W.: "Aerated Lagoons—A Report on the State of the Art," *Journal WPCF,* V 43, No 3, Pg 699, April, 1971.

2. Benefield, L.D., Randall, C.W.: *Biological Process Design for Wastewater Treatment,* Prentice-Hall, Inc., 1980.

3. Canter, L.W., Englande, A.J. Jr., Mauldin, A.F. Jr.: *"Loading Rates on Waste Stabilization Ponds,"* Sanitary Engineering Division, Proceedings ASCE, 95, SA6, Pg 1117, December, 1969.

4. Finney, B.A., Middlebrooks, E.J.: "Facultative Waste Stabilization Pond Design," *Journal WPCF,* V 52, No 1, Pg 134, January, 1980.

5. Kormanik, R.A.: "Design of Two-Stage Aerated Lagoons," *Journal WPCF,* V 44, No 3, Pg 451, March, 1972.

6. Kouzell-Katsiri, A.: "Design Optimization for Dual Power Aerated Lagoons," *Journal WPCF,* V 59, No 9, Pg 825, September, 1987.

7. Lawrence, A.W., McCarty, P.L.: "Unified Basis for Biological Treatment Design and Operation," *Journal San Eng Div.,* Proceedings ASCE, V 96, No 6, Pg 757, June, 1970.

8. Medcalf & Eddy, Inc.: *Wastewater Engineering-Treatment, Disposal, Reuse,* McGraw-Hill, 1991, Third Edition.

9. O'Connor, J.T. (Ed.): *"Environmental Engineering Unit Operations and Unit Processes Laboratory Manual,"* Association of Environmental Engineering Professors, July, 1972.

10. Oswald, W.J.: "Fundamental Factors in Stabilization Pond Design," *Advances in Biological Waste Treatment,* Pergamon Press, 1963, Pg 357.

11. Wang, L.K., Pereira, N.C.: *Handbook of Environmental Engineering,* V 3, Humana Press, 1986; [Poon, C.P.C., Wang, L.K., Wang, M.H.S., "Waste Stabilization Ponds and Lagoons," Chapter 7].

12. Rich, L.G.: "Design Approach to Dual-Power Aerated Lagoons," *Journal Environmental Engineering,* Proceedings ASCE, 108, No EE3, p 532, June, 1982.

13. Sawyer, C.N.: "New Concepts in Aerated Lagoon Design and Operation," In *Advances in Water Quality Improvement,* Gloyna, E.F., Eckenfelder, W.W., University of Texas Press, Pg 325, 1968.

14. Thirumurthi, D.: "Design Criteria for Waste Stabilization Ponds," *Journal WPCF,* V 46, No 9, Pg 2094, September, 1974.

15. Thirumurthi, D.: "Principles of Stabilization Ponds," *Journal Sanitary Engineering,* Proceedings ASCE, 95, No SA2, Pg 311, April, 1969.

16. Uhlmann, D.: "BOD Removal Rates of Waste Stabilization

Ponds as a Function of Loading, Retention Time, Temperature, and Hydraulic Flow Pattern," *Water Research*, V 13, Pg 193, 1979.

17. Uhlmann, D., et al: "A New Design Procedure for Waste Stabilization Ponds," *Journal WPCF*, V 55, No 10, Pg 1252, October, 1983.

18. U.S. Environmental Protection Agency: *Treatability Manual*, EPA-600/8-80-042a, 1980.

19. U.S. Environmental Protection Agency: *Handbook for Identification and Correction of Typical Design Deficiencies at Municipal Wastewater Treatment Facilities*, EPA-625/6-82-007, 1982.

20. U.S. Environmental Protection Agency: *Design Manual; Municipal Wastewater Stabilization Ponds*, EPA-625/1-83-015, 1983.

21. WEF Manual of Practice: *Design of Municipal Wastewater Treatment Plants*, Water Environment Federation, 1992.

22. Thirumurthi, D.: "Design Criteria for Aerated Lagoons," *Journal Environmental Engineering*, Proceedings ASCE, 105, No EE1, Pg 135, February, 1979.

Biological Oxidation: Fixed-Film Processes

Fixed-film processes are employed to destroy biodegradable organics; the applied loading depends on the specific process oxygen limitations.

FIXED-BED media include a variety of configurations in which wastewater-containing organic and nitrogen substrates are brought into contact with microorganisms attached to the media. Possible reactor configurations include trickling filters, combined systems with trickling filters, rotating biological contactors (RBCs), sparged columns, and submerged beds. *Trickling filters* are the most prominent systems, commonly used in municipal treatment and industrial applications and for the most part the processes discussed in this chapter. RBC process design criteria are explicitly identified so as not to confuse fix bed principles with those of a rotating media. Special bed column designs involve emerging technologies, where general criteria are sparse, requiring that the Process Engineer develop the design *and the process* for the specific application.

Figure 5.1 illustrates the basic trickling filter system. Waste enters a primary clarifier for solids removal prior to being fed to a trickling filter. The trickling filter contains a packed bed media covered with biological slime over which wastewater flows. Oxygen and organic matter diffuse into the film where oxidation and synthesis occur. Recirculation can be employed to improve the hydraulics of the system and thereby enhance the bed contact and mass transfer efficiency.

BASIC CONCEPTS

Fixed bed basic concepts will be discussed in terms of trickling filters because the vast amount of available operating experience facilitates their adoption to industrial application, and the principles discussed can be applied to other fixed bed configurations. One of the challenges in applying trickling filter technology to industrial waste treatment is adapting municipal data and historical models for industrial application.

Models employed in trickling filter design are a result of curve-fitting available municipal or select laboratory study data. This limits their use when conditions being considered deviate greatly from the original data base. As will be discussed, there are some inconsistencies in establishing the significant variables governing trickling filter design. When individual studies and resulting process models are reviewed, it will soon become obvious that any data supporting the significance of any purported process variable can be countered with equally impressive data refuting the claim. Specific trickling data sources, and the resulting data, emphasize or discount operating variables, depending on the conditions evaluated. Applying a trickling filter to aerobic biological oxidation requires an understanding of the operating and process variables governing the process and their appropriateness to specific industrial wastes.

A significant start in defining trickling filter technology is to evaluate current municipal treatment application, evolving from historical models or commonly accepted design criteria. This "state-of-the-art review" will enable a development of a general consensus of commonly employed municipal design parameters, evaluating their possible application to industrial waste systems. It should be strongly emphasized that evaluating means selecting significant variables for specific *pilot testing,* with specific wastes and under applicable operating conditions. Finally, trickling filters have been historically classified on the basis of unit loading, as indicated in Table 5.1. These classifications are seldom used as the sole process definition in present-day applications.

TRICKLING FILTER MODELS

The design criteria for trickling filters was developed from extensive analysis of sewage plant evaluations and operating

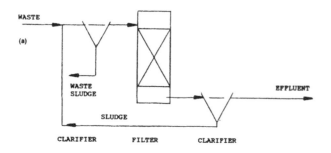

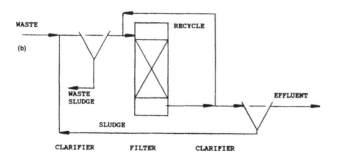

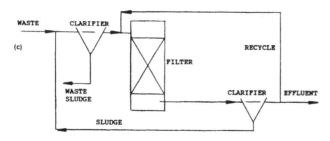

Figure 5.1 Trickling filter configurations.

TABLE 5.1. Trickling Filter Classifications (adapted from Reference [9,21]).

	Loading Rates	
	Hydraulic m³/m²/day (GPD/ft²)	Organic kg/d/100 m³ (lb BOD/day/1000 ft³)
Low rate	1–4 (25–90)	8–25 (5–15)
Intermediate rate	4–9 (90–230)	25–50 (15–30)
High rate	9–40 (230–925)	50–100 (30–60)
Super high rate (plastic)	12–90 (300–2100)	50–160 (30–100)
Roughing	45–190 (1100–4600)	160–800 (100–500)
Two-stage	9–40 (230–925)	100–200 (60–120)

$$dS = k\,S\,dT \qquad (5.1)$$

$$S_e/S_o = e^a \qquad (5.1a)$$

where $a = -kT$, S_e = effluent concentration, S_o = influent concentration, and K = reaction kinetics constant.

This formula is commonly referred to as Velez equation, which happens to be the equation for an "ideal" plug flow reactor. It can be modified by the Schultz definition for reaction or residence time, as defined by Equation (5.2).

$$T = C\,D\,/\,Q^n \qquad (5.2)$$

where T = time, C = constant, D = filter depth, and Q = hydraulic loading.

The Veltz, Schulze, Germain, and Eckenfelder equations use reaction or residence time as a major process variable. The media volume and area, the organic volumetric loading rate, and the recycle rate are not directly applied to the model; except for the Eckenfelder formula, which includes media surface area.

The *Schulze formula* can be represented by the formula:

$$L_e/L_o = e^{(-kD/Q^n)} \qquad (5.3)$$

where L_e is the filter effluent, mg/L, L_o is the *applied* influent, including recycle, mg/L, D is the depth, m, Q is the *applied* loading, Q is the hydraulic loading, (m³/min)/m², (gpm/ft²), and k and n are constants.

The following temperature correction is applicable:

$$k = k_{20} \cdot [1.035^{(t-20)}] \qquad (5.3a)$$

where t is the liquid temperature, °C.

data, applied in formulating direct guidelines or kinetic models. These models or procedures are commonly reported in standard municipal design texts as guidelines such as the "Ten State Standard Design Guidelines" and performance models such as the National Research Council (NRC) formula, the British Manual of Practice (BMP), the Velz formula, the Schulze formula, the Germain formula, the Eckenfelder formula, the Galer and Gotaas formula, and the Kincannon and Stover formula [9,21]. Data and theory are constantly being reviewed, evaluated, and upgraded to new approaches such as the Logan models. Models that are directly correlated from municipal operating plant measurements will not be discussed because it is difficult to apply them to industrial wastes, especially if their characteristics are not similar to domestic wastes.

Many of the models that are not a direct correlation of analyzed municipal plant data assume that the trickling filter biological reaction adheres to a first order reaction. The organic concentration (S) is related to the reaction kinetic factor (k) and the reaction time (T), as denoted by Equation (5.1).

The *Germain formula* is identical to the Schulze formula, except that S_o is the feed concentration, *excluding recycle,* and Q is the feed loading, excluding recycle. Equation (5.3b) defines the applicable temperature correction. It is important to repeat that recycle is not included in the formula because it was not considered a significant factor in the waste studies conducted with deep bed plastic media. The Germain formula is frequently used for industrial wastes employing tall towers and plastic media.

The *Eckenfelder formula* is similar to the Schulze formula, with the time exponent adjusted for the clean surface media area (As), so that:

$$L_e/L_o = e^{(-a\,D/Q^n)} \qquad (5.3b)$$

where $a = k \cdot As^{1+m}$

Assuming that the surface area and microbial film are constant, a coefficient K', equal to the A_s and k product, can be substituted, so that a relationship similar to the Schulze formula results. The applied temperature correction is similar to the Schulze formula, and the *applied organic concentrations* and *hydraulic loading* are used in the formula.

All of the previous equations have equivalent relationships for two-stage performance, employing the first-stage effluent as the second-stage influent concentration.

Table 5.2 illustrates how the variables discussed are included or excluded in available models. It is out of the scope of this book to evaluate the merits of these models because each of these no doubt can be verified for the conditions developed. However, it is obvious that not all models consider the same operating variables in defining the biofilter process. Significantly, most do not include kinetic constants that could be utilized to extrapolate filter performance from substrates whose biodegradability differ from municipal wastes. The Process Engineer should not assume that model constants are process variables and can be used for any waste by adjusting the k value. Most models have been developed from tests of specific substrates under limited operating conditions. They should only be used as a basis for initiating specific industrial evaluations.

This general model discussion demonstrates the difficulty in developing design criteria for trickling filters. As illustrated in Table 5.2, it is difficult to establish from the models the importance of operating variables commonly associated with tower performance; such as the

(1) Quantity of organics fed to the tower
(2) Bed volume
(3) Recirculation rate
(4) Bed depth
(5) Residence time
(6) *Applied* hydraulic loading
(7) *Applied* feed concentration
(8) Feed hydraulic loading (excluding recirculation)
(9) Feed concentration (excluding recirculation)
(10) Waste biodegradability in the constants
(11) Media effectiveness, as expressed in the constants

These variables as applied in developing design criteria will be discussed in the section Process Engineering Design Criteria.

FIXED FILM SYSTEMS

TRICKLING FILTERS

Trickling filters can be classified in terms of *conventional biofiltration systems* common in municipal treatment. In such cases the influent can contain both dissolved and suspended biodegradable influent components, incorporating a primary clarification step to remove excessive solid loadings. All wasting is from the primary clarifier.

Figure 5.1(a) illustrates a low rate trickling filter with primary clarification for bulk solids removal, followed by

TABLE 5.2. Trickling Filter Model Variables.

Model name	Media				Loading,		Rc	Comments
	Type	D	V	A	Org	Hyd		
Velz	All	x						First-order kinetics
Schultz	All	x				x		Residence time defined for Velz model
Germain	P	x				x		Schultz model plastic media
Eckenfelder	All	x	x	x		x	x	Follow first kinetics with an area term
Gottas	S	x				x	x	Based on data analysis
Kincannon and Stover	All			x		x		Based Monod kinetics

S = stone media; All = all media; P = plastic media; D = depth; V = volume; A = area; Org = Organic loading rate; Hyd = Hydraulic loading rate; Rc = Recycle.

a trickling filter and final clarification. Such systems could be subjected to considerable flow and concentration variations because no provision is made for waste equalization. Figures 5.1(b) and (c) illustrate a high- or superrate system applying recirculation to maintain the system hydraulics and dampen influent variations by either recirculating tower or final clarifier effluent. Recirculation is essential in stone media to improve removal efficiency and in plastic media to improve hydraulics and prevent dry spots. Final clarifier waste sludge is returned to the primary clarifier, where it serves as a flocculating aid to remove colloidal solids and is discharged with the primary excess sludge. Filters can be staged to improve effluent quality, with any of the configurations discussed used for either or both filters. In such cases, the first stage is designed as a high-rate system for bulk organic removal, with a second stage to polish the effluent to required limits. In addition, the aeration capabilities of the entire system are increased by providing two draft systems to satisfy the feed biological oxygen demand (BOD). In industrial systems where waste solids are frequently low, primary clarification may not be required, with excess sludge from the final clarifier directly processed and disposed and some final effluent recycled for hydraulic control.

COMBINED TREATMENT SYSTEMS

Combined treatment systems incorporate sequential two-stage treatment. Large bulk reductions are achieved in the first-stage fixed-film process, whereas the second-stage suspended growth system is designed with provisions for increased operating control to "fine tune" the process and achieve maximum effluent quality. In this manner system performance is improved by (1) incorporating the effective bulk removal and toxic resistent qualities of fixed-film systems with the high effluent quality characteristics of suspended growth systems; (2) providing more operator controllable variables to react to process variability; (3) providing better solids removal by incorporating a flocculation step consisting of solids contact between the clarifier recycle and influent colloidal solids; and (4) separating the carbonaceous and nitrification processes. However, the process loadings have to be balanced to avoid low feed concentrations to the second stage, which could result in an unstable filamentous growth process.

The simple activated biofilter (ABF) illustrated in Figure 5.2 uses sludge recycle to increase the filter food-to-microorganism (F/M) loading. This system tends to make the sludge more "viable," similar to a selector design in an activated sludge system. Process loadings are generally maintained at 96 to 160 kg BOD/100 m^3 packing volume per day (60 to 100 lb/day 1000 ft^3), to achieve less than 30 mg/L BOD and TSS [21]. Plastic or redwood media are the media of choice because of the return sludge. Municipal system experiences indicate poor cold weather performance. A special system referred to as a trickling filter/sludge contact system (TF/SC) employs a TF for bulk organic removals and a contact

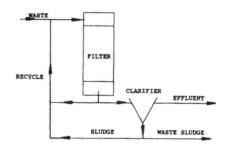

Figure 5.2 Activated biofilter (ABF).

tank to polish the TF effluent. The major elements of a TF/SC system is a TF which removes the bulk of the soluble BOD, sludge recycle, flocculation time, and a sludge contact stage to polish TF effluent or as an enhanced flocculation step. The specific configuration (Figures 5.3 to 5.5) depends on the TF effluent characteristics.

An aerated solids contact system (Figure 5.3) can be used to polish TF filter effluent to remove moderate SBOD quantities and small amounts of colloidal solids, a return sludge aeration tank to remove (and oxidize) high colloidal matter from the TF effluent (Figure 5.4), or both an aerated solids contact and return sludge contact tank for combined SBOD and colloidal removal, as illustrated in Figure 5.5 The filters can be configured to recycle effluent, sludge, or both. The concept of combined treatment can be expanded to "whatever works," by staging biofilters with common secondary treatment schemes such as (1) a trickling filter followed by an activated sludge system, (2) an activated sludge system followed by a trickling filter, (3) a trickling filter followed by a rotating biological contactor, or (4) a RBC followed by a trickling filter. In such cases the secondary system is not a polishing step, as in the TF/SC systems, but a staged treatment process employed to remove significant quantities of organics or as a separate nitrification step.

Some specific criteria recommended for combined wastewater municipal treatment systems are indicated in Table 5.3 [21]. Generally, the system design involves measures common for the individual components. Some important considerations applicable to combined treatment systems include

(1) Careful attention must be given to the selected trickling filter media when sludge is recycled to the filter. Large void space and vertical flow media must be used.

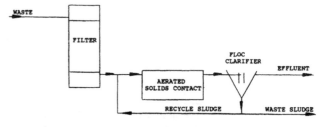

Figure 5.3 Combined treatment system with aerated solids contact.

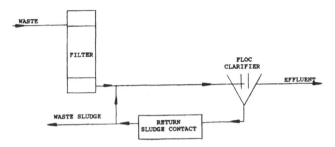

Figure 5.4 Combined treatment system with return sludge contact.

TABLE 5.3. Design Criteria for Combined Treatment Systems (adapted from Reference [21]).

Configuration	Loading kg BOD/day/ 100 m³ (lb/day/1000 ft³)	SRT Aeration Basin, days
1. Activated sludge	None	5–15
2. Trickling filter	20–50 (12–30)	None
3. ABF	32–120 (20–75)	None
3. Combined TF and solids contact	32–120 (20–75)	0.5–1.5
4. Combined TF and activated sludge	120–320 (75–200)	2–6

(2) The oxygen distribution between the two systems must be carefully allocated, with the worst design conditions deferred to the more flexible second stage.

(3) The carbonaceous reduction between the two stages must consider the potential for both overloading and underloading conditions. The second stage must be carefully sized to overcome high first-stage influent variability and possible poor performance. However, the first stage must not be oversized to "starve" the second stage, producing the very unstable conditions that the combined system is designed to overcome.

(4) The nitrification reactions should be deferred to the second stage.

(5) Suspended growth secondary treatment systems are commonly designed on the basis of MCRT, excluding the solids in the primary stage.

ROTATING BIOLOGICAL CONTACTOR

RBCs, also referred to as rotating biological discs (RBD), are appropriate for both suspended and dissolved organics removal. Figure 5.6 illustrates the basic RBC system. Feed (Q, S_o) enters the basin containing a bank of revolving discs with an active surface area. The disc, acting as an inverted trickling filter, rotates through the liquid, resulting in a growth of active biomass on the disc surfaces. Organic wastes biodegrade by wetting the disc containing the biomass. Retention time in the unit is short, but the feed-to-mass ratios developed are large, with the quantity of biomass accumulating on the discs dependent on its rotating speed and its holding capacity. Effluent from the basin is discharged to a final clarifier for solids removal.

RBC media are circular discs stacked in series, each series of discs considered a stage and the combined stages considered a unit. As with the trickling filter, a primary consideration is the total active surface area. This consists of the disc sides plus the adjoining edges, all of which can act as a source for microorganism growth. The unique characteristic of the RBC is the rotating discs' speed, which establishes the media exposure time to atmospheric oxygen, media sludge contact with the wastewater organics, disc cell maintenance (sludge sloughing), and mixing level in the vessel. The organic loading is limited by the unit's oxygen generating capacity. The basic substrate reduction mechanisms are similar to those of other biological systems.

The simplicity of these systems make them worth considering in some cases. Piloting of RBC systems is a challenge because reported prototype substrate loadings are generally higher than those achieved in full-scale operating systems, making pilot scale-up in other than *commercial size disks* difficult. The inability of some commercial units to meet design criteria, as well as mechanical problems associated with early units, have resulted in a diminishing interest in the RBC as an aerobic treatment system. However, with increasing experience, adequate site-specific data, and appropriate application

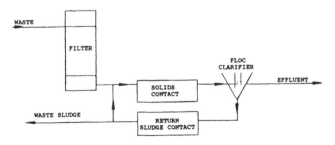

Figure 5.5 Combined treatment system with aerated solids and return sludge contact.

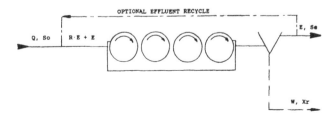

Figure 5.6 RBC treatment system.

of the available technology, there is no reason why these units cannot be evaluated as suitable alternatives to industrial waste treatment, alone or as part of a complex system.

EMERGING TECHNOLOGY

Requirements for more effective fixed film-bed systems has resulted in considerable *research* and *development* of packed-bed reactor technology in the laboratory, pilot plant, and some full-scale facilities. The basic drives for these systems are improved energy utilization, high treatment performance, and removal of specific waste substrate components. Design of these systems has not reached the stage where general criteria can be accurately cited. In many cases such data are exclusive to proprietary fixed bed systems and purchased as part of the technology package. In all cases where these systems are considered for industrial waste treatment specific evaluation is mandatory. The U.S. Environmental Protection Agency (EPA) cites some technologies for nitrification and denitrification application, classifying them as packed bed reactors (PBRs) and submerged beds [16]. Specific design criteria cited are results from theoretical evaluations, limited in scope, and in many cases based on limited performance data. Design data must be generated for each specific application, working closely with the prominent suppliers.

Packed-Bed Reactors (PBR): Nitrification

These units are developed for *nitrification*, generally operated with wastewater in an upflow, although they could be operated in a downflow configuration. They use a variety of media ranging from common sand filter to specialized plastic packing. Air or, if feasible, oxygen is sparged into the system from the tower bottom. Backwashing is usually necessary, the frequency of which depends on the media selected and the allowable pressure drop. As with all nitrification systems the efficiency depends on process conditions, especially the waste liquid temperature. Reported performance data indicate secondary effluents containing 6–20 mg/L ammonia nitrogen can be treated to 1–7 mg/L. Removal efficiencies range from 40 to 93% at loading of 8.8 to 88 m^3/m^2/day (0.15–1.5 gpm/sf) [16].

Submerged Low-Porosity Fine Media Packed Bed: Denitrification

These systems are commonly developed by sand filter manufacturers, utilizing sand filtration equipment technology. Some are designed as combined denitrification/final filtration systems employing 92 centimeters (36 inches) or more of mixed filter media. They are operated as downflow filters, with loading rates ranging from 0.7 to 2 L/m^2/s (1 to 3 gpm/sf), depending on the temperature. They employ backwash cycles similar to sand filters involving water or a combination of air and water. Reported removals of 95%

were achieved treating influent nitrate concentrations of 20 mg/L or less [16].

Submerged High-Porosity Fine Media Fluidized Beds: Denitrification

These systems are essentially fluidized denitrification beds, using small media similar to that employed in sand filters or activated carbon systems. The media provides significant biomass growth surface area, whereas fluidization is used to maintain an active bed. The system is fluidized in an upflow configuration with as much as 100% bed expansion. Nitrification rates range from 6 to 19 kg/m^2d (400 to 1200 lbs/d/1000 cf) of nitrogen at 10 to 20°C. Hydraulic loadings as high as 10 $L/s/m^3$ (15 gpm/sf) were reported for a 1.8 m (6 ft) sand and gravel bed fluidized to 3.7 m (12 ft) [16]. Theoretically, backwash may not be needed, with bed maintenance dependent on fluidized media shear action.

Submerged High-Porosity Packed Beds: Denitrification

These systems are operated in an upflow or downflow configuration, with backwash requirements minimized because high media void space is used. The specific media employed is as varied as with the other systems discussed, except that increased void space and larger media sizes result in lower denitrification rates. Reported surface denitrification rates range from 2–5 kg of nitrate nitrogen/day/m^2 at 10 to 25°C (0.4–1 lb/day/sf) [16].

Nitrogen Gas Fill Denitrification Packed Bed

These systems have specific characteristics distinguishing them from all systems previously mentioned. They are not submerged in liquid, but rather the void spaces are fill with nitrogen gas, assuring an anoxic denitrification environment. The media is selected to assure that the void space is adequate to prevent backup or clogging, allowing a thin film of liquid to form as a result of the waste downflow into the system.

PROCESS ENGINEERING DESIGN

General process limitations and defining operating characteristics of fixed-film systems are summarized in Table 5.4. *Oxygen capacity* and *tower temperature* are dominant factors governing tower performance, both of which are not operating controls. Unlike a dispersed biological system the fixed-film processes have limited oxygen capacity. In the case of filter towers the oxygen capacity depends on the available draft, which depends on the difference in the ambient and tower conditions. This limitation impacts allowable influent loading and could restrict tower packing depths to heights optimizing tower draft conditions. Oxygen capacity control can be enhanced by using supplementary forced air.

TABLE 5.4. Fixed Film Systems Operating Characteristics.

Variable	Operator Controllable	Critical
Waste Characteristics		
Waste generated	No	Yes
Composition	No	Yes
Concentration	No	Yes
Biodegradability	No	Yes
Toxicity	No	Yes
Operating Characteristics		
Flow rate	No*	No
Hydraulic loading	Yes*	Yes
Organic loading	No	Yes
Influent concentration	No	No
Nitrification	No	Yes
Recirculation	Yes*	No
Sloughing and sludge buildup	No	Yes
Required oxygen and ventilation	No	Yes
Process hydraulics	Yes*	Yes
Nutrients	Yes	Yes
Alkalinity	Yes	Yes
Temperature	No	Yes

*Related variable.

Tower temperature is difficult to control, primarily dependent on waste temperature and ambient conditions, although it is a major factor affecting both tower draft and reaction kinetics. System *hydraulics* can affect performance by impacting sludge sloughing, media wetting, and contact between reactants and sludge. Recirculation or dosing enables the operator to control the system hydraulics, although optimum recycle may have to be compromised to achieve required influent dilution or increase dissolved oxygen.

REPORTED PERFORMANCE DATA

Data presented in Tables 5.5 and 5.6 are from the EPA Treatability Series, summarizing operating performance for the facilities investigated [18]. Based on historical operating experience with municipal trickling filters and combined treatment facilities, achievable effluent quality is summarized in Table 5.7 [7]. For some industrial waste treatment facilities these expected effluent quality limits may be diffi-

TABLE 5.5. Trickling Filter Performance Data (adapted from Reference [18]).

	Effluent, mg/L		Removal, %	
	BOD$_5$	COD	BOD$_5$	COD
Minimum	4	290	76	0
Maximum	137	709	98	77
Median	27	623	92	23
Mean	39	541	90	33

TABLE 5.6. Trickling Filter Industrial Performance Data (adapted from Reference [18]).

	Influent, mg/L		Effluent, mg/L	
	BOD$_5$	COD	BOD$_5$	COD
Hospital	183–400		4–56	
Tanning	150–860		30–80	
Rubber Mfg.		379		290
Timber Products	1700	3110	137	709

cult to achieve because the influent may consist of fewer biodegradable components.

REQUIRED PROCESS DESIGN DATA

Prior to starting a fixed-film design the Process Engineer must have adequate process and performance data to (1) assure that the industrial waste is biodegradable over the range of influent characteristics encountered, (2) project achievable effluent quality, (3) have specific design parameters to establish an effective operating range, and (4) determine the required oxygen and excess sludge generated.

Because of the interdependence of the physical and process aspects of fixed-bed systems, performance is difficult to extrapolate from bench-scale studies. They are best evaluated applying pilot prototypes, which can be rented from some suppliers. Studies should include establishing oxygen limitation, bed depth (or disc stage), hydraulic loading, organic loading, and achievable effluent quality. For the media type and depth selected, operating variables of concern include

(1) Hydraulic loading
(2) Organic loading
(3) Limiting influent conditions
(4) Recycle requirements

Resulting pilot data can be tabulated or integrated into an applicable form, establishing the required design data listed in Table 5.8.

WASTE EVALUATION

Important influent characteristics include flow rate, concentration, and biodegradability. Flow rate is critical if recycle is not provided because it must be adequate to optimize system hydraulics and dissolved oxygen capacity. Recirculation provides a means to dampen potential feed toxicity and excessive waste variability as well as control tower hydraulics. However, high organic loading or waste toxicity encountered in some industrial facilities may not be effectively alleviated by recirculation and may eliminate a fixed-film process as a viable treatment method.

TABLE 5.7. Combined Treatment Performance Data (adapted from Reference [7]).

	Number of Facilities	Influent, mg/L		Effluent, mg/L		
		BOD$_5$	TSS	BOD$_5$	TSS	
		(Range in Reported Monthly Values)				
TF	13	95–508	39–179	11–149	13–71	Reported
ABF	4	115–295	38–115	16–62	18–41	Decreasing
BF/AS	14	52–2836	40–703	7–24	3–56	Average
TF/TF	4	68–148	56–94	4–27	8–17	Effluent
TF/AS	3	122–629	53–140	8–16	6–17	BOD
RF/AS	7	96–382	54–185	4–20	4–17	|
TF/SC	1	53	76	8	11	|
TF/TF/SC	1	190	150	6	5	V

(1) Data are from 43 operating plants as reported in Reference [7]; with data combined for the configurations indicated. The number of facilities indicate the number of individual or combined units evaluated.

(2) The plant data represent monthly average values.

(3) Data are reported on the basis of influent to the first unit and effluent from the final unit.

(4) Data are reported on the basis of the range of influent concentrations, and corresponding effluent quality. The reader should review References [7] for complete performance details.

(5) As indicated, the units are listed according to reported average monthly effluent BOD value, in descending order. Suspended effluent concentrations for the units generally follow the same order. In some cases the difference in average effluent quality between lower ranked facilities is not significantly different, and well within commonly required effluent quality. In addition, there is no assurance that all facilities were operating at optimum conditions: and it should be noted that the influent range is not consistent for all the facilities.

TF = trickling filter, ABF = activated biofilter, SC = solids contact, RF = roughing filter, AS = activated sludge, and BF = biofilter.

Feed Concentration

Tower feed concentration must be controlled to achieve an acceptable *applied* organic loading, maintain aerobic conditions, and prevent excessive sludge formation. Concentration as a viable operating variable is not evident in all fixed bed models, probably because of the limited feed concentrations evaluated in developing the models. Oxygen availability limits applied feed concentration to approximately 400 mg/L BOD$_5$, unless force air systems or emerging technologies are employed.

Treatability

As in all biological systems, feed biodegradability or toxicity affects system stability and performance. In fixed-film systems, biological sludge formed on the media is sensitive to the feed constituents, which under extreme conditions could terminate biological activity. As discussed in Chapter II-2, waste biodegradability significantly affects biological performance and can be related to specific compound characteristics. The biological effectiveness of any reactor configuration is commonly denoted in terms of the reaction rate, expressed in terms of kinetic constants. In processes such as rotating discs, constant removal rates are frequently assumed over the recommended influent range, with an "experience factor" employed to "fine tune" the design. Considerable operating data, and extrapolated treatability design data, are available for trickling filters, which is fortunate because it is the most frequently applied fixed film process and the data can be extrapolated to establish potential treatability effectiveness for other fixed film processes.

In trickling filters models reaction kinetics are "buried" in the many variables that physically and hydraulically define the tower performance. Because trickling filter models were established for municipal systems, definitive biode-

TABLE 5.8. Trickling Filter Required Design Data.

Critical pilot plant treatability data specific to the waste
(1) Design temperature
(2) Selected media
(3) Hydraulic loading
(4) organic loading
(5) Y, cell yield, mg/mg
(6) Achievable effluent quality
(7) Nitrification effectiveness
Waste solids characteristics that should be obtained from laboratory studies but can be estimated
(8) Fraction of effluent solids biodegradable
(9) Ultimate solids BOD per g solids
(10) Ratio of SBOD to ultimate BOD
(11) Ratio MLVSS/MLSS
SELECTED operating characteristics
(12) Recycle capabilities
Operating characteristics that should be obtained from pilot studies but can be estimated from treatability data
(13) Summer temperature effects
(14) Winter temperature effects
(15) Maximum influent concentration
(16) Required oxygen loading
(17) Waste sludge generated
(18) Nitrogen required per pound SBOD removed
(19) Phosphorus required per pound SBOD removed

gradability constants was not a primary concern relative to the other filter operating variables. In fact, within the accuracy of most models, unless a constituent was introduced to radically affect performance, its significance would be lost (for domestic waste). However, industrial waste characteristics cannot be assumed to be as accommodating.

For the most part, available trickling filter performance data are correlated in terms of "*k* and *n*" model constants. These constants combine the effects of specific waste biodegradability and concentration, as well as organic loading, filter media, media depth, and tower hydraulics. Based on other biological system models the *k* value would appear to be a removal constant and *n* a hydraulic loading exponent. In reality, the indiscriminate use of published constants is meaningless because removals are the result of combined kinetic and hydraulic tower effects as well as the tower physical characteristics. In some cases, available oxygen may be the controlling variable. In addition, published performance data could be skewed by the selected media or results reported as total or soluble BOD. Attempting to isolate specific waste characteristics from all the other interrelated variables is impossible. For all of these reasons, and probably many more, design data must reflect specific waste and configuration characteristics.

Biodegradability of some industrial wastes was evaluated by the Dow Chemical Company at the pilot scale, the resulting data are commonly cited in the literature [9,21]. The tests were conducted using VFC plastic media bed, 6.6 m (21.5 ft) deep. Most of the data were collected at hydraulic loadings ranging from 0.3 to 2 L/m^2-s (0.5 to 3 gpm/sf). Data were correlated using the Germain model, with the hydraulic exponent *n* set at 0.5. The interested reader should review the literature cited for specific details; the reported results can be summarized as follows:

(1) Removal efficiencies of 60 to 80% were reported at organic loading rates less than 4 kg/day/m^3 (250 lbs/day/1000 cubic foot) of packing. Removals ranging from 50 to 70% were reported at loadings from 4 to 8 (250 to 500) and generally lower at higher loadings. The 4 to 8 organic loadings may define the start of the oxygen limiting range.

(2) The measured *k* values at 20°C were normalized to a depth of 6 m (20 ft) and a 150 mg/L influent concentration using the relation:

$$\text{Normalized } k_{20} = (D/6)^{0.5} \, (S_e/150)^{0.5} \qquad (5.4)$$

These data are commonly presented by waste source, as indicated in Table 5.9 [9].

The data presented are not intended to define process biodegradability but (1) to emphasize the unique characteristics of some wastes, (2) the need to correct pilot data to applicable bed depth, influent concentration, and temperature, and (3) demonstrate the difficulty in arbitrarily utilizing reported trickling filter data.

TABLE 5.9. Trickling Filter Industrial *k* Values (adapted from Reference [21]).

Wastewater	Normalized k_{20}	
	$L/sec^{0.5}/m^2$	$gpm^{0.5}/ft^2$
Domestic	0.11–0.24	0.04–0.09
Domestic and industry	0.16–0.27	0.06–0.10
Kraft mills	0.08–0.11	0.03–0.04
Food and dairy	0.11–0.38	0.04–0.14
Meat packing	0.05–0.35	0.02–0.13
Tannery	0.19	0.07*
Pharmaceutical	2.22	0.82*
Fruit canning	0.16–0.19	0.06–0.07
Refinery	0.05	0.02*
Textile Mill	0.11–0.14	0.04–0.05

*Data from single test.
Reported *k* values in terms of $gpm^{0.5}/ft^2$ can be converted to $L/sec^{0.5}/m^2$ by multiplying by the factor 2.705, provided *n* is at 0.5.

Pretreatment

Generally, the waste preparation or pretreatment criteria discussed for activated sludge apply to fixed-film media systems. Influent conditioning optimizing biodegradable properties, nutrient addition, and alkalinity control must be considered in the process design. Specific to fixed-film systems, toxicity due to excessive organic loading, heavy metals, inorganic salts, ammonia, and chlorine is of concern. Chlorine is of particular concern because traditionally it is the compound used to control common biological growth problems.

PROCESS DESIGN VARIABLES: TRICKLING FILTERS

Fixed-film design considerations involve subjectively reviewing process basics to develop achievable effluent quality. Specific requirements for *trickling filters* will be outlined, which will serve as a basis for evaluating all fixed bed systems; specifics for rotating biological discs will be discussed separately. Trickling filter process design involves evaluating the following process variables:

(1) System configuration
(2) Carbonaceous oxidation
(3) Nitrification
(4) Media
(5) Packing depth
(6) Tower hydraulics
(7) Design loadings
(8) Production requirements
(9) Oxygen limitations
(10) Denitrification
(11) Fate of contaminants

System Configuration

Establishing a fixed-film process configuration involves considering both carbonaceous oxidation and nitrification requirements, using one of the systems discussed in the Fixed-Film Filter Systems Section. Frequently, the trickling filter is the initial configuration considered because of the vast amount of available operating experience. Combined systems are employed to improve operating flexibility and effluent quality. Innovative technologies are applied for special waste characteristics. Evaluating possible configurations involves the following sequence, assuming a trickling filter for the initial evaluation.

Trickling Filter Limitations

Selection of a trickling filter as a viable treatment system should be based on the following criteria:

(1) Is the waste biodegradable?
(2) Is the concentration less than 500 mg/L BOD$_5$?
(3) Are the nonbiodegradable suspended solids level low?
(4) Are there any significant waste toxic components that could prevent biofilm formation?
(5) Can the waste characteristics be defined and a design basis established?
(6) Can the oxygen requirements be met by a natural ventilation system? Is forced ventilation required?
(7) Is the influent SBOD too low to be effectively treated biologically?

Equalization Requirements

Is the plant waste flow, concentration, or composition highly variable? Waste storage or equalization, as well as recirculation, is a viable method of minimizing influent variations. Dampening influent variations allows design at an "average condition," within a reasonable operating range, preventing oversizing for peak loads.

High Waste Solids Content

Waste solids content should be completely analyzed, identifying suspended and colloidal solids level. The configuration should be adjusted for specific characteristics.

(1) Does the influent suspended solids concentration warrant the use of a primary clarifier?
(2) Does the influent contain colloidal solids? Evaluate configurations illustrated in Figures 5.1.
(3) Is the TF effluent colloidal solids content high requiring a polishing step as illustrated in the Figure 5.4 TF/SC system?

SBOD

A fixed-film system will be affected by very low or high soluble BOD levels. A low soluble organic content may not allow adequate biomass growth to assure a viable treatment process. This is equivalent to low growth in a dispersed growth system in which the biomass growth is less than the effluent solids concentration. In such cases the ABF configurations illustrated in Figures 5.2 or 5.3 may be applicable. A high influent organic concentration could result in a prohibitive tower loading, as discussed subsequently. If the influent SBOD is within an acceptable operating range, a trickling filter can be evaluated as a sole treatment or as a TF/SC system illustrated in Figure 5.3 to reduce the TF effluent SBOD content.

Enhanced Polishing Step

If the TF effluent requires a moderate polishing step to reduce SBOD and colloidal solids content a TF/SC system similar to that illustrated in Figure 5.5 should be evaluated.

Combined Treatment

If a single-stage filter or TF/SC polishing step does not provide the process stability and operating flexibility to consistently meet stringent effluent criteria, two-stage or a *combined treatment* system may be required. This can include a first stage roughing filter with a second stage filter, RBC, or a suspended growth system to improve carbonaceous oxidation, nitrification, or both.

Loading Limitations

Trickling filter hydraulic and organic loading limitations in single or combined treatment configurations must be studied, with the resulting stability and flexibility of multistage towers appraised as follows:

(1) Parallel towers should be considered to maintain hydraulic or organic loading design criteria and aerobic conditions for the expected influent range.
(2) Series operation should be considered to achieve required effluent quality for carbonaceous oxidation, nitrification, or any required combination, as well as to maintaining aerobic conditions for the entire operating range.
(3) The number of installed units will be directly related to the required total filter area or performance.
   (a) Additional tower area (or number of towers) may result from effluent recycle to dilute influent concentrations to less than 500 mg/L.
   (b) Multiple units will be required if the resulting total filter area results in excessive tower diameters.
   (c) Area requirements may be influenced by applicable organic loading criteria, resulting in increased tower diameter or multiple units.

Operating Flexibility

Evaluate the need for multistage operation based on operating flexibility. Operating flexibility can be obtained by installing two units, each capable of treating 75% of the total load, operating either in series or parallel. The resulting system will provide the following advantages:

(1) With seasonal or production variations the units can be operated in parallel, with one running at optimum conditions and the other taking the varying load.
(2) If toxic surges are common, and unavoidable, series operation allows the second unit an opportunity to recover while the problem is being remedied, possibly avoiding a complete shutdown.
(3) Two units permit maximum treatment efficiency during cold weather months, where loading rates are reduced and downtime maximum.

Care should be taken in sizing units in series because if the primary unit is sized too conservatively, the second unit will be starved and ineffective. Size the primary unit for high rate loading and the second unit for polishing rates. Series units should be configured so that the towers can be alternated, allowing biological growth in the trailing system.

Enhancements

Evaluate process enhancements to improve operating flexibility and system performance. Such enhancement could include the following:

(1) Forced air ventilation to assure tower stability at all ambient conditions.
(2) Effluent recirculation to enhance oxygen input to the tower top, reduce the effect of toxicity or concentration, and to allow operator flexibility in maintaining hydraulic stability.
(3) Sludge recirculation to improve system performance, primarily by promoting improved sludge activity, similar to selector design in dispersed systems.

Emerging Technology

Packed beds, sparged air, or submerged bed systems should be evaluated where nitrification (or denitrification) is a significant consideration in the overall treatment scheme. A separate or combined nitrification step, or a separate denitrification step, should be consistent with the carbonaceous oxidation requirements discussed in this section. If this technology proves viable for improved carbonaceous treatment or nitrogen control, site-specific testing will be required to upgrade the technology to full-scale application.

Carbonaceous Oxidation

Carbonaceous oxidation for industrial wastes is frequently evaluated using the Germain or similar relations applicable to plastic media. Test data are used to establish relevant kinetic constants, corrected to the design concentration and depth, and applied to design, winter, and summer conditions. Appropriate hydraulic loadings can be established using Equation (5.5).

$$Q_v = \left[\frac{k \cdot D}{ln \, (S_o/S_e)} \right]^2 \qquad (5.5)$$

where Q_v, hydraulic load, m^3/m/SM, (gpm/sf); S_o, feed concentration, BOD_5 mg/L; S_e, settled effluent, BOD_5, mg/L; D, depth, m, (ft); and k, coefficient, units to suit.

It is essential that the appropriate influent flow rate and concentration are employed, consistent with the model and study data collected, employing or *excluding* recirculation in establishing the applied influent conditions. The applied hydraulic loading must be consistent with the requirements of the media employed and if too high adjusted accordingly. The organic loading is estimated by using Equation (5.6), and the design hydraulic loading. The resulting organic loading is compared to the recommended system loading range and if too high adjusted accordingly.

$$Q_o = \left[\frac{Q_v \cdot S_o}{D} \right] \qquad (5.6)$$

where Q_o, organic loading, kgs/min/m^3 media (lbs/min/sf). Q_v is expressed as volume/area/min, and S_o in compatible weight/volume units.

Based on the *Germain model,* the tower area is estimated by dividing the feed flow, *excluding the recirculation,* by the design hydraulic loading. The media volume is estimated by dividing the organic mass feed rate by the design organic loading. Although recirculation does not enter into the Germain model calculation, recirculation is a critical design consideration to

(1) Maintain hydraulically effective bed conditions
(2) Dilute the feed to acceptable concentrations
(3) Maintain aerobic conditions

Nitrification

Basic stoichiometric and kinetic expressions applicable to nitrification are discussed in Chapter II-2. The organic carbon to ammonia nitrogen ratio (or organic nitrogen) establishes whether a combined or separate nitrification system is applicable. The primary mechanism governing the competitive nature of the two oxidation processes is the slower nitrification cell growth relative to that of carbonaceous microorganisms. Some basic principles governing nitrification include:

(1) Because nitrifying cell growth is slower than the carbonaceous cell rate, nitrification will initiate only at low BOD levels.

(2) When BOD levels are high any nitrogen ammonia removal is governed by the synthesis reaction of the carbonaceous oxidation, i.e., the nutrient requirement.

(3) Filter oxygen limitations must be carefully considered for industrial wastes where ammonia/nitrogen concentrations could be much greater than 20 to 30 mg/L.

(4) At high ammonia/nitrogen influent levels, alternate treatment systems should be investigated.

(5) Where final ammonia nitrogen effluent concentrations lower than 1 or 2 mg/L are required, chemical treatment (such as breakpoint chlorination) may be required.

Specific design procedures are frequently based on criteria obtained from municipal systems, or laboratory studies using domestic wastes. Collectively these studies offer some insight as to the variables and limitations governing biological filter nitrification.

Combined Oxidation/Nitrification Treatment Systems

The bulk of information frequently quoted for combined carbon oxidation nitrification systems are from the EPA Nitrogen Design Manual, often presented as shown in Figure 5.7 [16]. An overriding parameter in applying EPA data is the *allowable* BOD *loading* at which nitrification commences. This is the prime design criteria cited in these stud-

ies, directly incorporating (or ignoring) hydraulic loading, temperature, oxygen limits, applied nitrogen concentration, and applied nitrogen loading. Data from eight full-scale and pilot facilities studied indicate that a rock media filter treating *municipal wastes* can achieve *75% nitrification (or better) at a BOD₅ loading of 0.16 to 0.19 kg/m³/day* (10 to 12 lb/day/1000ft³). The nitrification rate diminishes with increasing organic loading, being insignificant at 0.48 to 0.64 (30 to 40). No data were cited for nitrification at temperatures less than 20°C. Other studies emphasize the effect of BOD level on combined treatment, confirming that BOD levels of less than 30, or SBOD less than 20 mg/L, are necessary to initiate nitrification [2,6,12]. It is important to differentiate between biological oxidation aimed at *ammonia/nitrogen* removal, which is relatively effective, and *organic nitrogen* removal for some industrial requirements, which for the EPA evaluated facilities varied from 20 to 80%.

An intensive investigation by Obey, whose data is commonly cited for combined nitrification municipal design, proposes a direct correlation between removal rate and influent loading (BOD₅/TKN ratio), corrected for operating temperature [21]. Figure 5.8 represents a correlation of removal rates at three temperature ranges, 9 to 20°C, 20 to 25°C, and 25 to 31°C. Significantly, the data indicate reduced nitrification rates at increased temperature, attributed to reduced oxygen capacity. The best nitrification rates were achieved at the 9 to 20°C range, as illustrated in Figure 5.9.

In summary, current combined treatment design criteria are frequently based on operating limits observed in the municipal systems evaluated, which is primarily based on the requirement that the organic content be lowered substantially so that the nitrifying microorganisms can survive. This has been observed at soluble BOD₅ levels of less than 20 mg/L and/or an organic loading of less 0.16 to 0.19 kg/m³/day. Although some pilot data may indicate that effective nitrification might be achieved at higher organic loadings, the possibility of (high density) media plugging for extended

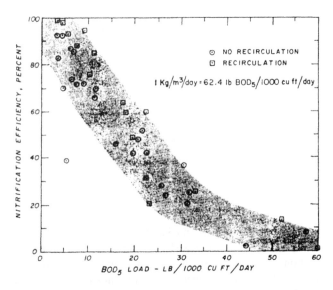

Figure 5.7 Combined nitrification design criteria (adapted from Reference [16]).

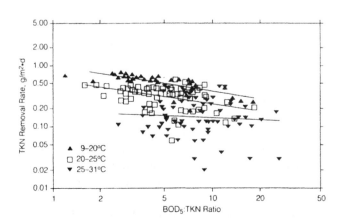

Figure 5.8 Modified combined nitrification design criteria, varying temperature range. Copyright © Water Environment Federation and American Society of Civil Engineers, reprinted with permission.

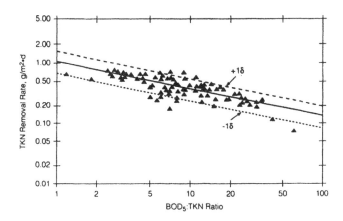

Figure 5.9 Modified combined nitrification design criteria, 9–20°C. Copyright © Water Environment Federation and American Society of Civil Engineers, reprinted with permission.

full-scale operation should deter such design. Besides the loading criteria mentioned, industrial waste nitrification designs must include investigating the affect of media, temperature, and nitrogen composition differences from those in the municipal study data cited.

Separate Stage Nitrification

Separate nitrification biofiltration criteria, as with combined systems, are frequently based on the collective results of municipal system evaluations. These include data reported in the EPA Nitrogen control manual [16], supplemented by studies conducted by individual investigators. EPA data for separate nitrification of *treated municipal effluent* have been reported on the basis of temperature, media, ammonia nitrogen loading, and required effluent ammonia nitrogen concentration [16]. These data are often cited as base study data and as such reproduced for reference purposes in Figures 5.10 and 5.11. The data presented indicates 372 m² (4000 ft²) of surface plastic packing surface area is required for each pound of ammonia oxidized. This loading produces a municipal effluent of 3 to 6 mg/L nitrogen/ammonia at a BOD₅/TKN ratio ranging from 0.36 to 1.1 and at a temperature range from 13 to 19°C.

Gullicks and Cleasby proposed modification of the base data design curves to include temperature and applied ammonia nitrogen concentration and loading as well as hydraulic loading as design variables [4]. Figures 5.12 and 5.13 are based on modification of the EPA data cited, and applicable to settled secondary treatment *municipal* effluent. The data include influent ammonia nitrogen concentrations of less than 18 mg/L, producing effluent ammonia nitrogen levels of less than 4 mg/L. Although constant rate nitrification is implied, they concluded that nitrification rates are nitrogen controlled at ammonia nitrogen concentrations less than 4 mg/L, at a temperature of 20°C, and a corresponding oxygen

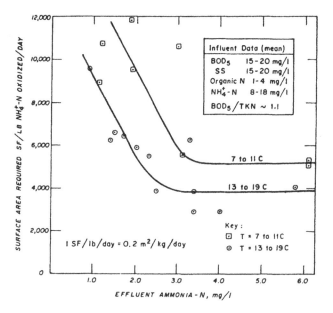

Figure 5.10 Separate nitrification design criteria, EPA Facility 1 (copied from EPA Reference [16]).

saturated conditions of 9.1 mg/L. In addition, the potential for reduced nitrification as the liquid approaches the tower bottom occurs as a result of an alkalinity loss, and lowered pH, from the nitrification process.

In separate pilot studies involving nitrification of municipal secondary effluent in a biofilter at low temperatures (10°C), Gullicks and Cleasby concluded that at winter conditions [5]

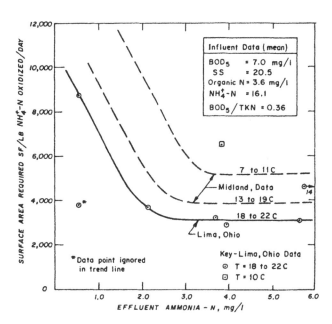

Figure 5.11 Separate nitrification design criteria, EPA Facility 2 (copied from EPA Reference [16]).

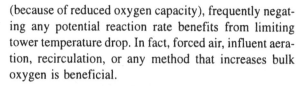

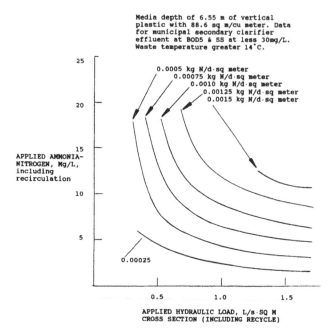

Figure 5.12 Modified separate nitrification design criteria, >14°C. Copyright © Water Environment Federation, reprinted with permission.

(because of reduced oxygen capacity), frequently negating any potential reaction rate benefits from limiting tower temperature drop. In fact, forced air, influent aeration, recirculation, or any method that increases bulk oxygen is beneficial.

(2) Filterable chemical oxygen demand (COD) levels greater than 60 mg/L adversely affected nitrification, with performance improving below this concentration.

(3) Total hydraulic loadings greater than 0.8 L/m²/s (1.2 gpm/sf), when combined with low DO, adversely affected nitrification.

Performance curves developed by Gullicks and Cleasby for a 10°C operating temperature are shown in Figure 5.14.

All the correlations cited assume a constant nitrification rate, representing apparent overall reduction rates based on analyzing influent and effluent conditions. Other investigators proposed that these correlations do not describe all the tower mechanisms affecting nitrification. If that is the case, a constant rate throughout the tower may not be an appropriate design procedure, especially if low ammonia effluent concentrations are required. Investigators have found a significant reduction in nitrification rate at low hydraulic loadings, being further affected by oxygen, depth, and nitrogen concentration [3,6,10,13]. These investigators have proposed that total tower nitrification be corrected for these parameters. The suggested corrective correlations are not presented because they are difficult to apply outside the study limits and the required design constants for general application are sparse. However, the Process Engineer should review the

(1) Bulk dissolved oxygen (DO) concentration is critical to nitrification, with nitrification diminishing with oxygen levels at 60 to 65% of the saturation level and may be totally eliminated below 45 to 50% saturation. The DO level is so critical that restricting ventilation to minimize temperature drop is detrimental to nitrification efficiency

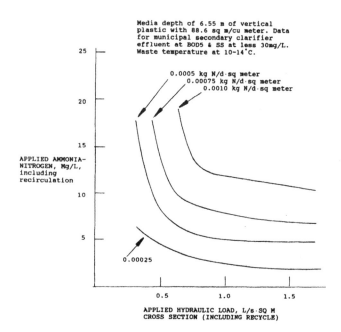

Figure 5.13 Modified separate nitrification design criteria, 10–14°C. Copyright © Water Environment Federation, reprinted with permission.

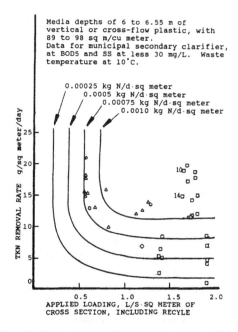

Figure 5.14 Modified separate nitrification design criteria, 10°C. Copyright © Water Environment Federation, reprinted with permission.

cited references to establish whether these parameters affect selected design criteria.

Industrial Nitrification System Design

Process Engineers can easily misinterpret the data cited, approximate constants and obtain a *definitive* industrial nitrification design. The underlining theme of this book that reliable industrial design criteria can only be obtained from site specific evaluations is especially critical for nitrification. The data and criteria cited should be put in prospective.

(1) Municipal nitrification involves secondary effluents with ammonia levels well below 50 mg/L. Industrial levels can be well above this level, and the component may be a combined nitrogen, not the "ammonia/nitrogen" species common in municipal effluents.

(2) Definitive design methods are not available for biofilter nitrification. The data discussed, as well as similar studies, are for the most part limited "correlated observations," specific to the facility or experimental conditions evaluated. Reported data sometimes conflict with other study results and should always be used cautiously outside the conditions evaluated.

(3) Nitrification does not reduce effluent nitrogen content, it converts ammonia to nitrite. Therefore, total effluent quality must be established and a complete nitrogen control strategy defined.

With this understanding, the studies discussed do indicate some significant criteria applicable to nitrification design:

(1) Total tower kinetics are not adequately defined by simple zero-order kinetics. Nitrification at levels above about 4 mg/L will probably adhere to zero-order kinetics, represented by constant nitrification rates. Low effluent concentrations may adhere to first-order kinetics, affected by oxygen availability, depth, nitrogen concentration, and operating conditions. Chemical treatment may have to be included for low effluent requirements.

(2) Nitrification will not initiate until conditions favoring carbonaceous oxidation are eliminated. This is generally cited as a BOD_5 loading of 0.16 to 0.19 kg/m^3/day or less, a BOD level of less than 30, or SBOD less than 20 mg/L.

(3) Available oxygen is critical to achieving significant nitrification, making a forced air ventilation system a required consideration for industrial systems.

(4) As discussed for both combined or separate treatment, hydraulic and organic loading is a significant design variable.

(5) Cold weather conditions could lead to unstable or ineffective tower performance.

Media

Media Type

Media characteristics are not a direct input in most trickling filter models but indirectly reflected in the model constants. Media characteristics are primarily related to shape and size, governing the specific surface area (surface area per unit volume) and percent bed void space. The surface area provides the opportunity for slime growth to promote the biological process and can be related to the F/M ratio in a suspended growth biological system. A large surface area provides the opportunity for high cell mass accumulation, equivalent to increased MLVSS in suspended growth systems. Increased void space affords the opportunity for higher hydraulic loading, enhanced aeration capacity (forced or draft), and reduced plugging potential.

Total bed contact area is a product of the media specific surface area and the bed volume. Theoretically, the greater the bed contact area the greater the performance efficiency. Although the total area can be increased by increasing bed volume and the tower size, media selection is a more effective and economical method of controlling performance. By selecting a small shape with a high area to volume ratio, the total transfer area is increased for equal bed volume. Again a balance must be reached because media with high area per volume can also result in small void space; resulting in higher energy requirements and higher maintenance due to media plugging and breaking.

Available Media

Early trickling filter media consisted primarily of *rock* and *horizontal (wood) slats,* with the significant design parameter being size and orientation. The introduction of *random* plastic packing and prepackaged *bundles* extended tower bed use and improved performance. Bundled packings are fabricated with *vertical* or *cross-flow* patterns, the vertical flow pattern most simulating plug flow and cross-flow a complete mix system. Vertical bundled packings are constructed in the *semicorrugated* and *fully corrugated* mode.

Cross-flow bundled packing are constructed as 60 and 45 degree flow patterns. Characteristics and general applications associated with different media types are summarized in Table 5.10 [21]. Selection of tower media requires an evaluation of both physical and process performance characteristics.

Media Physical Performance

The effect of media in reported studies or operating data is not readily apparent because most process studies record performance on the basis of a single selected bed. In addition, related hydraulics or other process variables commonly expressed in trickling filter models are obscured in the equa-

TABLE 5.10. Media Characteristics (adapted from Reference [19,21]).

Media Type	Size in.	Surface Area, ft²/ft³	Void Space %	Application			
				R	B	N	T
Bundle	24 · 24 · 48	27–32	>95	x	x	x	
	24 · 24 · 48	42–45	>94				x
Rock, small	1–3	17–21	40–50	a	x	x	
Rock, large	4–5	12–50	50–60		x	x	
Random	Varies	less 40			x	x	
Random	Varies	more 40					x
Wood	48 · 48 · 1.875	14	70–80	x	x	x	

In. · 2.54 = cm ft²/ft³ · 3.28 = m²/m³.
R, roughing filter; B, BOD removal application; N, BOD and nitrification application; T, tertiary treatment application; and a, coarse media only.

tion, usually in the model constants. Harrison and Daigger [8] conducted a study using municipal waste, evaluating six different media in six parallel systems. The data obtained were correlated using the modified Velz equation. The resulting average n and k constant values established for the media evaluated are tabulated in Table 5.11.

The significance of the constants are explained in terms of the media capabilities, with the constant n signifying a quantitative measure of the media hydraulic characteristics. A high n value indicates media sensitivity to hydraulic loading. Because all the towers were tested with the same domestic waste, k can be assumed to be a measure of the biological reaction efficiency for the specific waste considered, as affected by the media sludge growth, oxygen availability, and waste sludge contact.

As a result of test results and observations, Harrison and Daigger concluded [8]

(1) Oxygen transfer characteristics of rock media were inferior to others, random media were superior, and no significant difference were observed among the others.
(2) Visual observations indicated improved wetting and less potential for filter flies (dry areas) or plugging with the vertical and 60 degree cross-flow media, relative to the others. Richards and Reinhart [14] observed less potential for plugging problems with vertical packing interfaces than with random or cross flow packing.

Media Process Performance

Media performance must account not only for the physical characteristics discussed, but also BOD and nitrification removal efficiency. Much of the reported performance data is from pilot studies, with little assurance that pilot results can always be produced in full-scale operation. Some pilot investigations suggest that plastic media outperform standard rock or horizontal media, with prefabricated flow oriented bundles outperforming the random packed plastic media [8,11,14]. The Harrison and Daigger study [8] provides an indication of plastic media performance based on selected test conditions. They concluded that at low loadings (0.59 TBOD and 0.27 SBOD kg/m³-day) a 60-degree cross-flow cross-flow produced an effluent with lower soluble BOD and COD than the 45-degree cross-flow and vertical media. However, at medium (2.64 TBOD and 1.68 SBOD kg/m³-day) and high (5.95 TBOD and 3.54 SBOD kg/m³-day) loading the vertical flow media effluent quality was superior. Other studies at equivalent loadings have also demonstrated excellent performance with plastic media, with the potential for cross-flow media to outperform vertical at the low ranges evaluated [11,14].

General Guidelines

It should be observed that the pilot media test results discussed are offered as a *guide* and optimizing system hy-

TABLE 5.11. Media Performance (adapted from Reference [8]).

Media	Description	Area*	Void, %	n	k
Vertical	PVC corrugated	88	97	0.34	0.35
45°	PVC corrugated	98	95	0.41	0.32
60°	PVC corrugated	98	95	0.44	0.35
Horizontal	Redwood slats	46	94	0.44	0.27
Random	24 × 12 cm cylinder	105	92	0.62	0.35
Rock	8 cm	50	35	0.70	0.38

*Area in m²/m³.

draulics (as discussed below) may alter the results cited. Although the only total valid media testing is full scale, this is not practical in the early evaluation stage and too late during operation if the wrong media is selected. Therefore, pilot studies and supplier data can establish not only design criteria but also suitable media. Utilizing the information discussed, specific evaluations should include

(1) Investigating the potential wetting characteristics and required liquid distribution for selected media, realizing that optimum hydraulics is specific to the packing employed
(2) Sizing and configuring the pilot facilities to minimize the impact of short circuiting and obtain realistic full scale operation impact

Based on the testing data cited previously, some general design guidelines can be offered:

(1) Synthetic media, especially bundled media, are frequently selected for industrial application because of performance and ease of installation.
(2) Fouling is enhanced with media that redistribute flow (such as rock, random, and cross-flow). This is because of reduced flushing effectiveness, and a higher tendency to retain solids [21].
(3) High concentrated industrial wastes produce greater biomass concentration, favoring wood or vertical media. Cross-flow patterns are generally considered for low loadings. Vertical flow patterns are generally effective for high loadings, minimizing the potential for increased media plugging. However, improvements in flushing and hydraulic design may warrant evaluation of cross-flow patterns.
(4) Because there is no significant evidence that high density media substantial improves carbonaceous oxidation performance, media with surface areas greater than 100 m²/m³ is discouraged because of greater plugging potential [21].

Packing Depth

Packing depth is a direct input in most trickling filter models, implying a direct relationship between bed depth and performance, the assumption being that the bed depth governs the system mass transfer characteristics and contact time, in turn governing reaction efficiency. However, the significance of bed depth as related to substrate reduction is complicated because of two factors. First, system effectiveness is not exclusively related to the bed depth, but to other independent (or interrelated) factors such as media type, wetting efficiency, hydraulic or organic loading, and the dosing mechanism. As with all biological systems, reaction efficiency is related to *solids retention time* (SRT), a measure of the effective biomass growth, which in biofilter is difficult to control or determine. Second, a (natural draft) biofilter

is an oxygen-limited system capable of treating a maximum feed concentration of 400 mg/L BOD and under the best of conditions discharging an effluent containing 5 to 40 mg/L of BOD (utilizing traditional configurations). In fact increased depth, and thereby increased draft resistance, tends to decrease air flow (oxygen availability) to the tower. Therefore, oxygen availability and transfer could be significantly affected by the selected tower depth and a more meaningful factor to the performance than the depth itself.

Model and study results are not decisive as to the significance of depth, or volume irrespective of depth, as a vital design criteria. Some municipal system data suggest improved removals with increased depth, and other data suggest that any significant improved performance attributed to increased depth was probably due to other process variables. Investigator's conclusions as to the significance of bed depth to system performance can be summarized as follows [21]:

(1) Some municipal system operating data for taller towers appear to indicate improved performance with depth. However, this may be the result of increased hydraulic loading commonly applied to deep beds or the inherent result of better wetting efficiency. Deep beds improve distribution from the top to the effective lower packing sections, thereby reducing the required minimum flow and recirculation.
(2) Poorer performances of some shallow towers may be impacted by the fact that they must be adequately irrigated in the surface levels or risk the possibility of poor performance due to low wetting efficiency.
(3) Some study results are further complicated by the occurrence of anaerobic conditions at the tower top sections, making the system reaction kinetics oxygen controlling and thereby seemingly minimizing the significance of additional retention time.
(4) Some of the effectiveness of taller towers may be because they tend to be designed with a configuration exposing less surface areas to wind and climatic exposure, reducing temperature loss. This could improve the natural draft and ventilation capacity as well as maintain a higher waste temperature to improve removals.
(5) The results of most studies involving higher towers, indeed any size tower, are difficult to analyze on the basis of isolated variables. Because higher tower depths tend to complicate the internal hydraulics, performance data become exceeding more difficult to interpret on a common basis.

Discussions of bed depth effects would appear to assume that bed depths are installed to the nearest centimeter. In reality most full-scale systems and pilot studies are operating with bed depths of 1.8, 2.4, 3.7, 6.6 m (6, 8, 12, or 21.5 ft). Some highly loaded roughing filters employing depths as low as 1 m (3 ft) and some plastic media are constructed as high as 12.8 m (42 ft). Significantly, most pilot studies and historical operating data were obtained from towers 1.8

and 2.4 m in depth, covering the tower heights commonly employed in municipal biological filters. Because of a lack of consensus on the significance of bed depth, the Process Engineer should establish the applicable bed depth range based on overall system requirements. This includes considering

(1) Whether the filter is employed for maximum organic removals, high effluent quality, or as component in a combined treatment system
(2) The system oxygen limitations, evaluating available tower ventilation with depth, and the need for forced ventilation
(3) Tower hydraulics to maximize effectiveness at lower depths
(4) Limitations of the media selected

As a guide,

(1) If stone or random media are employed, 2 to 2.5 m (6 to 8 ft) maximum depths should be considered because of ease of construction and the oxygen limits. If increased capacity is required, separate units in series would provide additional oxygen.
(2) If stone or random media is used in a roughing operation, or as the primary treatment in a combined treatment system, shallower depths down to 1 m (3 ft) could be considered. However, the shell and hydraulics should be evaluated for up to 2.5 m (8-ft bed) depth to allow future enhancements.
(3) If plastic media is selected, 4 or 7 m (12 or 21.5 ft) depth (based on prepackaged bundle standards) should be evaluated. If a trickling filter is the primary or only treatment system, the tower should be designed to allow adjustments to facilitate future treatment demands. If greater than 7 m of packing is warranted, the ventilation limitations should be thoroughly evaluated and either forced ventilation or units in series considered.

Tower Hydraulics

Tower hydraulic effects are a direct input in some models as the hydraulic loading rate raised to some exponential factor n, directly impacting substrate reduction. However, tower performance can be affected by hydraulic parameters such as media wetting, oxygen and reactant transfer rates, and media solids buildup control. The hydraulics being affected by volumetric loading, wetting efficiency, recirculation, and dosing.

Hydraulic Loading and Wetting Efficiency

In fluid bed design the reactor wetting efficiency is defined as the fraction of the packing surface area covered with liquid. As with any contact process it depends on physical

variables such as liquid distribution, air to liquid relative rates, and physical packing characteristics; and waste water temperature, viscosity, and surface tension. This is commonly related to the media type as the allowable hydraulic loading, as indicated in Table 5.12 [9]. As discussed in the Dosing Section, conditions improving the wetting efficiency of a system are believed to optimize removal efficiency and along with organic loading significantly impact the system performance.

Hydraulic loading is commonly reported as the volume per unit of time per cross sectional area of filter, i.e., $L/m^2/s$ or $gal/min/ft^2$. Sometimes the filter area is expressed as acres. Depending on the data base or model, it may or may not include recirculation. Loadings applied to the Germain Model and plastic media commonly do not include recirculation.

Recirculation

Recirculation appears as a process variable in many of the trickling filter models, but the direct contribution to tower performance is still questioned by some investigators [21]. Recirculation appears to improve tower performance as a result of (1) dilution of high strength wastes to the aeration capacity of the tower, (2) increase oxygen capacity to the system, and (3) improvement of the tower *hydraulics*. Where recirculation has proven operationally significant it has been assumed that a major contribution was improved tower hydraulics [9,21]. Because adequate hydraulics can not be assured by the waste flow volumes, recirculation capabilities or dosing frequency should be evaluated as a means of providing operating flexibility. In addition, recirculation produces a stabilizing effect by dampening sudden volumetric and concentration surges to the system. If shallow towers are selected, recirculation must be included in the design to assure adequate tower hydraulics.

However, recirculation could produce some adverse effects. In winter months tower temperature will be reduced when cooler tower bottoms are recycled. If nitrification oc-

TABLE 5.12. Hydraulic Loadings (adapted from Reference [9]).

Media	$L/m^2/s$	GPM/ft^2
Rock filter		
Slag, Low rate	0.01–0.04	0.02–0.06
Slag, intermediate	0.04–0.11	0.06–0.16
Rock, high rate	0.11–0.43	0.16–0.64
Two-stage	0.11–0.43	0.16–0.64
Wood filter		
Roughing	0.54–2.17	0.80–3.20
Plastic		
Roughing	0.54–2.17	0.80–3.20
Super high rate	0.14–0.81	0.20–1.20
Two-stage	0.11–0.43	0.16–0.64

curs in the tower, low pH fluid recycled back to the highest nitrification section (inlet) could reduce nitrification rates.

Recirculation ratios are commonly expressed as the percent or ratio of effluent recycled relative to the forward feed. Recycle ratios from 0.5 to 4 are generally applied. In some industrial applications ratios as high as 10 are applied for high strength wastes, although ratios greater than 4 are not economical, and should not significantly alter tower efficiency [21].

Dosing

Great emphasis has been placed on effective media wetting impacting filter efficiency by optimizing mass transfer and biological reaction mechanisms, flushing excess solids, improving contact time, and stabilizing the biological processes. An important consideration in achieving optimum hydraulic conditions is liquid distribution at the surface. Liquid is sprayed over the surface with a rotating arm, one diameter in length, containing spray nozzles throughout its length, and rotating around the filter center. Therefore, a single arm sprays or doses a specific point once every revolution at a frequency equal to the rotation of the arm. The total dosing rate equals the total feed (influent plus recirculation) rate and the nozzle rate equals the total rate divided by the number of nozzles on the single arm. Accordingly, dosing effectiveness of a specific area is a function of the arm speed, the total nozzle capacity and the nozzle spray pattern. If only a single arm were used, a specific area would be affected once every revolution. In actual practice, two and four arms are generally used, so that any specific area is dosed two or four times per complete arm revolution.

The dosing intensity is a product of the number of sprays per arm and the hydraulic rate of each nozzle. A low dosing intensity could result in poor wetting efficiency and unstable process conditions, whereas an excessive dosing intensity could result in flooding conditions. Conventionally, trickling filter distributors operate at less than 1.5 min per revolution, utilizing 2 to four arms, at continuous dosing rates.

A considerable amount of research and operating data support the theory that trickling filters are more effective utilizing intermittent dosing, at increased flushing intensities and at slower distribution arm speeds [1,9,21]. The effectiveness of this method has been explained on the basis of improved wetting, allowance for better promotion of aerobic and oxygen transfer conditions, improved solids flushing conditions, or a combination of any of these and other reasons. The basis for this theory is that in the trickling filter tower, continuous dosing does not produce consistent performance or effluent quality because of the inherent trickling filter solids control mechanism. Under optimum and stable tower conditions, solids accumulation is cyclical, self-adjusting to sustain an operating system. The filter bed cycle evolves around biomass accumulation, storage, and sloughing. Therefore, although there is a steady volumetric

feed rate, and even if the feed characteristics were constant, the filter performance is cyclical. This results in varying effluent quality! Under nonideal conditions, poor performance can result in excessive solids buildup, reducing available media surface area, and increasing resistance to oxygen transfer. Not only is excessive buildup not beneficial, but at extreme conditions detrimental to the filter performance.

The Spulkraft flushing intensity (*SK*) [1], as defined by Equation (5.7) is a proposed tool to measure the effective dosing intensity.

$$SK = \frac{(Q \cdot r) \cdot (1000 \text{ mm/m})}{(a) \cdot (n) \cdot (60 \text{ min/hr})} \qquad (5.7)$$

where *SK* is the flushing intensity, mm per pass of arm, $Q + r$ is the average hydraulic rate, $m^3/m^2/hr$, a is the number of arms, and n is the rotation speed in rpm.

Current systems in the United States are believed to operate at too high a rotational speed, resulting in a *SK* of 2 to 6 mm/pass, although *SK* values from 50 to 500 are practical and beneficial. The dosing procedure and resulting distribution affect media "wetting," impacting tower performance as previously discussed. The Process Engineer must carefully consider the following criteria influencing the system hydraulics [1,9,21].

(1) The distributor system should use a rotating or reciprocating distributor based on whether a circular or rectangular filter is employed.

(2) The distributor system should be selected for optimum design and flushing rates, and the rotation speed and number of arms chosen to achieve the desired dosing rate.

(3) The selected hydraulic rate should assure proper wetting. A minimum 0.5 L/m²-s (0.75 gal/min/ft²) should be applied to VFC filter media and 0.11 to 0.30 (0.16 to 0.44) for cross-flow packing, especially if shallow heights are employed. The specific value depends on the packing characteristics, configuration, and flow pattern.

All of the parameters mentioned should be evaluated on the basis of the *SK*, using the design and flushing values recommended in Table 5.13 as a guide. To assure maximum operational flexibility remote controlled electrically driven rotary distributors are recommended, operating independently of flow, and with the ability to program the cycles with either timers or microprocessors.

Organic Loadings

Some researchers [21] conclude that organic volumetric loading is the most significant filter bed design criteria, related to F/M or SRT in suspended growth systems, the packed volume being substituted for mass weight because the biomass is difficult to establish. An analogy being that hydraulic load-

TABLE 5.13. Dosing Rate Guidelines (adapted from Reference [1]).

BOD$_5$ loading						
kg/m3-day	<0.4	0.8	1.2	1.6	2.4	3.2
lbs/day-1000 ft^2	25	50	75	100	150	200
Design SK						
mm/pass	15–40	25–75	40–120	50–150	75–225	100–300
in./pass	0.6–1.6	1.0–3.0	1.6–4.7	2.0–5.9	3.0–8.9	3.9–11.8
Flushing SK						
mm/pass	100	100	150	200	300	500
in./pass	3.9	3.9	5.9	7.9	11.9	19.9

ing is equivalent to HRT, and can be subordinated to SRT. Organic loading is commonly expressed as mass of either COD or BOD$_5$ per unit time per *volume* of packing bed. As in activated sludge systems, excessive organic loading will result in limited treatment efficiency and possible anaerobic conditions. Low loads will result in difficulty in maintaining a biomass population. Two critical operating concerns affect the selection of an organic design loading. First, feed concentrations should be limited to a maximum 400 mg/L BOD$_5$ based on oxygen availability. In addition, the organic substrate must be applied to the active media surface at a rate which will optimize cell growth and substrate oxidation, and assure a viable microorganisms population.

The range applied to domestic treatment systems depends on the filter classification [17,21]. Low loadings range from 8 to 25 kg/day per 100 m^3 (5 to 15 lb/day/1000 cf), intermediate loadings from 25 to 50 (15 to 30), high loadings up to 100 (60), and super loading using plastic media up to 480 (300). Typical industrial loading are indicated in Table 5.14 [20,21].

TABLE 5.14. Trickling Filter Loadings (adapted from Reference [20]).

Industry	Applied BOD, kg/day/100 m^3	Loading lb/day/1000 ft^3
Domestic (DOM)	64–280	40–175
DOM + Canning	104–320	65–200
DOM + Cereal	34–384	21–240*
DOM + Meat Packing	157–357	98–223
	151–2025	94–1264**
DOM + Industrial	194–622	121–388
	109–344	68–215
Sugar	232–1065	145–665
Meat Packing	248–8329	155–5200
	43–322	27–201
Canning	74–1052	46–657
Fruit Canning	506–2419	316–1510
Pharmaceutical	250–820	156–512
Kraft Mill	16–449	10–280
Paper	27–798	17–498
Chemical	136–783	85–489

*Indicates 42 feet of packing.
**31.5 feet.
All data based a 21.5 foot bed of American Surfpac Corporation media.

Oxygen Requirements and Sludge Production

Basic *oxygen requirements* and *sludge production* rate calculations are similar to other biological systems and covered in Chapter II-2, Biological Aerobic Oxidation Section.

Oxygen Limitations

Aerobic tower conditions are achieved by restricting the organic loading to balance available ventilation. Unless a forced draft system is considered, oxygen available for a trickling filter depends on the air flow induced by a natural draft, a result of the difference between ambient and filter orifice air conditions. In heavy loaded towers forced air may be advantageous or necessary. By whatever means the air flow is selected, a means of hydraulically balancing the tower must be provided to assure tower fluid stability and adequate contact between the waste and air flows.

A system's natural draft depends on (1) design of the tower bed so that adequate ventilation area is available, (2) media selection, (3) restricting the organic loading to minimize flow *resistance* from solids plugging, and (4) environmental conditions favorable to maintain a natural draft.

Air flow is a product of *area, resistance,* and *driving force (draft)*.

Providing adequate bed flow *area* to maintain a natural tower draft requires having sufficient air passages and maintaining them free to assure sufficient air flow. Primary inlet air transits include underdrain and effluent channels; with secondary air inlets from vent stacks peripheral to the filter, underdrain extensions, ventilating manholes, underdrain sidewall louvers, and any other open channels to the influent or effluent lines. Reported recommendations for rock filters are [9,21]

(1) The unsubmerged underdrain area shall be equal to at least 15% of the filter surface area.
(2) Less than 50% of the under drainage system, effluent, channels and pipe should be submerged at the design hydraulic loading.
(3) One square meter ventilating area for each 250 m^2 of filter area.

Reported plastic media manufacturers recommendations are [21]

(1) A ventilating area of 0.1 m² (1 ft²) for each 3 to 4.6 m (10 to 15 ft) of trickling foot tower periphery
(2) A ventilating area of 1 to 2 square meters per 1000 m³ (0.3 to 0.6 ft²/1000 cf) of media

Actual media void space directly impacts air flow *resistance* and therefore vital in controlling available oxygen. As a result, consideration must be given to the effective void space resulting from liquid holdup and sludge buildup, which is affected by the selected packing size and the hydraulic and organic loadings.

Driving force conditions in a draft system depends on the buoyancy forces created by a density difference through the tower. Density differences result from ambient and tower temperature and humidity differences. In essence, the air flow through the system is proportional to the natural draft, which can be estimated as follows:

$$D = (po - p)(0.1923) \qquad (5.8)$$

where D is the draft in inches of water per foot of bed, po is the outside air density at the ambient humidity and temperature, lb/ft³, and p is the saturated air density inside the tower at the liquid temperature, lb/ft³.

This can be converted to mm of water per meter of bed by multiplying the result by 83.3.

The draft through a system is subject to ambient conditions, varying seasonally and diurnal. Figure 5.15 illustrates the draft variation that could occur during a day.

Several ambient conditions can affect tower performance. First, ambient conditions reducing or eliminating a tower draft, resulting in stagnant or oxygen deficient conditions, will obviously result in a "shut down" of the biological process. In addition, cold winter conditions which result in excessive air-water temperature differences can produce large air flows, reducing the liquid temperature, and thereby decrease the reaction efficiency. In such cases, the tower draft must be optimized to assure adequate oxygen without excessive waste temperature drops.

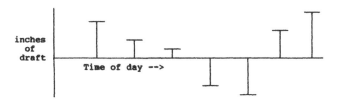

Figure 5.15 Daily draft profile.

Denitrification

Denitrification is not feasible in a trickling filter because of the difficulty in eliminating air from the system. Instead one of the submerged packed tower systems indicated in the *fixed-film filter systems* may be considered.

Fate of Contaminants

Like other biological systems the *fate of the substrate contaminants* must be evaluated in biological towers, more specifically the pathways by which the substrate contaminants are separated and their final deposition. In general terms the route the contaminants can take will be one, a combination, or all of the following:

(1) Stripping from the bulk of the waste
(2) Volatilization from the waste surface
(3) Sorption by the biological solids
(4) Biodegradation in the treatment system.

Volatilization from the waste surface is not a serious concern in biological towers, but the other routes are definite substrate pathways. The fate of contaminants is best evaluated as part of the pilot study by analyzing exit air for organic content and analyzing the waste sludge. General concepts of fate model components for biological systems are discussed in Chapter II-2, Biological Aerobic Oxidation section.

SPECIAL PROCESS CONDITIONS FOR ROTATING BIOLOGICAL CONTACTORS

RBC performance is governed by (1) the available oxygen level, (2) disc rotation, (3) the organic concentration, (4) flow rate, (5) wastewater temperature, and (6) the system configuration. These parameters control the system loading rates and retention time, the excess sludge generated, sustaining aerobic conditions, and organic reduction. RBC design procedures are based on applying generally accepted criteria, with a judgment factor applied for specific waste characteristics and expected variations. Available performance data do not always translate to rational design criteria which the engineer can verify, whereas proposed theoretical approaches can not always be translated into proprietary system design. A considerable amount of municipal system operating experience has been evaluated, correlated, and reported by the EPA in two separate reports [23,24].

Based on reported operating experience, critical RBC design considerations include

(1) Preventing first stage overloading as well as in the succeeding stages
(2) Preventing shaft loping

(3) Providing adequate surface area to meet effluent quality

(4) Maintaining aerobic condition throughout the RBC system

(5) Providing adequate sludge growth and sloughing conditions

Design Loadings

The dominant RBC performance parameters are retention time and influent concentration (organic feed rate), which are managed by controlling the hydraulic and organic loadings. In turn, influent concentration restrictions reflect limited available oxygen capacity. RBC removal efficiency normally increases with increasing concentration as long as aerobic conditions are maintained, although this does not always result in decreased effluent concentration because the higher *removal rate* may not always offset higher initial concentration. Retention time is related to influent flow rate, usually expressed as hydraulic loading. Increased retention time results in increased treatment efficiency.

Hydraulic Loading

Allowable RBC flow rate depends on two basic criteria, hydraulic loading and retention time. Like any biological reactor large flow variations will tend to upset, and under extreme conditions washout, the system. Process stability depends on avoiding frequent system shocks, upsetting or destroying the disk biogrowth. Generally, RBC units are designed for a hydraulic residence time of approximately 40 to 90 min, allowing time for the substrate entering the system to be absorbed by the rotating media floc. High flow rates result in low retention times, reducing substrate absorption and internal equalization. This could result in immediate effluent concentration increases due to unreacted, or partially reacted, substrate.

Hydraulic loading is defined as the flow rate per media unit area, and therefore directly related to the flow rate. Hydraulic loading rate affects the system dynamics as follows:

(1) Low hydraulic rate not only increases the total detention time, but the detention time per stage. Because large removal rates are achieved in the initial stages, concentration in subsequent stages is reduced. This decreases latter stage cell growth, the concentration driving force, and the removal rate. Although improved effluent quality is generally achieved, the later stages are underutilized and ineffective.

(2) Conversely, high hydraulic loadings reduce treatment efficiency in the initial stages, increasing substrate concentration, mass transfer driving force, and removal rate in the later stages. Although a poorer effluent quality could be produced with lower retention time, the final stages are more effectively utilized.

Ideally, required effluent quality should be achieved by optimizing the surface area by judiciously selecting the design hydraulic and organic loadings and the retention time. Municipal systems are commonly designed for a carbonaceous oxidation hydraulic loadings ranging from less than 40 and up to 200 L/day/m^2 (1.0 to 5.0 gpd/sf) and 40 to 90 minutes vessel retention time and nitrification loadings of 12 to 82 (0.3 to 2), with 90 to 230 min retention time. The hydraulic loading must be adjusted for the industrial characteristics and the applied concentration. The configuration can be optimized by selecting a volume-to-surface area ratio of 4.9 L/m^2 (0.12 gallons per square foot of media). Reported pilot studies indicate that basin volumes corresponding to greater than 4.9 L/m^2 of media did not improve BOD removals [22,23]. Based on a hydraulic loading ranging from 40 to 200 L/day/m^2, this corresponds to 35 to 175 minutes retention time.

Organic Loading

Initial and interstage RBC organic concentration is a key parameter governing overall process performance. High organic concentrations can result in low dissolved oxygen and excessive biomass generation, both of which can destabilize a system. In turn, low organic concentrations reduces both biogrowth and concentration driving force, resulting in low removal rates. Concentration can be defined in terms of soluble BOD or total BOD, the difference between the two being suspended or colloidal organic matter. Where suspended biodegradable matter is significant, the system must either include a preliminary settling step or allow additional retention time for solids degradation.

Concentration gradient is an important parameter closely associated with the system configuration. In a plug flow system minimal reactor mixing occurs. The result is a bulk concentration gradient, measured by the feed concentration (high value) and the effluent concentration (low value). In contrast, in a completely mixed system the effluent concentration is the same as the bulk reactor concentration. Bulk concentration is significant in either configuration because it determines substrate transport and removal rate. A common RBC design assumption is that zero-order kinetics govern most of the treatment process, resulting in an implied constant rate removal rate applicable through a significant segment of the unit. In lower concentrations, first-order kinetics are frequently assumed, resulting in lower removal rates and larger units, acting as later stage(s) polishing sections.

Organic loading is defined as the mass feed divided by the media area, so that increased influent concentration results in increased organic loading. As with any biological system the organic loading must be adequate to support the biological population and consistent with the available DO. Because industrial waste concentrations can be high, although volumes frequently low, hydraulic and organic loading rates must be balanced to operating levels. This can be accomplished by utilizing parallel units to accommodate organic loads and recycle for dilution or hydraulic balancing.

Although design curves are commonly expressed in terms of hydraulic loading and required effluent concentration, influent organic concentration and loading is a key design parameter defining overall performance. Organic limits are applied to the first stage(s) to avoid overloading, excessive biofilm accumulation, and anaerobic conditions. Overloading can cause septic conditions and reduced treatment efficiency as well as structure failure. The *maximum* municipal total BOD loading based on oxygen transfer is usually limited to 31 kg/day/1000 m² (6.4 lb/day/1000 sf), which is equivalent to 12.7 to 18.6 SBOD (2.6 to 3.8) [23]. This is applied to the first stage, and therefore a limitation on every stage. The applied design criteria is generally more conservative and limited to a first stage loading of 12 kg/day/1000 m² SBOD (2.5 lb/day/1000 sf) for mechanical drives, and 14 (3) for air-driven units [23,25]. Finally, establishing industrial waste *performance* solely on municipal design criteria can be risky, unless the industrial waste effects are accurately assessed or allowance made for add-on modules.

System Configuration

Developing an RBC system involves establishing the applicable configuration based on loading, constructed to optimize the total unit area to establish cost-effective physical dimensions.

General Configuration Consideration

The first step in establishing an RBC configuration is evaluating the need for *pretreatment, parallel flow,* and *staging (series flow).*

Pretreatment (Figure 5.16) involves waste preparation to control stage loadings and to eliminate waste components detrimental to biological activity. This includes limiting first stage loading, determining the effects of soluble and colloidal feed constituents, reviewing the waste characteristics for toxic constituents, establishing solids removal or similar pretreatment requirements, and adjusting reactor pH and nutrient levels.

Next an appropriate *parallel* and *series* configuration must be selected to achieve a desired effluent quality. Current design methods are based on bulk organic removal at a constant rate adhering to zero-order kinetics, with the initial stage loading limited to assure aerobic conditions and deter overloading. Effluent concentration in latter stages could be reduced as a result of first order (or other than zero order)

Figure 5.17 RBC parallel configuration.

kinetics. The number of final stages installed to assure required effluent quality is frequently based on "operating experience."

The number of *parallel flow* (Figure 5.17) units are a result of applied organic and hydraulic loadings to each (parallel) system. This is based on

(1) Each shaft containing approximately 100,000 (standard) to 150,000 (high density) ft² of media
(2) An oxygen capacity of approximately 64 to 68 kg/day (140 to 150 lb/day) for a 9,290 m² (100,000 sf) shaft and 95 to 102 kg/day (210 to 225 lb/day) for a 13,935 m² (150,000 sf) shaft
(3) Applying the hydraulic and organic loadings discussed as defined in the nitrification and carbonaceous loading sections

Parallel systems provide the flexibility to distribute hydraulic and organic loading, allowing operating response to changing influent conditions and achieve performance requirements.

Series flow (Figure 5.18) involves unit *staging* to achieve plug flow configuration, minimizing influent leakage and maximizing treatment efficiency. Staging can be achieved by physical separation of media sections within a tank using baffles separating single shaft systems with perpendicular baffles, or separation between multishafts in the same tank. The organic loading to a unit should be carefully evaluated to avoid starved and ineffective final stages.

It is important to understand series staging effects. First, each successive stage experiences an organic loading level lower than the previous. In addition, a different environment may be established with predominant strains of microorganisms developing in response to that environment. The removal mechanism is related to the substrate concentration, with constant removal rate occurring at high concentrations, changing to first (or another) order at lower concentrations. If the reaction order is first order or higher, substrate removal efficiency is greater in a multistage unit than with an equivalent capacity single stage. Most units are believed to operate as zero-order reactors in the initial stages, with subsequent

Figure 5.16 RBC pretreatment.

Figure 5.18 RBC series configuration.

TABLE 5.15. **RBC Staging Guidelines (adapted from Reference [23,24]).**

Autotrol		Lyco		
Effluent	min. stages	BOD Removal, %	Stages	Clow
>25	1	Up to 40	1	At least four stages
15–25	1 or 2	35–65	2	per flow path
10–15	2 or 3	60–85	3	recommended
<10	3 or 4	80–95	4	
		Min. 4 stages for combined BOD$_5$ and NH3N removal		

stages acting as a polishing step, and the final stages promoting nitrification.

Generally, staging is based on manufacturers' experience with municipal systems. Available data reported in the EPA design manual are summarized in Table 5.15 [23].

Series or parallel flow configurations can be combined into a *mixed-flow operation* or as part of a *combined system.*

Mixed-flow operation (Figure 5.19) combines both parallel and series flow, including parallel flow as required for loading restrictions and series units for performance requirements. Mixed-flow operation is a vital consideration when influent conditions involve high substrate or flow quantities, nitrification is part of the treatment, and final denitrification is required.

The combination of industrial waste variability, poor recovery from upset conditions, and limited process flexibility or control makes single unit RBC unreliable and impractical. However, because of their simplicity multiple units are effective when configured as mixed-flow systems or when enhanced and combined with other processes. *Combined system,* coupling an RBC unit with other biological units, are a viable alternative when a large influent loading is applied and a high effluent quality is required.

Interstate clarification may be required if large quantities of sludge are generated in the initial (or any) stage. Although this is possible, basic process limitations would suggest that the quantities of soluble substrate required to produce adequate sludge to necessitate clarification would exceed the aeration capacities of the stage. However, excess sludge can be generated as a result of high influent solids, in which case primary clarification is a better consideration. In any event, large sludge inventories will result in septic conditions, oxygen depletion by endogenous respiration and premature erosion of the biomass on the disk surfaces.

Temperature

Design criteria commonly do not predict any improved treatment performance above 12°C (55°F), probably because reduced oxygen levels dominates performance [23,24]. Below 12°C the reaction rate, as expressed by the reaction rate constant *k,* is significantly affected by temperature, commonly expressed in terms of the Arrhenius relation,

$$Kt \, / \, K_{20} = \Phi^{T-20} \qquad (5.9)$$

Reported values for the temperature factor (Φ) range from 1.010 to 1.025 with definitive data applicable to RBC performance sparse [24]. Reduced reaction efficiency at low temperatures may be compensated by increased saturated dissolved oxygen capacity in sections where performance is oxygen controlled. Manufacturer's design guidelines frequently express temperature corrections as correction factors to the estimated media area.

Nitrification

Basic nitrification chemistry and kinetic expressions are discussed in Chapter II-2. The staged or plug flow configuration of RBC units makes effective nitrification dependent on soluble organic and nitrogen concentrations at individual stages. First stage loading is especially critical because oxygen capacity must not exceed total carbonaceous oxidation and *nitrification* demand. In that regard, the total nitrogen concentration (TKN) is a significant measurement of feed ammonia and the potential conversion of combined organic nitrogen constituents to ammonia. With adequate oxygen, nitrification will occur with diminishing carbonaceous growth cell rate because nitrification cell growth is considerably slower than the carbonaceous cell rate.

Nitrification starts when soluble BOD$_5$ declines to approximately 15 mg/L and is accelerated when below 10 mg/L. Municipal system nitrification occurs in the third or fourth stage. At ammonium nitrogen concentrations greater than 5 mg/L and 13°C (55°F), a maximum zero-order design rate of approximately 1.5 kg NH$^-$N/day/1000 m^2 (0.3 lb/d/1000 sf) is a common design basis at 13°C (55°F) and a pH range

Figure 5.19 RBC mixed flow configuration.

from 7 to 9 [23]. This represents an equivalent oxygen demand of 6.8 kg/day/1000 m² (1.4 lb/day/1000 sf), approximately equal to the 7.3 (1.5) available at a rotational speed of 1.6 rpm.

An RBC system is effective at wastewater temperatures between to 4 to 13°C (40 to 55°F), governed either by reaction or oxygen transfer rates [23]. Below 4°C the reaction rate starts to diminish, although the oxygen transfer rates increases as a result of increased oxygen solubility. Because of controlling oxygen transfer limitations, nitrification rates do not increase at temperatures higher than 13°C (55°F). Required media area is usually adjusted for anticipated adverse temperature conditions.

Separate Oxidation/Nitrification Systems

Separate systems are usually employed to segregate and protect nitrifying microorganisms or to upgrade an existing system. System upgrade can include a separate add-on to an activated sludge system, a trickling filter, or to another RBC. The number of separate stages selected are an important consideration in achieving system performance.

The following nitrification design criteria attributed to three manufacturers (Clow, Autorol, Lyco) were reported by the EPA for municipal systems [23]:

(1) *Clow* design procedures recommended a minimum of four stages, with reduction removals increasing approximately five percent with two additional stages.
(2) *Autorol* design procedures state that staging is ineffective for zero-order kinetics (greater than 5 mg/L ammonia nitrogen), with staging recommended for concentrations less than 5 mg/L.
(3) *Lyco design procedures estimate individual stage performance as follows:*

    Up to 40% NH₃N removal   1 stage
    35 to 65%   2
    60 to 85%   3
    80 to 95%   4 stage
    Lyco, combined oxidation/nitrification procedures, recommends a minimum of four stages.
(4) If peak to average flow is less than 2.5 the use of average flow is recommended by *Clow* and *Autorol.*
(5) All manufacturers apply nitrification at 15 mg/L SBOD or a total BOD of 30 mg/L or lower.
(6) All manufacturers recommend temperature correction factors to increase the design surface area for operating conditions below 13°C (55°F).

Combined Carbon Oxidation and Nitrification

Combined oxidation/nitrification requirements are identical to those for separate nitrification, with adequate "pretreat-

ment" area requirements included to achieve 15 mg/L SBOD or 30 mg/L total BOD in the proceeding stages.

GENERAL ENGINEERING CRITERIA

Figure 5.20 illustrates a Trickling Filter Preliminary Concept Flowsheet. A complete *trickling filter system* could involve many components which are discussed in other chapters, where specific unit process design criteria are described. Trickling filter components could include the following:

Equalization (optional)   Chapter I-4
Waste feed system
Trickling filter system
Sludge clarification   Chapter II-8
Sludge recycle
pH control (optional)   Chapter I-8
Nutrient addition (optional)   Chapter I-6
Sludge treatment system   Chapters II-8 and III-7

CRITICAL EQUIPMENT SELECTION

The elements of a trickling filter include a floor upon which the filter rests, a containing wall, media, a feed distribution system, and an underdrain to collect effluent and allow ventilation into the tower.

Floors

Floors are the base of the filter unit, upon which the bed and ancillary equipment rest. They are generally constructed of reinforced concrete, sufficient to support the tower and hydraulic load. The floor is basically the underdrain system, sloped up to 5% to a drainage channel, which discharges the effluent from the system.

Walls

Filter walls are constructed to house the bed media and tower internals, protecting it from the elements. A primary consideration is whether the walls support the media and liquid contents, common with "loose" media such as stone and plastic. In the other alternative, the wall contains self-supporting, prefabricated, plastic packs. Construction and indeed the use of construction materials other than reinforced concrete depend on how much supporting force the wall must retain. Walls should be extended, with enough freeboard to assure adequate protection of the system from weather and wind conditions.

The walls can be constructed of a variety of materials provided that corrosion and structural support requirements are adequately considered. Materials of construction include reinforced concrete, reinforced masonry, structural steel frame supported from precast concrete, or a steel tank. Reinforced concrete is commonly employed; reinforced-fiber-

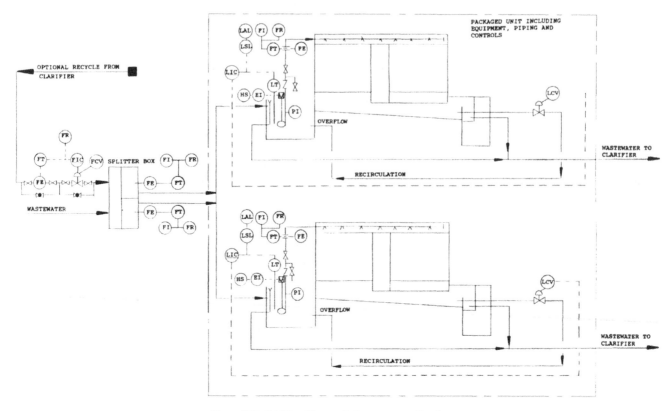

Figure 5.20 Trickling filter preliminary concept flowsheet.

plastic is frequently used for self-supporting, prefabricated, plastic pack. Theoretically, walls can be any shape but are almost always circular to confirm with common rotary distributor mechanism designs.

Special wall design conditions include ventilation ports, air dump gates and flooding provisions. Ventilation ports located at the wall bottom or an air vent system improve natural draft through the tower. Air dump gates are openings at the top of the wall to drain or flush the arms. The wall construction may be designed to be able to flood the bed to control ponding or filter flies. In that case the wall construction must be tight, the drainage channel must contain a gate valve, and provision must be made to discharge the flood water to a point beyond the gate.

Media

Generally, media are available as rock or plastic. Common filter media include the following:

(1) Crushed stone and crushed slag
(2) Random-dumped type plastic media, commonly polypropylene
(3) Self-supporting, stacked, modular type media

Plastic media or modular packing are generally employed in new installations because of lighter construction, easily adaptable to elevated heights. In addition, they provide excel-

lent oxygen mass transfer characteristics and are available in a wide range of configurations. Whatever media is selected, media type and size must be carefully balanced to achieve high mass transfer efficiency without promoting excessive clogging or poor tower hydraulics.

Media Support

Media support must be adequate to not only support the bed but to separate the media from the under drainage system, allowing drainage from the bed and air to the media. For stone, or similar small random media, concrete blocks (or similar materials) are used for the support grill. The blocks are constructed with the large surface openings, allowing waste flow to the underdrain channel and air to the bed. Beam column supports or grating are used for self-supporting plastic media.

Distribution System

An influent pipe, connected to a center column, injects waste to a center well. The center well is designed to provide adequate liquid head above the distributor arms to maintain equal flows to them. A distributor base rests on the center column and supports the distributor arms. Distributor bearings allow smooth rotation of the distributor arms. Distributor arms rotate around the distributor base, conveying equal waste flows from the center well through outlet orifices on the hollow arms. Multiple arms can be installed for increased media coverage. Each rotating arm length is approximately equal to the radius of a circular filter, containing spray nozzles throughout its length. Two and four arms are generally used, so that any specific area is dosed two or four times per complete arm revolution.

Outlet orifices control flow to the media, and are adjustable to balance flow distribution. In hydraulic distributors the jet action from speed regulating orifices controls the rotation of the arms. In newer facilities electric drives are recommended to separately control dosing rate and rotating arm speed. The arm rotation is commonly at less than 1.5 min per revolution, with new data suggesting decreasing speeds to maintain recommended *SK* dosing values, using electrically driven distributors to optimize dosing rate. Arm splash plates evenly distribute the flow from the orifices to the media.

Underdrain System

Sloped floors receive treated wastewater from the media, through the support grill, conveying it to the underdrain system. The underdrain system is part of the support grill, consisting of openings allowing treated water to discharge from the media (through the support grill), and air to flow to the media. The underdrain channel conveys collected water to the outlet box and along with the ventilation ports lets air flow into the system. The outlet box collects water from the underdrain channel, discharging it to the outlet pipe and out of the filter system, when the outlet valve is open. The (closed) outlet valve is used to flood the filter or regulate the flow from the channel to the outlet pipe.

Pumps

Wastewater recirculation is commonly achieved using low head dry or wet wall centrifugal pumps. Solids will never be a problem if open-faced pumps similar to sludge pumps are used.

Ventilation System

Draft systems are based on port openings in the filter floor walls, allowing air flow counter to the liquid waste through the underdrain system, and through the media. Liquid submergence in the underdrain system must be controlled to allow free air flow through the system. In some industrial systems forced ventilation systems are used.

PROCESS CONTROLS

The following filter controls should be considered for system monitoring:

(1) Splitter box or forward feed valves to control flow to individual trickling filters

(2) Splitter box or recirculation valves to control recirculation to the filter bed

(3) Damper control for forced air ventilation systems

(4) Electric drive control system to control distributor speed

(5) Running lights to indicate that critical equipment is operational and alarms to indicate that critical parameters are not at acceptable limits

(6) High torque alarms to alert and shut down the distributor

(7) Flow recording of influent, effluent, and recycle streams

(8) Provision for continuously sampling the influent and effluent streams

RBC units are proprietary systems with configurations purchased as complete packaged units. *Other fixed-film bed* systems must be developed from emerging research to fit specific process requirements.

COMMON TRICKLING FILTER DESIGN DEFICIENCIES

Some design deficiencies associated with municipal facilities have been investigated by the EPA and are identified as a checklist of process and mechanical design considerations [19]. The cited reference should be reviewed for more details.

(1) Poor distributor arm design results in clogging and rotation problems caused by (a) inadequate port sizes, (b) an inability to flush the system, (c) an inability to prescreen grit and rags, or (d) an inadequate feed flow or recirculation to maintain correct distributor arm movement.

(2) Inadequate tower sidewall height provided to prevent splashing and mist fog.

(3) Lack of piping and valving flexibility prohibits filter flooding.

(4) The filter cannot be chlorinated.

(5) Oxygen deficiencies result from poorly operated and/ or inadequate tower ventilation.

(6) Piping and pumping design does not permit multitower operation in parallel or series flow.

(7) Filter performance does not meet effluent quality because of (a) an inadequate system, (b) an inability to operate towers in series operation, or (c) poor media selection and/or size.

(8) Tower is hydraulically unable to handle recirculation, resulting in tower overloads.

(9) No provision was made for flushing the underdrains.

(10) Limited waste inventory and no provisions for recycle result in periods of inadequate flow to tower.

(11) Freeze protection of exposed piping and the pumping system is inadequate, further compounded by "icing" of media surface areas from resulting tower mist.

(12) Feed and recirculation flows cannot be monitored or controlled.

(13) The dosing feed system is inadequate.

(14) The recirculation capacity is inadequate.

COMMON RBC DESIGN DEFICIENCIES

(1) Bearings are susceptible to flooding because of their location.

(2) Excessive temperature reduction results from excessive heat loss from unhoused units or housing poorly insulated.

(3) High solids in the influent results in media plugging and anaerobic conditions in the vessel.

(4) Feed becomes septic because of excessive upstream transfer or storage detention times as well as poor hydraulic conditions in the RBC channels.

(5) Hydraulic overloading results from excessive feed volumes because of unaccounted process or waste treatment plant side streams.

(6) RBC media plugging results from a lack of waste (physical) pretreatment.

(7) RBC unit dead spots and solids deposition results from poor mixing and improper design of overflow baffles.

(8) Secondary clarification is ineffective, with no provision for coagulation and flocculation chemicals to improve the performance.

(9) No provision was included to sample or measure waste sludge flows.

CASE STUDY NUMBER 16

Develop a trickling filter system based on the influent conditions and effluent requirements detailed in Case Study 10.

REQUIRED DESIGN DATA

The results of treatability studies conducted on pretreated plant wastes using cross-flow plastic media are as follows.

(1) Temperature, °C	25
(2) Feed concentration, mg/L	500
(3) Required effluent, mg/L	7.45
(4) Reaction kinetic constant	0.09 @ 25°C
(5) Depth, ft	20
(6) Hydraulic loading, gpm/sf	0.18
(7) Nitrification rate, lbs/cu ft/day	0.0025
(8) Required ammonia-nitrogen reduction	75%

The treatability data *for this study* was correlated assuming first order kinetics, with the effects of recirculation neglected and the influent concentration not corrected for recirculation.

PROCESS CALCULATIONS: GENERAL

(1) Establish biokinetic coefficients

    (1) Selected depth, 30 ft

    (2) Applied influent concentration, 500 mg/L

    (3) Required effluent concentration, 7.45 mg/L

    (4) Waste flow, 1 MGD

    (5) Design temperature, 68°F (20°C)

    (6) Winter temperature, 50°F (10°C)

    (7) Design temperature, 80°F (26.6°C)

    (8) Exponent n 0.5

$$K = \text{test } k \cdot \left[\frac{\text{test } D}{D}\right]^{\frac{1}{2}} \left[\frac{S_a \text{ test}}{S_a}\right]^{\frac{1}{2}} \cdot 1.035^{(t-\text{test})}$$

$$K_{\text{design}} = 0.09 \cdot (20/30)^{\frac{1}{2}} \cdot (500/500)^{\frac{1}{2}} \cdot 1.035^{(20-25)} = 0.062$$

$$K_{\text{summer}} = 0.062 \cdot 1.035^{(26.7-20)} = 0.078$$

$$K_{\text{winter}} = 0.062 \cdot 1.035^{(10-20)} = 0.044$$

(2) Kinetic related hydraulic loading

$$Q_v = [K \cdot D/LN(S_o/S_e)]^2$$

$$Q_{\text{design}} = [0.062 \cdot 30/LN\ (500/7.45)]^2 = 0.20 \text{ gpm/sf}$$

$$Q_{\text{summer}} = [0.078 \cdot 30/LN\ (500/7.45)]^2 = 0.31 \text{ gpm/sf}$$

$$Q_{\text{winter}} = [0.044 \cdot 30/LN\ (500/7.45)]^2$$

$$= 0.10 \text{ gpm/sf (this is low for packing)}$$

(3) Select Tower Recirculation
Minimum recirculation required to dilute the influent to 400 mg/L:

$$\text{recirculation} = \frac{\text{influent BOD} - \text{applied BOD}}{\text{applied BOD} - \text{effluent BOD}}$$

$$\text{recirculation} = \frac{500 - 400}{400 - 7.45} = 0.25$$

The system will be designed for 100% recirculation to allow the operator flexibility in controlling tower hydraulics. The diameter will be based on the feed flow; however, the effects of recirculation will be evaluated.

(4) Calculate tower organic loading

$$\text{Organic load} = Q_v \cdot 8.34 \cdot 1440 \cdot (So/1,000,000) \cdot 1000$$
$$\cdot \text{cf/depth} = \text{lbs/dy/1000 cf}$$

Design (within acceptable range):

$O_d = 0.20 \cdot 1440 \cdot (500/1,000,000) \cdot 1000 \cdot 8.34/30$
$= 40$ lb/day/1000 cf

Summer conditions (within acceptable range):

$O_s = 0.31 \cdot 1440 \cdot (500/1,000,000) \cdot 1000 \cdot 8.34/30$
$= 62$ lb/day/1000 cf

Winter conditions (within acceptable range):

$O_w = 0.10 \cdot 1440 \cdot (500/1,000,000) \cdot 1000 \cdot 8.34/30$
$= 20$ lb/day/1000 cf

(5) Select loadings governing tower sizing
Winter conditions will be selected as the design basis, using 0.10 gpm/sf, an organic limiting loading of 20 lb/day/1000 ft³, and an effluent of 7.45 mg/L SBOD.

(6) Estimate required area

$$\text{area} = \frac{\dfrac{\text{MGD} \cdot 1,000,000}{1440}}{Q_v}$$

$$\text{area} = \frac{1 \text{ MGD} \cdot 1,000,000}{1440 \cdot 0.10 \text{ gpm/sf}} = 6944 \text{ ft}^2$$

(7) Establish filter parameters.
  (8) Tower recirculation: provide for 100% recycle.
  (9) Select number of units: three
  (10) Select filter diameter: 55 ft
  (11) Unit filter area: 2376 ft²
  (12) Unit volume: 71,280 ft³
  (13) Total filter area: 7128 ft²
  (14) Unit tower depth: 30 ft
  (15) Check organic loading rate

1 MGD $\cdot$ 8.34 $\cdot$ 500 mg/L/(7,128 ft² $\cdot$ 30 ft) $\cdot$ 1000
$= 20$ lb/1000 ft³; (acceptable for all conditions).

(16) Calculate the effects of recirculation
Estimate effects for winter conditions, with $k = 0.044$

$$\frac{S_e}{S_i} = \exp\left[-K \cdot D \left[\frac{A}{(1+R) \cdot Q}\right]^n\right]$$

S_e = Effluent concentration, mg/L
S_i = Influent (diluted) concentration, mg/L
$1 - Se/Si$ = Fraction removed based on *applied* (diluted) concentration.

$$\frac{S_e}{S_b} = 1 \Big/ \left[(1+R) \cdot \frac{S_i}{S_e} - R\right]$$

S_o = Waste concentration, mg/L
$1 - Se/So$ = Fraction removed based on raw concentration.

	Recycle Ratio			
	0	0.5	1	4
S_e/S_i	0.02	0.03	0.05	0.15
removal, %	98	97	95	85
S_e/S_o	0.02	0.02	0.03	0.04
removal, %	98	98	97	96
Q_v	0.10	0.15	0.20	0.50

Based on the *raw influent* concentration and a recycle ratio ranging from 0 to 4, the organic reduction can range from 96 to 98%, not significantly affected by the recycle, within the range of in which recycle is applied (up to 100%) and the accuracy of the calculation. The selected recycle ratio of 1 allow results in a hydraulic loading of 0.20 gpm/ft², adequate for cross-flow vertical packing. A higher recycle ratio may be required if vertical flow packing is employed. This must be checked with the supplier, and conform to the selected configuration.

(17) Nutrient requirements
Based on the required BOD removed, an excess ammonia nitrogen of 212 lb/day (25.4 mg/L) results. See Case Study 10 for detail calculations.

(18) Calculate nitrification requirements for combined treatment

(19) Check BOD₅ loading at exit

$O_s = 0.31 \cdot 1440 \cdot 8.34 \cdot (7.45/1,000,000) \cdot 1000 \text{ ft}^3 \cdot /30 \text{ ft}$
$\equiv 1$ lb BOD/day/1000 ft³
lb/day feed   ppm BOD

The *maximum SBOD* is well below the recommended 25 to 36 loading, near the tower exit, where nitrification should initiate.

(20) Required removal rate: 75% ammonia nitrogen removal.

(21) Quantity ammonia nitrogen to be removed:

$0.75 \cdot 212 = 159$ lb/day

(22) Nitrification rate, lb/ft³/day: 0.0025

(23) Required packing: 159/0.0025 = 63,600 ft³

62,600/total of 7128 ft² $\equiv$ 9 additional ft.

(24) Check tower draft conditions

	Tower	Ambient	
Temp, °F	70	60	
Humidity	0.0158	0.0066	lb/lb dry air
lb water	0.0158	0.0066	lb water
lb dry air	1.0000	1.0000	
Total weight	1.0158	1.0066	
Moles water	0.0009	0.00037	M_w air = 28.84
Moles air	0.0347	0.0347	M_w water = 18
Total moles	0.0356	0.0350	
ft³ air + water	13.7675	13.3180	
Density air + water	0.07378	0.07558	

dP, in./ft packing = draft developed = (0.07558 − 0.07378)
· 12 in/ft/ 62.4 lb/ft³ = 0.00035 in/ft

total pressure drop = 0.00035 in/ft · 30 (or 39)
feet pack = 0.01 in. of water

The effect of ambient temperature change during summer days is illustrated below; assuming the wastewater temperature remains at 70°F and the ambient at 60% saturated humidity.

Time	°F	dP, in./ft	Time	°F	dP, in./ft
0700	60	0.00035	1400	80	Negative
0800	65	0.00020	1500	80	Negative
0900	70	0.00005	1600	80	Negative
1000	70	0.00005	1700	70	0.00005
1100	75	Negative	1800	70	0.00005
1200	80	Negative	1900	65	0.00020
1300	80	Negative	2000	60	0.00035

DISCUSSION

The trickling filter design is driven by the high removal rate specified, 98% based on a 500 mg/L feed, to achieve a 7.45 mg/L SBOD effluent. These conditions were imposed as comparison of performance relative to the other biological treatment options, especially activated sludge as described in Case Study 10. It is doubtful that this effluent could be achieved! An effluent of 50 to 125 mg/L is more realistic, based on a 75 to 95% reduction of the *raw feed*. Designing for a 125 mg/L effluent would dramatically reduce the required tower characteristics. When a trickling filter is applied for a heavy industrial load, it is best applied as a roughing filter and part of a combined treatment system.

CASE STUDY NUMBER 17

Develop a RBC system based on Case Study 10 influent (1 MGD @ 500 mg/L BOD, 7.45 effluent). Assume that the effluent concentration can be *approximated* using a second-order equation, with the K value of 0.080 at 20°C.

The RBC unit has the following physical characteristics:

(1) Rotation, 1.5 rpm
(2) Hydraulic loading of 2.5 gpm/ft² of media
(3) Organic loading of 2.5 lbs SBOD/day/1000 sq media
(4) Volume to surface ratio, 0.12 gals per ft² media.
(5) An acceptable 75% ammonia nitrogen removal can be obtained with a loading of 0.3 lbs ammonia-N/day/1000 ft² ammonia
(6) Each disk contains 250 ft² of media.
(7) Each shaft contains 100,000 ft² of media.
(8) The oxygen capacity of the unit is equivalent to 1.5 pounds of oxygen per day per 1000 ft² of media.
(9) Approximately 0.35 net VSS is produced per pound of SBOD removed.

PROCESS CALCULATIONS: GENERAL

(1) Establish biokinetic coefficients and required effluent quality

(1) Second-order reaction rate coefficient for test conditions: 0.080 @ 20°C

(2) Reaction rate coefficient for design conditions: 0.080 @ 20°C

(3) Reaction rate coefficient for winter conditions:
$K_w = 0.080 \cdot 1.028^{(4.44 - 20)} = 0.052$

(2) Establish theoretical feed limit

(1) Select hydraulic loading rate: 2.5 gpm/ft²

(2) Allowable feed limit: see equation (5.11).

Q, gpm/ft² · TBOD, mg/L · 0.00834 = 6.4 lb/day/1000 ft²
Q, gpm/ft² · SBOD, mg/L · 0.00834 = 3.8 lb/day/1000 ft²

$$Q = 2 \text{ gpm/ft}^2$$

$$\text{max SBOD} = 3.8 \text{ lb/day/ft}^2$$

$$\text{SBOD} = 3.8/(0.00834 \cdot 2.5) = 182 \text{ mg/L}$$

(3) Estimate required recycle

(1) Basis: $S_o = 500$ mg/L, $S_e = 7.45$ mg/L, and an applied soluble loading of 182 mg/L max.

(2) Required recycle ratio

$$R = (S_o - S_a)/(S_a - S_e)$$

$$R = (500 - 182)/(182 - 7.45) = 1.82$$

Assume an recycle ratio of 2, total flow 3 MGD.

(4) Check required disk area based on hydraulic and organic loadings

(5) Calculate applied hydraulic loading

(1) Base on a recycle ratio of 2, total flow equals 3 MGD.

(2) Applied hydraulic loading of 2.5 gpm/ft²

(6) Calculate required disk area

$$\text{area} = 3 \text{ MGD} \cdot 1,000,000 / 2.5 \text{ gpm/ft}^2 = 1,200,000 \text{ ft}^2$$

(7) Estimate organic loading

(1) Organic feed = 1 MGD · 8.34 · 500 = 4170 lb/day
(2) Organic loading at 2.5 lb/day/1000 ft²

(8) Verify disk area based on organic loading

$$[4170/2.5] \cdot 1000 = 1,668,000 \text{ ft}^2$$

(9) Specify required disk area

A disk area of 1,700,000 ft² will be selected. In this case the recycle acts to dilute the influent concentration, which is another way of providing sufficient area to meet organic loading requirements.

(10) *Approximation* of effluent concentration

Although a constant removal rate is assumed in estimating the disc area, second-order kinetics will be assumed to approximate effluent concentration.

(11) Theoretical retention time

   (1) Theoretical volume, gals based on selected area

$$0.12 \text{ gal/ft}^2 \text{ media} \cdot 1.7 \text{ million ft}^2 = 0.204 \text{ mg}$$

   (2) Residence time

$$[0.204 \text{ MG} / 3 \text{ MGD}] \cdot 1440 \text{ min/day} = 98 \text{ min}$$

   (3) Applied influent concentration
   Based on a waste of 500 mg/L, an assumed effluent of 7.45 mg/L, and a recycle ratio of 2:

$$S_a = [2 \text{ MGD} \cdot 7.45 \text{ mg/L} + 1 \text{ MGD} \cdot 500 \text{ mg/L}]/[1 + 2]$$
$$= 172 \text{ mg/L}$$

(12) Design conditions

$$C_e = \frac{-1 + (1 + 4Kt\,S_a)^{\frac{1}{2}}}{2\,K\,t} \quad [26]$$

$$C_e = \frac{-1 + (1 + 4 \cdot 0.080 \cdot 98 \cdot 172)^{\frac{1}{2}}}{2 \cdot 0.080 \cdot 98} = 5 \text{ mg/L}$$

(13) Winter conditions

$$C_e = \frac{-1 + (1 + 4 \cdot 0.037 \cdot 98 \cdot 172)^{\frac{1}{2}}}{2 \cdot 0.037 \cdot 98} = 6 \text{ mg/L}$$

Base rest of calculation on 7.45 mg/L effluent!

(14) BOD removed

forward flow $\cdot$ 8.34 $\cdot$ (feed BOD $-$ effluent BOD)

$$1 \text{ MGD} \cdot 8.34 \cdot (500 - 7.45) = 4108 \text{ lb BOD/day}$$

(15) Estimate nitrification requirements
Based on the required BOD removed, an excess ammonia nitrogen of 212 lbs per day (25.4 mg/L) results. See Case Study 10 for detail calculations.

(16) Nitrification area

   (1) Loading of 0.3 lb nitrogen/day/1000 ft² media

   (2) $[212/0.3] \cdot 1000 = 707{,}000$ ft² of media

(17) MLVSS generated per day

4108 lb BOD removed/day $\cdot$ 0.35 lbs MLVSS/BODr
$$= 1438 \text{ lbs/day}$$

1438/(1 MGD discharge $\cdot$ 8.34)
$$= 172 \text{ mg/L of MLVSS}$$

(18) Oxygen required, lbs per day

(19) Carbonaceous oxidation:

$$\text{BOD}_5/\text{BOD}_u \text{ ratio} = 0.68$$

$$\text{BOD}_u/\text{VSS} = 1.42$$

$$4108 \text{ BOD}_r/ 0.68 - 1438 \cdot 1.42 = 3999 \text{ lb/day}$$

(20) Nitrification:
Assume 75% removal.

212 lb/day nitrogen $\cdot$ 0.75 = 159 lbs removed per day

159 lb/day ammonia nitrogen $\cdot$ 4.57 Ox/N = 727 lb/day Ox

(21) Total oxygen demand:

$$3{,}999 + 727 = 4726 \text{ lb/day oxygen}$$

(22) Oxygen available, lbs per day

(SBOD) 1,700,000 + (N) 707,000
$$= 2{,}407{,}000 \text{ ft}^2 \text{ disc area}$$

2,407,000 ft² media $\cdot$ 1.5 lb/day/1000 ft²
$$= 3611 \text{ lbs/day oxygen available.}$$

not enough oxygen supplied, supplementary source required, or additional disk area supplied. Install 3,200,000 ft² of disk area, 3,200,000 ft² media $\cdot$ 1.5 lb/day/1000 sq ft = 4800 lb/day available.

(23) RBC configuration

(24) Total area installed: 3,200,000 ft²

(25) Number of shafts

$$3{,}200{,}000 \text{ ft}^2 / 100{,}000 \text{ ft}^2 \text{ per shaft} = 32 \text{ shafts}$$

DISCUSSION

In this case the RBC design is driven by the high removal rate specified, 98%+ based on a 500 mg/L feed and a 7.45 mg/L effluent. These conditions were imposed as a comparison of performance relative to the other biological treatment options, especially activated sludge. It is doubtful that this effluent could be achieved! Once the influent concentration exceeds 180 mg/L and the removal exceeds 75 to 80%, the design is driven by oxygen requirements! In such cases, an RBC units is best employed as part of a combined treatment configuration.

REFERENCES

1. Albertson, O.E.: "Slowdown That Trickling Filter," *Operations Forum,* Pg 15, January, 1989.
2. Bruce, A.M., et al.: "Pilot Studies on Treatment of Domestic Sewage by Two-Stage Biological Filtration-With Special Reference to Nitrification," *Water Pollution Control,* Pg 80, 1975.
3. Gujer,W., Boller, M.: "Nitrification in Tertiary Trickling Filters Followed by Deep-Bed Filters," *Water Research,* V 20, No 11, Pg 1363, 1986.
4. Gullicks, H.A., Cleasby, J.L.: "Design of Trickling Filter Nitrification Towers," *Journal WPCF,* V 58, No 1, Pg 60, January, 1986.
5. Gullicks, H.A., Cleasby, J.L.: "Nitrification Performance of a Pilot-Scale Trickling Filter," *Research Journal WPCF,* V 62, No 1, Pg 40, January/February, 1990.
6. Harremoes, P.: "Criteria for Nitrification in Fixed Film Reactors," *Water Science Technology,* 14, 167, 1982.

7. Harrison, J.R., Daigger, G.T., Filbert, J.W.: "A Survey of Combined Trickling Filter and Activated Sludge Processes," *Journal WPCF,* V 56, No 10, Pg 1073, October, 1984.

8. Harrison, J.R., Daigger, G.T.: "A Comparison of Trickling Filter Media," *Journal WPCF,* V 59, No 7, Pg 679, July, 1987.

9. Medcalf & Eddy, Inc.: *Wastewater Engineering-Treatment, Disposal, Reuse,* McGraw-Hill, 1991, Third Edition.

10. Okey, W.O., Albertson, O.E.: "Evidence for Oxygen- Limiting Conditions During Tertiary Fixed-Film Nitrification," *Journal WPCF,* V 61, No 4, Pg 510, April, 1989.

11. Parker, D.S., Merrill, D.T.: "Effect of Plastic Media on Trickling Filter Performance," *Journal WPCF,* V 56, No 8, Pg 955, August, 1984.

12. Parker, D.S., Richards, T.: "Nitrification in Trickling Filters," *Journal WPCF,* V 58, No 9, Pg 896, September, 1986.

13. Parker, D., et al.: "Enhancing Reaction Rates in Nitrifying Trickling Filters through Biofilm Control," *Journal WPCF,* V 61, No 5, Pg 618, May, 1989.

14. Richards, T., Reinhart, D.: "Evaluation of Plastic Media in Trickling Filters," *Journal WPCF,* V 58, No 7, Pg 774, July, 1986.

15. U.S. Environmental Protection Agency: *Trickling Filter/Solids Contact Process; Full-Scale Studies,* EPA-600/2-86-046, 1986.

16. U.S. Environmental Protection Agency: *Process Design for Nitrogen Control,* PB-259-149/38A, October 1975.

17. U.S. Environmental Protection Agency: *Process Design Manual for Upgrading Existing Wastewater Treatment Plants,* EPA-625/1-71-004a, 1974.

18. U.S. Environmental Protection Agency: *Treatability Manual,* EPA-600/8-80-042a, 1980.

19. U.S. Environmental Protection Agency: *Handbook for Identification and Correction of Typical Design Deficiencies at Municipal Wastewater Treatment Facilities,* EPA-625/6-82-007, 1982.

20. WEF Manual of Practice No 8: *Wastewater Treatment Plant Design,* Water Pollution Control Federation, 1982, 2nd Edition.

21. WEF Manual of Practice: *Design of Municipal Wastewater Treatment Plants,* Water Environment Federation, 1992.

22. Antonie, R.L.: *Fixed Biological Surfaces—Wastewater Treatment; The Rotating Biological Contactor,* CRC Press, Inc., 1975.

23. U.S. Environmental Protection Agency: *Design Information on Rotating Biological Contactors,* EPA-600/2-84-106, 1984.

24. U.S. Environmental Protection Agency: *Review of Current Rotating Biological Contactor Performance and Design Procedures,* EPA-600/2-85-033, 1985.

25. Evans, F.L.: "Consideration of First-Stage Organic Overloading in Rotating Biological Contactor Design," *Journal WPCF,* V 57, No 11, Pg 1094, November, 1985.

26. Opatken, E.J.: "An Alternative RBC Design—Second-Order Kinetics," *Environmental Progress,* V 5, No 1, Pg 51, February, 1986.

Aerobic Digestion

Aerobic digestion is primarily employed to stabilize excess biological waste solids.

AEROBIC digestion is employed to stabilize sludge and to lower its solids content. It can be considered a protraction of the aerobic biological process to the extended aeration phase, encountered in activated sludge processes and aerated lagoons operating at long retention times. Endogenous digestion involves volatile solids destruction to sustain the process, reducing the mass that must be further processed or disposed, and stabilizing the sludge to minimize further biological action. When biological cells are aerated under endogenous (food-limited conditions) the microorganisms utilize the food within the cells, causing the eventual collapse of aged cells, releasing organic matter to sustain the living organisms. This catabolism results in volatile solids reduction to carbon dioxide, water, and nondegradable components. Under these conditions nitrification can occur. Applicable biological principles are discussed in Chapters II-2 and II-3. The basic aerobic batch digestion process is illustrated in Figure 6.1. Sludge is fed intermittently to an aerobic digester; concentrated sludge and decant are discharged at regular intervals. Figure 6.2 illustrates a continuous aerobic digester, with optional sludge separation and recycle.

Aerobic digestion is a valuable process for industrial waste application, when convenience and ease of operation overshadow any concern for recoverable energy. In such cases, sludge quantities are usually small, making any economical disadvantage with competing processes minimal. It is an alternative to anaerobic digestion, involving a simpler process and operation, lower capital costs, and eliminating the potential for releasing odorous or hazardous gaseous byproducts. Its disadvantages are higher net energy costs resulting from aeration and mixing power, and not producing recoverable energy (methane).

Two significant criteria may limit the viability of aerobic sludge digestion in industrial facilities when employed solely for solids reduction. First, because of the relatively small quantities of sludge generated dewatering and/or disposal costs may not be significantly reduced and may not offset the digester capital and operating costs. In fact, sludge thickening with intermittent air injections (to prevent septic conditions) may be equally effective. Second, in some cases the dewatering qualities of aerobically digested sludge is poor, requiring enhanced sludge conditioning or increased dewatering capacity, which may offset the potential savings from reduced sludge volumes. Sludge digestion must be economically evaluated, along with any process advantages, as part of a complete sludge management program.

BASIC CONCEPTS

STOICHIOMETRIC RELATIONS

Aerobic digestion is best described as a continuation of the basic biological system involving first the destruction of remaining substrate organics and formation of cells, followed by the endogenous respiration of the cells. In simple terms these processes can be expressed as follows:

(1) Removal of remaining organic substrate:

organic substrate + oxygen + nutrients = new cells
+ CO_2 + H_2O + energy for new cells

(2) Destruction of cells at low substrate concentration:

sludge cells + O_2 = CO_2 + H_2O + NH_3
+ inactive material

Assuming a cell formula of $C_5H_7NO_2$, the endogenous

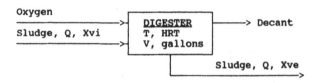

Figure 6.1 Aerobic batch digester.

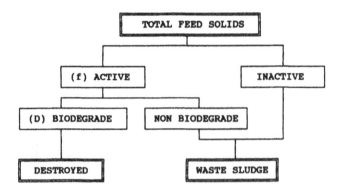

Figure 6.3 Aerobic digester solids balance.

respiration phase occurring in aerobic digestion under *limited* substrate can be represented as follows [6,9]:

$$C_5H_7NO_2 + 5O_2 = 5CO_2 + 2H_2O + NH_3 + energy$$

Under suitable conditions the process proceeds toward nitrification as follows:

$$C_5H_7NO_2 + 7O_2 = 5CO_2 + 3H_2O + H^+ NO_3^- + energy$$

Accordingly, 1.42 kg of oxygen is required for oxidation of the cell, and 1.98 kg is required for the process to complete to nitrification.

The previous stoichiometric relations are based on destruction of the *active* portion of the feed solids, which in the case of biological treatment waste sludge is composed of 60 to 80% oxidizable cell tissue, the remaining being inert components or nonbiodegradable, as illustrated in Figure 6.3 [1].

Net solids production in aerobic digestion must be based on the destruction of the volatile (active) and accumulation of the inactive portions of the sludge.

REACTION KINETICS MODEL

Theoretical aerobic digestion models are usually based on first-order kinetics, with the final form dependent on the process configuration [1].

Basic System Balance

A steady-state material balance, based on consistent VSS characteristics and all VSS being biodegradable, can be rep-

resented by the continuous or intermittent configurations illustrated in Figure 6.4. An intermittent system operates with periodical feed and discharge, a continuous, completely mixed system with continual decant and sludge withdrawal.

In any configuration the process can be depicted using three defining flows: the (1) influent sludge flow, (2) decant discharge, and (3) processed sludge discharge. Where sludge recycle is included, sludge wasting and decant are discharged from the final separation equipment. These flows, along with the reaction, can be represented as follows:

$$Q_o \cdot X_o = Q_w \cdot X_w + Q_d \cdot X_d + V \cdot K_d \cdot X \qquad (6.1)$$

where Q_o is the feed sludge flow, L/day (GPD), X_o is the biodegradable feed concentration (VSS), mg/L, Q_w is the waste sludge flow, L/d (GPD), X_w is the biodegradable waste sludge concentration (VSS), mg/L, Q_d is the decant flow, L/day (GPD), X_d is the decant biodegradable concentration (VSS), mg/L, V is the reactor volume, L (gals), and K_d is the decay coefficient, 1/day.

When combined with the definition of SRT, Equations (6.2) and (6.3) define the digester basin volume and system performance.

$$V = \frac{Q_o \cdot X_o \cdot SRT}{X_w [1 + K_d \cdot SRT]} \qquad (6.2)$$

$$\frac{X_w}{X_o} = \frac{1}{1 + K_d \cdot SRT} \cdot \frac{SRT}{HRT} \qquad (6.3)$$

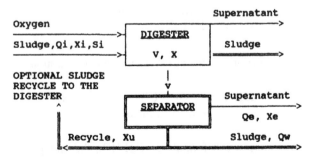

Figure 6.2 Continuous aerobic digester.

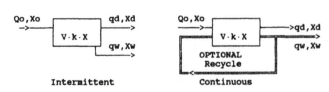

Figure 6.4 Steady state material balance.

Completely Mixed System

Based on solids VSS concentration, the above equation simplifies to Equation (6.4) for a completely mixed, *flow-through (no recycle)*, single-stage system.

$$\frac{X_w}{X_o} = \frac{1}{1 + K_d \cdot SRT} \tag{6.4}$$

Expressed in terms of digester retention time, the basic single stage equation can be rearranged to Equation (6.5).

$$SRT = \frac{X_o - X_w}{K_d X_w} \tag{6.5}$$

These equations can be expanded to further distinguish among total feed solids, volatile feed solids, inactive solids, active solids, and the reaction conversion efficiencies. Defining influent solids on the basis of total feed solids (S_o), containing a constant nondegradable component (X_n), so that

$$X_o = S_o - X_n \tag{6.6}$$

$$X_w = S_w - X_n \tag{6.6a}$$

where S_o is the total feed VSS concentration, mg/L, S_w is the total waste VSS concentration, mg/L, and X_n is the feed nonbiodegradable VSS concentration, mg/L.

Assuming that X_n is inactive, passing through the reactor, Equation (6.5) can be expressed as follows [1]:

$$SRT = \frac{S_o - S_w}{K_d (S_w - X_n)} \tag{6.7}$$

$$S_w = \frac{S_o + SRT \cdot K_d \cdot X_n}{1 + SRT \cdot K_d} \tag{6.8}$$

Further defining reaction solids as illustrated in Figure 6.3, with the fraction of biodegradable solids defined as f and the fraction degraded D, Equation (6.5) can be represented as follows:

$$SRT = \frac{S_o - S_e}{K_d \cdot f \cdot D \cdot S_o} \tag{6.9}$$

Commonly cited decay coefficients are indicated in Table 6.1 [1,5]. Digester temperature must be selected to maximize reduction and avoid extreme temperatures that could inhibit digestion. Studies suggest that digester reduction is maximized at 30°C [2]. Whether 30°C is a critical temperature governing reaction kinetics or not may be overshadowed by the economics favoring anaerobic digestion as the reactor temperature increases. Investigators have recommended

TABLE 6.1. Decay Coefficients (adapted from References [1,5]).

Sludge Type	K_d, 1/day	Properties
Waste activated	0.05	VSS @ 15°C
Waste activated	0.10–0.12	VSS @ 20°C
Waste activated	0.14	VSS @ 25°C
Extended aeration	0.18	VSS @ 20°C
Trickling filter	0.05	VSS @ 20°C
Primary and TF	0.04	VSS @ 20°C

temperature corrections to the decay rate coefficient based on the relation [1,9]:

$$(Kd)_t = (Kd)_{20} \cdot \Phi^{(t-20)} \tag{6.10}$$

where Φ ranges from 1.02 to 1.11, with 1.072 representing a doubling of the reaction rate for every 20°C rise, a common rule-of-thumb for biological systems.

AEROBIC TREATMENT SYSTEMS

There are three configurations generally employed in aerobic digestion: *batch, continuous with no recycle,* and *continuous with recycle.*

BATCH AEROBIC DIGESTER

Batch aerobic digesters are employed for small sludge quantities, operating in sequential stages similar to an SBR treatment system. Operating steps include

(1) Sludge feed
(2) Aeration of the full digester, initiating the VSS reduction process
(3) Shut down and settling when the reaction is complete
(4) Supernatant withdrawal
(5) Waste sludge withdrawal
(6) Batch processing is complete and the cycle is repeated

CONTINUOUS (CONVENTIONAL) DIGESTER

Digester volumes can be reduced by employing a *continuous system* with *no recycle*, operating similar to an aerated lagoon. Sludge is fed continuously or at regular intervals to the digester, with equivalent decant and concentrated sludge volumes withdrawn from the vessel. The system is controlled by adjusting the flow rate and therefore the system hydraulic retention time (HRT). Typical design criteria are cited in Table 6.2.

TABLE 6.2. Typical Continuous Aerobic Digester
Design Criteria for Municipal Sludge
(adapted from References [6]).

| Solids Retention Times, Days | VSS Removal | |
	40%	55%
At 4.4°C (40°F)	108	386
At 15.6°C (60°F)	31	109
At 26.7°C (80°F)	18	64
Oxygen, kg O₂/kg VS destroyed:	2 @ less 45°C (113°F)	
	1.45 greater 45°C (113°F)	
Oxygen for soluble BOD: Specific to the influent		
Oxygen nitrification: Specific to the influent		
Maximum concentration: 2.5 to 3.5% (depending on sludge)		
Mixing, mechanical, watts/m³: 13–105		
hp/1000 ft³: (0.5–4)		
Diffuser, m³/min/1000 m³: 20–40		
Dewatering characteristics: Generally poor		

CONTINUOUS DIGESTER WITH RECYCLE

The conventional process can be upgraded to include sludge separation and recycle to separately control the system SRT, and thereby reduce the digester volume. In capacities commonly encountered in industrial facilities, the sludge quantity generated does not warrant the additional capital expenditures for recycle equipment. If large sludge quantities are generated, aerobic treatment advantages may be overshadowed by anaerobic treatment economics. In addition, poor sludge quality common with endogenous digester conditions often result in poor sludge settling characteristics, possibly making digester control difficult.

EMERGING TECHNOLOGIES

Digester performance can be improved by utilizing pure oxygen or more effective temperature ranges. Economic justification for these processes for industrial size facilities must be evaluated on a site-by-site basis. They are briefly discussed to review available advanced technologies.

Pure Oxygen Aerobic Digestion

Pure oxygen aerobic digestion is an emerging technology in which pure oxygen replaces air as the oxidant, possibly applicable for large installations. The system operates as a closed tank to conserve oxygen usage, thereby retaining much of the heat released in the process and making the system especially applicable in cold climates. Costs dictate the practical application of this process, balancing reduced capital costs resulting from a lower reactor volume with increased oxygen generation costs.

Thermophilic (autothermal) Aerobic Digestion

Thermophilic aerobic digestion or autothermal aerobic digestion is an emerging technology developed to operate in the thermophilic zone (above 42°C), utilizing heat energy released from the reaction to sustain process conditions. The process has been evaluated for both ambient air and pure oxygen aeration and proven effective for feed VSS concentrations of 2.5 to 5%, resulting in biodegradable reductions up to 70% at the pilot stage [5,9].

Cryophilic Digestion

Cryophilic digestion involves low-temperature operation (less than 10°C), applicable to small treatment plants at cold climatic conditions. Research has been conducted at elevated SRT values to compensate for low temperatures, with tests being conducted at SRT · temperature (°C) product ranges of 250 to 300 °C-days to achieve acceptable solids reduction performance. Full-scale operating data are sparse [9].

PROCESS ENGINEERING DESIGN

General process limitations and defining operating characteristics of aerobic digesters are summarized in Table 6.3. The digesters possess all the characteristics of their comparable suspended solids biological treatment systems: contrasting batch digestion to SBR, a once-through digester to an aerated lagoon, and a continuous digester with recycle to an activated sludge system. Great operating flexibility is not necessary; the design should be focused on the rather limited performance requirements of solids reduction, sludge

TABLE 6.3. Operating Characteristics.

| Variable | Waste Characteristics | |
	Operator controllable	Critical
Waste generated	No	Yes
Composition	No	Yes
Concentration	No	Yes
Biodegradability	No	Yes
Toxicity	No	Yes
Operating Characteristics		
Flow rate	Yes	Yes
SRT	When applied	Yes
HRT	Yes	Yes
VSS level	Yes	Yes
Sludge recirculation	When applied	Yes
Mixing	Yes	No
Oxygen capacity	Yes	Yes
Wasting	Yes	Yes
Nutrients	Yes	Yes
Alkalinity	Yes	Yes
Temperature	No	Yes

stabilization, and producing a sludge with good dewatering qualities. Large upstream secondary treatment capacity as well as the digester volume itself provide large dilution capabilities diminishing upset or process variability concerns. The available upstream storage capacity allows great flexibility in controlling flow rate and thereby reactor retention time. The major detriments to significant solids reduction are reduced temperatures during winter conditions and waste sludge properties. Process winterization counter measures are discussed in the Reactor Temperature Section. The digesting process is complicated when feed components other than volatile suspended solids are injected into the digester, such as dissolved solids, inert solids, toxic components, nitrogen, and excessive soluble BOD.

REPORTED PERFORMANCE DATA

Aerobic digester performance for municipal sludge has been reported in the EPA *Treatability Manual,* and are summarized in Table 6.4 [7].

REQUIRED PROCESS DESIGN DATA

A complete sludge analysis is paramount! Sludge characterization should establish the volatile suspended solid, soluble BOD, COD, nitrogen, alkalinity, and toxic content. Representative samples should be taken and analyzed until its characteristics and variability can be established. When the sludge is generated in a suspended growth biological system, the secondary treatment plant and any downstream operations such as thickening or storage should be investigated for their affect on sludge properties. At a minimum the operating SVI should be monitored, and its value for collected samples should be recorded. The relation between sludge concentration and digester performance should be established so that the advantages of thickening can be evaluated. In some cases *thickening with intermittent aeration*

may be adequate as a preparatory step for dewatering or direct disposal.

Digester design criteria can be obtained from batch laboratory testing using techniques similar to those proposed for activated sludge and once through aerated lagoons. Resulting data can be tabulated or integrated into first order, or similar applicable aerobic digestion models, to formulate design criteria over the applicable operating range. The required design data are listed in Table 6.5.

Important information not obtained from batch tests are the digested sludge dewatering characteristics, which is a significant property affecting the downstream dewatering equipment capacity. This information is commonly included as part of overall dewatering study covering not only sludge digestion and concentration, but its effects on specific final dewatering design parameters. Dewatering and settling tests discussed in Chapter III-7 (*dewatering*) may prove beneficial.

SLUDGE CHARACTERISTICS

Sludge characteristics affecting digestion include sludge age, nitrification potential, and specific composition. Theoretically, increased upstream sludge age should reduce the decay rate coefficient because in such cases endogenous respiration has initiated the digestion phase, resulting in less degradable sludge. Studies generally confirm this, although some contradictory results dispute that biological system sludge age has an effect on digestion [3,10].

Both influent soluble BOD and ammonia nitrogen concentrations affect digester design. High soluble waste BOD will

TABLE 6.4. **Reported Performance Data (adapted from Reference [7]).**

Sludge	
Influent total solids, %	2–7
Influent volatile solids, %	1–6
Effluent total solids, %	3–12
Volatile solids reduction, %	30–70
Typical vs reduction, %	35–45
Supernatant	
Suspended solids, mg/L	100–12,000
BOD_5, mg/L	50–1700
Soluble BOD_5, mg/L	4–200
COD, mg/L	200–8000
Kjeldahl nitrogen, mg/L	10–400
Total phosphorus, mg/L	20–250
Soluble phosphorus, mg/L	2–60
pH	5.5–7.7

TABLE 6.5. **Required Design Data.**

Critical pilot plant treatability data specific to the waste
  (1) Design temperature
  (2) first order constant
  (3) Y, excess cell yield, mg/mg
Waste solids characteristics that should be obtained from
  laboratory studies but can be estimated
  (4) Fraction of solids biodegradable
  (5) Ultimate solids BOD per g solids
  (6) Ratio of SBOD to ultimate BOD
  (7) Ratio MLVSS/MLSS
SELECTED operating characteristics
  (8) Required final sludge concentration
  (9) Mixed liquor volatile suspended solids
  (10) MART, calculated from treatability data
Operating characteristics that should be obtained from
  pilot studies, but can be estimated from treatability data.
  (11) Summer temperature effects
  (12) Winter temperature effects
  (13) Minimum aeration basin DO, mg/L
  (14) α factor, K_{la} waste/tap water
  (15) β, C_s waste/C_s tap water
  (16) Required oxygen loading
  (17) Final sludge quantity
  (18) Any deficient nutrient requirements

require additional digester volume to complete the degradation process and to further digest the resulting sludge. Relatively low dissolved BOD and high nitrogen concentration will promote nitrification, also requiring increased digester volume and aeration capacity. The amount of inorganic dissolved solids in the waste influent, passing through the secondary process, will concentrate in the sludge. These concentrated inorganics could exhibit toxic properties in the digester solids, although having exhibited little effect diluted in the upstream biological processes.

Sludge Concentration

The digester feed concentration is a significant process variable, with the digestion effectiveness dependent on the active solids concentration. The volatile solids concentration is a measure of the active or biodegradable sludge solids, and a primary design consideration. Inactive dissolved or settleable solids can significantly increase the final sludge quantity, making final disposal more difficult or expensive. As an example, if final disposal involves incineration, inactive solids can significantly impact air pollution emissions (and required controls) and primary combustion chamber slagging. If the inorganics are toxic or hazardous classified components, not only will incineration be more complex but landfill alternatives can be either banned or prohibitively expensive. Finally, significant quantities of inactive solids could retard (or terminate) the digestion reaction.

The sludge VSS or active concentration can impact the required digester volume and the reactor effectiveness, based on the kinetic and SRT definitions expressed by Equation (6.11).

$$\text{digester volume } (V) = \text{SRT} \cdot \text{Wasting quantity}/X \quad (6.11)$$

So that with all other variables being equal, the digester volume should decrease with increasing VSS concentration. If this is the case, waste prethickening could be effective and should be evaluated as part of the digester process design.

Toxicity

Toxic factors affecting aerobic digestion generally follow those cited for all biological systems, discussed in Chapter II-2. One important consideration is that manufacturing wastes could contain low level toxic elements not significantly affecting upstream treatment but could concentrated to critical levels in subsequent solids management systems prior to the digester. A similar concern is the release of "dormant" sulfides and metals into solution as a result of reduced alkalinity associated with accelerated nitrification, potentially being detrimental to the digestion process.

Sludge Preparation

As with all biological processes some feed preparation may be required to optimize reactor conditions, alkalinity control and nutrient addition being of primary concern. Generally, biological reactions are optimized at a pH in the 5 to 9 range, with the efficiency improving at the higher range [3]. Significantly, in aerobic digestion the process alkalinity is influenced not only by the feed pH but the degree of nitrification occurring, as indicated by the basic equation.

The waste sludge ammonia nitrogen content should be carefully monitored for nitrification potential because at excessive levels the nitrification process can result in significant reactor pH suppression. This will always be the case for high ammonia nitrogen content sludges discharged at neutral conditions. In such cases influent alkalinity may have to be supplemented to adjust the process pH.

Biological system nutrient requirements discussed in Chapter II-2 apply to aerobic digestion. However, if the sludge is a biological treatment product the required nutrients should be present.

PROCESS DESIGN VARIABLES

Critical aerobic digester design variables include the following:

(1) Reactor configuration
(2) Reactor temperature
(3) Reactor mixing level
(4) Hydraulic and solids retention time
(5) Oxygen requirements
(6) Fate of contaminants

Reactor Configuration

The primary consideration driving any process decision is to implement the simplest operating system to accomplish the intended overall sludge management program. Digestion, as with all sludge management processes, is economically driven because the goal is to dispose the smallest quantity of stable residue, whether that be raw, processed, concentrated sludge, or incinerator ash.

First, the benefits of aerobic digestion should be established:

(1) Does digestion improve downstream processing or disposal?
(2) Does the digested sludge dewater, or does the solids reduction complicate the dewatering process?
(3) Would sludge thickening be more beneficial that digestion and sludge reduction?
(4) Does the sludge volume warrant anaerobic digestion?

Next, a series of evaluations should be made to optimize the system, aimed at implementing a conventional treatment system.

(1) Should thickening precede digestion?

(2) If digestion proves beneficial does the volume generated warrant more than a conventional configuration? In most industrial facilities, the complexity of separate sludge control and recycle does not warrant the benefits of reactor volume reduction.

(3) Are there any significant benefits in operating at any temperature other than ambient? Again in industrial facilities the added complexity of operating at other than ambient temperatures is usually not warranted.

Where conventional treatment is not applicable, frequently as a result of sludge volume, the evaluation sequence in order of increased complexity would include

(1) Utilizing anaerobic treatment for large sludge quantities, where energy reduction and generation would be beneficial or where aerobic dewatering sludge quality is poor

(2) Sludge recycle, with separate HRT and SRT control, to reduce reactor volume

(3) Innovative technology involving pure oxygen or operating outside ambient temperatures to significantly improve the design and operation

Reactor Temperature

Conventional processes operate as unheated reactors over a wide range of ambient temperatures. As with all biological reactions, digester temperature controls the microorganisms population selection and reaction effectiveness, commonly expressed in terms of the decay constant (K_d) value. Thermophilic aerobic digestion processes are being developed which would operate at temperatures greater than 45°C (113°F). Similarly, cryophilic digestion process are being developed to operate at temperatures less than 20°C (68°F) [9]. Unless specific circumstances dictate otherwise, industrial aerobic sludge systems are designed to operate at ambient conditions, in the 5 to 25°C (41 to 77°F) range. If elevated temperatures are required anaerobic treatment will have to be evaluated as a viable alternative along with thermophilic aerobic digestion.

In the commonly operated temperature range, process efficiency is assumed to follow a first-order reaction, with the associated decay constant adhering to the Arrhenius relationship as defined by Equation (6.10).

The variation of decay constant with temperature is illustrated in Figure 6.5, based on pilot and full-scale data as presented in the EPA sludge treatment manual [6]. Two specific process conditions are defined by (1) an expected rapid decline in aerobic digestion at low temperatures, as encountered in winter conditions, and (2) an implied temper-

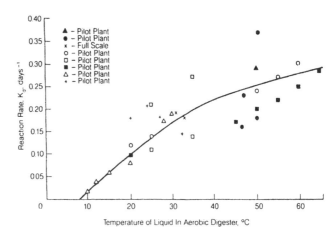

Figure 6.5 Affect of temperature on aerobic decay constant (copy of EPA diagram [6]).

ature related peak performance, after which a retarded digestion rate is suggested.

The significance of two operating variables, temperature and SRT, is illustrated in Figure 6.6, as reported in the EPA *Sludge Treatment Manual* [6]. Utilizing the product °C-(SRT) days, decreasing performance with decreasing product is indicated, implying that winter conditions can be off-set by increased solids inventory (SRT). This operating variable can *only* be accomplished if provision is made for digester concentration control because in a once-through system the SRT is equal to the HRT and solely dependent on the influent flow rate.

Anticipated adverse cold weather effects can be countered with increased SRT, applying an external heat source to the reactor to maintain a suitable temperature, housing the reactor, minimizing losses and conserving endogenous heat release, or a combination of these measures. Housing the reactor may be a viable alternative for the relatively small vessel sizes used for industrial facilities. If these measures are not adequate the configuration enhancements discussed in the Process Systems section to either conserve heat release

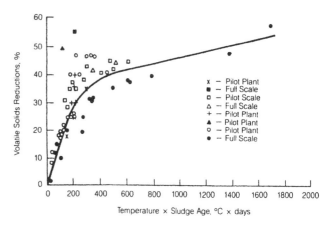

Figure 6.6 Aerobic digester performance curve (copy of EPA diagram [6]).

and minimize heat loss or maximize the system effectiveness at lower temperatures should be considered.

Reactor Mixing

Digester mixing criteria are identical to those governing other dispersed growth treatment systems, based on maintaining a uniform dissolved oxygen and solids content and providing intimate contact between the reactants. Traditionally, mechanical aerators are employed at levels ranging from 13 to 105 W per 1000 m³ of volume (0.5 to 4 hp/1000 ft³) or diffused air at 20 to 40 m³ per 1000 m³-min (20 to 40 ft³/1000 ft³ · min) [6,9].

Hydraulic or Solids Retention Time

The required reactor volume is a product of the *flow rate* and the hydraulic retention time (HRT), the reaction effectiveness is measured by the SRT. The HRT and SRT are equal in a once-through system. SRT or HRT selection, as with all biological reactions, is based on maintaining a suitable cell growth rate, a viable microorganism population and a stable process. Basic theoretical considerations are similar to those discussed for activated sludge systems (Chapter II-3). The SRT is defined by Equation (6.12), the HRT and SRT impact system performance as indicated by Equation (6.3)

$$\text{SRT} = \frac{X \cdot V}{Q_w \cdot X_u + Q_e \cdot X_e} \qquad (6.12)$$

where X_u is the thickened sludge concentration, mg/L, X is the digester suspended solids, mg/L, Q_w is the digester waste rate, m³/day (GPD), Q_e is the digester decant rate, m³/day (GPD), X_e is the effluent suspended solids, mg/L, and V is the digester operating volume, m³ (gal).

As illustrated in Figure 6.6, a 40% VSS reduction is projected at a temperature · SRT product of 400 °C · days, a 50% reduction requires 1400 °C · days, and a 55% reduction requires 1800 °C · days. At an operating temperature of 20°C, this equates to an SRT of 20 to 90 days. The impact of sludge recycle on reactor size is similar to any suspended growth system, with reactor vessel reduced significantly when the SRT can be independently controlled. However, unless the sludge volume is significant, batch or a once-through system provides a simpler, less expensive system.

Treatment requirements for municipal sludge are many times mandated by local or federal regulations, which could require aerobic digester volumes exceeding that established by theoretical considerations. Whether industrial waste sludge is subject to the same regulations or specific industrial regulations should be investigated *prior* to the process design completion.

Oxygen Requirements

Digester oxygen requirements include that needed to degrade the remaining BOD, destroy cell tissue, and nitrification. Based on stoichiometric relations, assuming a cell composition represented by the formula $C_5H_7NO_2$, 1.98 kg of oxygen per kg of cell are theoretically required for cell degradation and nitrification, 1.43 kg being required for the cell degradation alone. A biological system is considered viable if the reactor dissolve oxygen level is 1 to 2 mg/L. Aeration equipment sizing is a balance between mixing and aeration requirements, as discussed in Chapter II-1.

Fate of Contaminants

The *fate* of the *contaminants* depends on the sludge characteristics, which are a direct result of the upstream processes generating the sludge. Generally, the criteria discussed for activated sludge apply to aerobic digesters. The differences being that (1) the volatile organic level should be low because appreciable stripping would have occurred in the upstream aeration basin and (2) any inorganics fed into the biological system will be concentrated. Potentially, the most significant problem could be digested sludge disposal because of the concentrated inorganics. Digester air emissions could occur if anaerobic reactor conditions are encountered; sulfide release could be enhanced if nitrification is prominent. If volatile emissions are a nuisance or exceed air pollution control regulations, then a covered digestion system can be employed.

Supernatant from the aerobic digestion system will be returned to the front end of plant treatment system; as a result the plant must be capable of handling the additional hydraulic and organic loading. In addition, at some point the recycle supernatant stream may concentrate a specific inorganic component to intolerable levels, detrimental to the secondary treatment system. This is especially significant in industrial waste treatment, where manufacturing wastes can contain a variety of rejected chemicals.

Concentrated and stabilized sludge generation is often dewatered to final disposal. The dewatering process fixes the cake characteristics. Poor dewatering sludge may be a result of a failure of any of the upstream treatment processes or the aerobic digestion itself. Factors generating poor dewatering sludge in a biological system are essentially those that produce bulking sludge, and are covered in Chapters II-2 and II-3.

The final sludge composition determines the difficulty and cost of final disposal. Certain toxic components will result in a RCRA classification as a hazardous waste, prohibiting landfill disposal, or make it expensive. In addition, high inorganic concentration, either as innocuous solids or as specific components such as heavy metals, can make incineration difficult and expensive. Both can contribute to primary combustion chamber slagging or increase the air pollution control equipment costs.

GENERAL ENGINEERING CRITERIA

A complete aerobic sludge system could involve the following process components:

Waste feed system	Chapter II-1
Aerobic digester	
Aeration equipment	
Sludge recycle (optional)	Chapter I-8
pH control (optional)	
Nutrient addition (optional)	Chapter I-5

Each of these elements, except the aerobic digesters, are discussed elsewhere. Aerobic digester process engineering requirements are similar to those discussed for activated sludge (Chapter II-3) and aerated lagoon (Chapter II-4), and will not be repeated here. Some special considerations include

(1) Aerobic digesters, in the sizes commonly encountered in industrial facilities, can be purchased as complete packages, configured as a batch process consisting of the tank and all system components. They can also be designed as miniature biological treatment plants, similar to an (extended aeration) activated sludge system. They can be constructed as reinforced concrete basins or above-ground steel vessels.

(2) The number of batch tanks depends on capacity, although single units are common in industrial systems because (1) of the relatively small sludge volumes generated, (2) the high reliability of the equipment, and (3) the operation is not critical to the entire process, and temporary shutdowns can be tolerated. Where multiple vessels are provided piping should be included for complete flexibility to allow transfer from one vessel to another.

(3) Vessels should be constructed with provision for level indication, monitoring, and control. In addition, the vessel should have a bottom drain or provisions to completely empty for servicing and maintenance. Small units should be designed to allow support of mixing equipment; larger units may be constructed with separate mixer support.

(4) Provision should be made for digester pH control, especially if nitrification is expected. Although lime is commonly used for pH control, its use should be carefully evaluated for relatively small industrial systems. It is difficult to handle and increases the reactor inert concentration, which results in increased sludge ash content. Other neutralizing agents should be evaluated and their overall cost effectiveness carefully considered.

(5) Digesters can be covered to minimize heat loss, freeze protection, or preserve oxygen. Open tanks are more common, but where covered digesters are considered they should confirm with OSHA requirements for entrance to confined spaces.

(6) Batch systems should have multiple supernatant withdrawal points for flexibility and the ability to see and sample the effluent.

(7) Provision should be made to be able to flush sludge lines with plant effluent.

(8) Provision should be made for foam control and defoam addition.

(9) Odor control provisions should be considered.

(10) Provision should be made for freeze control, including submerged or diffused air systems rather than surface aerators.

(11) Vessel design in high groundwater areas should include provision for emptying the tank without foundation problems.

(12) Basins with common walls should be designed to allow emptying of one cell and maintaining a full hydraulic load in cells utilizing the same wall.

(13) Provision should be made for multiple ports to decant the digesters by gravity.

(14) Pumps should be selected for concentrated sludge characteristics, utilizing plunger type rather than centrifugal.

COMMON AEROBIC DIGESTER DESIGN DEFICIENCIES

Some design deficiencies common to municipal systems have been investigated by the EPA [8] and are identified for the reader as a checklist of process and mechanical design considerations. The cited reference should be reviewed for further details.

General

(1) No provision was made for sludge thickening prior to digesting.

(2) An inadequate common wall design results in structural failure during batch operations.

(3) Freeze protection measures are inadequate because above line pipes and pumps were not traced, tank covers were not provided, or the vessels were poorly insulated.

General

(1) Denitrification in the digester results in settling problems.

(2) Inadequate mixing results in solids deposition and poor DO distribution.

(3) Reactor has inadequate freeboard.

(4) No provision was made for removing supernatant at various heights.

(5) Supernatant quality cannot be observed upon withdrawal.

(6) No provision was made for pH control.

(7) Supernatant cannot be withdrawn at low vessel levels.

(8) The digester design did not consider lower efficiencies at winter temperatures.

(9) Level gages acre not provided to establish vessel operating depths.

(10) No provisions were made for defoaming.

Poor Pump Design

(1) No spare pumps were installed to allow for pump maintenance without a system shutdown.

(2) The pumping capacity is inadequate to allow for rapid supernatant removal.

(3) The wrong sludge pump was selected, and the head selected was not corrected for the solids concentration and viscosity.

Inadequate Aeration Design

(1) The blower air supply was inadequate for the installed diffused air systems.

(2) Consideration was not given to diffuser maintenance or clogging problems.

or

(1) The mechanical aeration capacity is inadequate.

(2) Fixed mechanical aerators cannot be used at levels other than when the tank is at full capacity.

CASE STUDY NUMBER 18

Develop an *aerobic sludge digester* for the sludge generated in Case Study 10, which is thickened to 2% (Case Study 25), having the following characteristics:

(1) 9658 gal/day of sludge @ 2%.

(2) 20,000 mg/L SS concentration, 1611 lb/day.

(3) 16,000 mg/L VSS concentration, 1289 lb/day

(4) 80% of feed SS volatile suspended solids

(5) 90% of feed VSS biodegradable

(6) First-order decay constant, 0.06 day^{-1} @ 20°C

(7) 40% degradable VSS destruction efficiency

(8) Dissolved BOD_u at 7.45 BOD_5/0.68 ≡ 11 mg/L

(9) Use a temperature correction coefficient of 1.04.

PROCESS CALCULATIONS

(1) Establish influent concentrations
  (2) Suspended solids: 20,000 mg/L
  (3) (S_o) volatile suspended solids (VSS): 16,000 mg/L
  (4) Fixed solids: 20,000 − 16,000 = 4000 mg/L
  (5) VSS biodegradable: 0.9 · 16,000 = 14,400 mg/L
  (6) (Xn) VSS nonbiodegradable: 0.1 · 16,000 = 1,600 mg/L
  (7) System material balance, lbs/day

(8) Influent

$$\text{feed TSS} = 9658 \text{ GPD} \cdot 8.34 \text{ lb/gal} \cdot 20,000 \text{ mg/L}/1,000,000$$
$$= 1611 \text{ lb/day}$$

$$\text{feed VSS} = 9658 \text{ GPD} \cdot 8.34 \text{ lb/gal} \cdot 16,000 \text{ mg/L}/1,000,000$$
$$= 1289 \text{ lb/day}$$

feed nonbio VSS
$$= 9658 \text{ GPD} \cdot 8.34 \text{ lb/gal} \cdot 1,600 \text{ mg/L}/1,000,000$$
$$= 129 \text{ lb/day}$$

feed bio VSS
$$= 9658 \text{ GPD} \cdot 8.34 \text{ lb/gal} \cdot 14,400 \text{ mg/L}/1,000,000$$
$$= 1160 \text{ lb/day}$$

fixed solids
$$= 9658 \text{ GPD} \cdot 8.34 \text{ lb/gal} \cdot 4,000 \text{ mg/L}/1,000,000$$
$$= 322 \text{ lb/day}$$

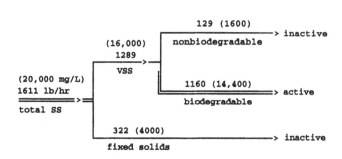

(9) Establish effluent concentrations
  (10) VSS Biodegradable: 14,400 · (60/100)
  $$= 8640 \text{ mg/L } (Xw)$$
  (11) VSS nonbiodegradable: 1600 mg/L (Xn)
  (12) (S_e) volatile suspended solids (VSS):

$$8640 + 1600 = 10,240 \text{ mg/L}$$

(13) Estimate digester retention time for design conditions

$$\text{HRT} = \frac{S_o - S_w}{K_d \cdot (S_e - X_n)} = \frac{16,000 - 10,240}{0.06 \cdot (10,240 - 1600)} = 11.1 \text{ days}$$

(14) Estimate digester retention time for summer conditions

(15) Corrected coefficient @ 80°F (26.7°C)

$$0.06 \cdot 1.04^{(26.7-20)} = 0.078 \ \text{day}^{-1}$$

$$\text{HRT} = \frac{16,000 - 10,240}{0.078 \cdot (10,240 - 1600)} = 8.5 \ \text{days}$$

(16) Estimate digester retention time for winter conditions
(17) Corrected coefficient @ 40°F (4.4°C)

$$0.06 \cdot 1.042^{(4.4-20)} = 0.032 \ \text{day}^{-1}$$

$$\text{HRT} = \frac{16,000 - 10,240}{0.05 \cdot (10,240 - 1600)} = 20.8 \ \text{days}$$

(18) Selected digester retention time 21 days
(19) Reactor volume, gal

$$21 \cdot 9658 \ \text{GPD} = 203,000 \ \text{gal}$$

(20) Selected digester volume 203,000 gal
(21) Performance
   (22) Design digested sludge concentration

$$S = \frac{S_o + (K_d \cdot \text{HRT} \cdot X_n)}{1 + K_d \cdot \text{HRT}}$$

$$= \frac{16,000 + 0.06 \cdot 21 \cdot 1600}{1 + 0.06 \cdot 21} = 7972 \ \text{VSS mg/L}$$

Effluent degradable VSS = 7972 − 1600 = 6372 mg/L

(23)   Summer digested sludge concentration

$$S_s = \frac{16,000 + 0.078 \cdot 21 \cdot 1600}{1 + 0.078 \cdot 21} = 7059 \ \text{VSS mg/L}$$

Effluent degradable VSS = 7059 − 1600 = 5459 mg/L

(24) Winter digested sludge concentration

$$S_e = \frac{16,000 + 0.032 \cdot 21 \cdot 1600}{1 + 0.032 \cdot 21} = 10,212 \ \text{VSS mg/L}$$

Effluent degradable VSS = 10,212 − 1600 = 8612 mg/L

(25) Maximum oxygen demand (summer), lb/day
   (26) Oxygen demand constants
      Effluent soluble BOD = 7.45 mg/L
      (see Case Study 10)
      Effluent ultimate BOD = 7.45/0.68 = 11 mg/L
      Wt oxygen required/Wt VSS destroyed: 1.42
      Wt nitrogen released/Wt VSS destroyed: 0.12
      Wt oxygen required/Wt nitrogen nitrified: 4.57
      Total oxygen required/Wt VSS destroyed,
      1.42 + 0.12 · 4.57 = 1.97

(27) SBOD oxygen demand per day

$$9658 \ \text{GPD} \cdot 8.34 \cdot 11 \ \text{g/L BOD}_u \ \text{feed} \ /1,000,000$$
$$\equiv 1 \ \text{lb/day}$$

(28) Maximum VSS daily oxygen demand

$$(\text{Feed VSS}_d - \text{Eff VSS}_d) \cdot 8.34 \cdot Q/1,000,000$$
$$\cdot 1.97 \ \text{Wt Ox/VSS}$$

$$(9685/1,000,000) \cdot 8.34 \cdot (14,400 - 5459)$$
$$\cdot 1.97 = 1418 \ \text{lb/day}$$

(29) Total maximum VSS daily oxygen demand

$$\text{SBOD} + \text{VSS oxygen demand}$$
$$1 + 1418 = 1419 \ \text{lb/day}$$

(30) Estimate digester sludge product
   (31) Nonbiodegradable feed VSS: 129 lb/day
   (32) Fixed feed solids: 322 lb/day
   (33) Maximum digester degradable VSS per day (winter):

$$9658 \ \text{GPD} \cdot 8.34 \cdot 8612 \ \text{mg/L}/1,000,000 = 694 \ \text{lb/day}$$

(34) Total (dry) sludge per day: 694 + 322 + 129
   = 1145 lb/day
(35) Determine digester physical design
   (36) Digester volume: 203,000 gal = 27,141 ft³
   (37) Number of digesters: 2
   (38) Working depth: 10 ft
   (39) Total projected area: 2714 ft²
   (40) Each basin:
      projected surface area: 1357 ft²
      26 × 52 ft
      freeboard: 2 ft
      total depth: 12 ft
(41) Determine digester mixing requirements
   (42) 14,000 ft³ each basin.
   (43) 1 hp per 1000 ft³
   (44) 14,000/1000 · 1 hp = 14 hp each basin

DISCUSSION

The 203,000, gal aerobic digester volume represents approximately 25% of the aeration basin volume (920,000 gal) and may not be economically justified for the volume of sludge generated. Alternatives include (1) operating the process at an elevated temperature and thereby reducing the digester volume, (2) increasing the size of the aeration basin, allowing the process to operate under endogenous conditions, (3) using a small sludge storage tank, providing minimum aeration to maintain aerobic conditions, serving to store and (batch) thicken the sludge, or (4) feeding the sludge directly to the dewatering device, providing (slightly) increased dewatering capacity.

REFERENCES

1. Benefield, L.D. and Randall, C.W.: *Biological Process Design for Wastewater Treatment,* Prentice-Hall, Inc., 1980.

2. Hartman, R.B., et al.: Sludge Stabilization Through Aerobic Digestion," *Journal WPCF,* V 51, No 10, Pg 2353, October, 1979.

3. Krishnamoorthy, R., Loehr, R.: "Aerobic Sludge Stabilization-Factors Affecting Kinetics" *Journal Environmental Engineering,* Proceedings ASCE, 115, No 2, Pg 283, April, 1989.

4. Mavinic, D.S., Koers, D.A.: "Fate of Nitrogen in Aerobic Sludge Digestion," *Journal WPCF,* V 54, No 4, Pg 352, April, 1982.

5. Medcalf & Eddy, Inc.: *Wastewater Engineering-Treatment, Disposal, Reuse,* McGraw-Hill, 1991, Third Edition.

6. U.S. Environmental Protection Agency: *Process Design Manual for Sludge Treatment and Disposal,* EPA-625/9-79-007, September, 1979.

7. U.S. Environmental Protection Agency: *Treatability Manual,* EPA-600/8-80-042a, 1980.

8. U.S. Environmental Protection Agency: *Handbook for Identification and Correction of Typical Design Deficiencies at Municipal Wastewater Treatment Facilities,* EPA-625/6-82-007, 1982.

9. WEF Manual of Practice: *Design of Municipal Wastewater Treatment Plants,* Water Environment Federation, 1992.

10. Reece, C.S., Roper, R.E., and Grady, C.P.L. Jr.: "Aerobic Digestion of Waste Activated Sludge," *Journal Environmental Engineering,* Proceedings ASCE, 105, No EE2, Pg 261, 1979.

Anaerobic Waste Treatment—Anaerobic Sludge Digestion

Anaerobic waste treatment is primarily employed to destroy large quantities of biodegradable organic solids, although it can be applied for the economical removal of mass quantities of soluble organics.

Anaerobic biological treatment systems are an effective method of treating two specific types of wastes:

- Biological waste sludge from secondary treatment systems, commonly referred to as *anaerobic sludge digestion.*
- *Anaerobic waste treatment* of high strength biodegradable wastes, containing dissolved organics, suspended organic matter, or both.

Although anaerobically treating concentrated suspended matter or dissolved organics have some mass transfer differences, the concepts are similar and will be covered as one subject.

Anaerobic treatment can be explained in terms of two basic reactions, after substrate hydrolysis is achieved. The *hydrolysis* step is a preparation step in which solids and complex dissolved substrate are hydrolyzed into simple organic components. The first reaction step uses acid-forming bacteria to convert the hydrolyzed organic material to volatile fatty acids (VFA), which are capable of being stabilized. The second reaction involves stabilization of these acids by converting them to methane and carbon dioxide. Substrate stabilization requires completion of the slower growing methane bacteria forming step because the initial steps do not remove the BOD or COD, rather they are converted to different species. Detention time and temperature are dominant process variables. Although a wide operating range has been reported for sustaining an anaerobic process, economical and performance considerations have generally limited the process to the 29 to 38°C (85 to 100°F) range and the production of mesophilic bacteria [2,9,10,12,15].

Because anaerobic treatment systems operate at elevated temperatures, they are effectively employed for high strength wastes with ultimate BOD values of at least 5000 mg/L, to generate adequate methane to heat an influent from 20°C (68°F) to 35°C (95°F). The required influent BOD concentration increases with decreasing influent temperature. Anaerobic systems allow the economical biological treatment of wastes whose loadings on aerobic systems could be prohibitive. An anaerobic system can stabilize a high percent of a substrate COD, converting it to methane gas which can be used as a fuel or energy primer; while utilizing no aeration energy, generating low solids production, and requiring minimum nutrient supplement.

These advantages are many times overshadowed by process stability and operability concerns in applications *other than sludge digestion.* The scarcity of industrial installations has resulted in less practical waste treatment operating experience than with aerobic systems. Applications have generally been limited to wastes similar to sludge, such as fermentation spent broths, food, and pulp/paper wastes. Application to other wastes are sometimes discouraged because (1) of the great concern about the sensitivity of the process variables to waste toxic conditions, (2) long start-up periods, and (3) the process will not consistently produce an effluent quality suitable for direct discharge. In fact, the supernatant from anaerobic treatment systems will many times require further treatment.

However, when the principles of anaerobic treatment are understood and satisfactorily applied, these systems are extremely competitive for their intended application. This includes (1) the economical destruction of high solid loadings for sludge reduction and stabilization, (2) treating high strength wastes to achieve an effluent that can be used for its nutrient value, or (3) as a pretreatment to reduce the load to secondary treatment systems. The response to the concern of large and possible unstable systems to industrial treatment has been continued improvements to the conventional digesters, to more effective high rate systems resembling activated sludge systems, and to new fixed bed technologies allowing treatment of lower concentrated wastes. Suspended growth and fixed film systems are illustrated in Figures 7.5 to 7.11.

BASIC CONCEPTS

Anaerobic treatment theory, limitations, and application has been thoroughly discussed in the literature, with considerable detail presented in a series of articles published by McCarty [2,9–12,15]. Although the actual process is complex, the chemistry is often represented as the sequential series of steps illustrated in Figures 7.1 and 7.2.

Sustaining an anaerobic treatment system requires maintaining the necessary bacteria for the individual reactions. After hydrolysis, acid forming bacteria initiate the reaction by forming organic fatty acids, releasing energy for continued growth, and forming additional cells. Organic acid production is a critical step because these acids are basic to the final stabilization step, methane gas and water production by methane forming bacteria. The methane bacteria developed are related to the variety of fatty acids formed, involving species compatible to the variety of organic acids present. Although complex organic breakdown can result in a variety of acids forming, the two most important acids formed, directly or through intermediate compounds, are acetic and propionic acids. The significance of acetic acid and propionic acid formation is illustrated in Figure 7.3, which tracks the probable route of 100 units of complex organic waste. These compounds are readily attacked by the methane forming bacteria, producing the primary waste stabilization mecha-

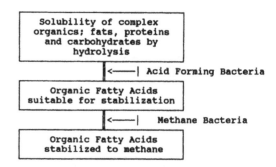

Figure 7.2 Anaerobic sequential steps.

nism. The system is in balance when the methane bacteria stabilize the intermediate acids as rapidly as formed.

REACTOR PROCESS KINETICS

Reactor process kinetics for anaerobic treatment follow the same logic as that discussed for aerobic treatment, with the design criteria being related to the cell growth rate and the solids retention time (SRT), expressed as follows:

$$dX/dT \cdot 1/X = \mu = 1/SRT \qquad (7.1)$$

If the Monod model is selected to represent the cell growth rate, substrate utilization and sludge generation can be represented by Equations (7.2) and (7.3), similar to those proposed for aerobic treatment [6].

$$\frac{dS}{dt} = \frac{k \cdot S \cdot X}{Ks + S} \qquad (7.2)$$

$$\frac{dX}{dt} = Y \cdot \frac{dS}{dT} - kd \cdot X \qquad (7.3)$$

where dS/dt equals the rate of organic removal, mass/volume-time, S equals the substrate concentration, mg/L, k equals the maximum substrate removal rate, mass per day-mass bacteria, K_s equals the half-velocity constant, mg/L, X equals the cell concentration, mg/L, dX/dt equals the cell

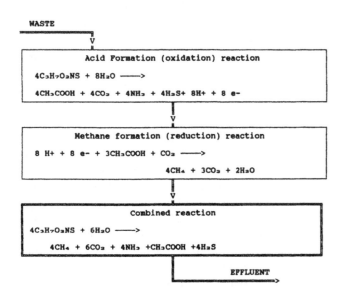

Figure 7.1 Anaerobic treatment reactions (adapted from Reference [25]).

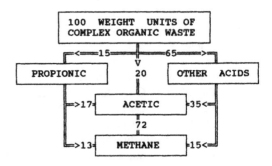

Figure 7.3 Acid utilization steps (adapted from Reference [9]).

growth rate, mass/volume-time, Y equals the cell yield, mg bacteria/mg substrate, and kd equals the decay rate, 1/time.

In turn, the cell growth rate can be related to the SRT as indicated by Equation (7.4).

$$\frac{dX}{dt \cdot X} = \frac{1}{\text{SRT}_{\text{eff}}} = \frac{Y \cdot K \cdot S_e}{K_s + S_e} - b \qquad (7.4)$$

where S_e is the required effluent concentration.

Equation (7.5) defines the minimum SRT (SRT_{min}), the "washout" condition where the influent (S_o) and effluent (S_e) concentrations are equal.

$$\frac{1}{\text{SRT}_{\text{min}}} = \frac{Y \cdot K \cdot S_o}{K_s + S_o} - k_d \qquad (7.5)$$

In the special case where $S_o >>> k_s$, a more conservative limiting condition is defined by Equation (7.6).

$$1/\text{SRT lim} = Y \cdot K - k_d \qquad (7.6)$$

Commonly cited kinetic constants are indicated in Table 7.1.

When anaerobic treatment is considered for high complex soluble organics, an analogy can be developed between aerobic and anaerobic treatment, relating the performance of each to the applied substrate loading. When the substrate concentration is small compared to the biomass, so that the biomass concentration is not limiting, the loading to an anaerobic system has been related to substrate removal efficiency S_e/S_o according to Equation (7.7) [7].

$$F_r/M_v = K\, S_e/S_o \qquad (7.7)$$

where F_r is the substrate removal rate, mass per day, M_v is the biomass quantity, mass, S_e is the substrate concentration, and S_o is the effluent concentration.

K is a constant which for paper and pulp anaerobic treatment varies from 2 to 5 (7 to 17 for the equivalent aerobic treatment), which is explicit for each industrial waste.

As in aerobic treatment, the utilized F_r/M_v can be expressed as the relationship:

$$F_r/M_v = Q\,(S_o - S_e)\,/\,X \cdot V \qquad (7.8)$$

or

$$F_r/M_v = (S_o - S_e)\,/\,X \cdot t \qquad (7.8a)$$

where t is the hydraulic retention time, days.

Equation (7.9) defines the relation between F_r/M loading, the quantity of organic removed per day per unit mass of biomass in contact with the substrate, and SRT.

$$1/\text{SRT} = a\, F_r/M_v - b \qquad (7.9)$$

TABLE 7.1. Anaerobic Treatment Kinetic Constants (adapted from References [2,7,15,20]).

Domestic Sludge
$K = 6.67$ g COD/g VSS-day $\cdot\ 1.035^{(T-35)}$
$K_s = 1.8$ g COD/L $\cdot\ 1.112^{(35-T)}$
$k_d = 0.03$ 1/day $\cdot\ 1.035^{(T-35)}$
$Y = 0.04$ to 0.054 g VSS/g COD

General Wastes General Values for Methane Fermentation[2]
$K = 6.67$ g COD/g VSS-day $\cdot\ 1.035^{(T-35)}$
$K_s = 2224$ mg COD/L $\cdot\ 1.112^{(35-T)}$
$k_d = 0.010$ to 0.040/day
$Y = 0.040$ to 0.054 mg VSS/mg COD

General Wastes
General Values for Organic Classes

	Y mg VSS$_p$/mg COD$_r$	kd mg VSS$_d$/mg VSS-day
Carbohydrates	0.01–0.18	0.02–0.03
Acetic acid	0.04–0.06	0.011–0.015
Mixed organic acids	0.04–0.06	0.015

VSS$_p$	is the volatile solids produced
COD$_r$	is the COD removed (influent minus effluent)
VSS$_d$	is the volatile solids destroyed
VSS	is the volatile solids in the reactor
t	Temperature, °C

The following related terms illustrate the similarities between Equations (7.4) and (7.9).

$$Y = a$$

$$F_r/M_v = k \cdot S_e/(K_s + S_e)$$

$$K_d = b$$

Researchers utilize these similar relations to emphasize an analogy between the kinetics of aerobic and anaerobic treatment, and to explain fixed film anaerobic reactor performance [7]. However, although the concept of SRT and cell growth is conventionally used for suspended growth anaerobic reactors, it is difficult to apply directly to fixed-film system design.

SOLIDS RETENTION TIME (RETENTION TIME)

SRT represents the conventional parameter selected for suspended growth anaerobic treatment system design. SRT is defined as the total weight of suspended solids in the treatment system divided by the suspended solids leaving the system, as indicated by Equation (7.10).

$$SRT = L_i/L_o \qquad (7.10)$$

where L_i represents all the solids in the reactor, transfer lines, and separator and L_o represents the waste sludge rate (excess sludge and supernatant).

A primary consideration in anaerobic treatment design is the slow growing microorganisms, relative to an aerobic system, and the corresponding hydraulic retention times (HRTs) needed to achieve the required treatment efficiency and prevent microorganism washout. The HRT equals the SRT when they cannot be separately controlled. The relation among the design (SRT_d), theoretical (SRT_e), washout (SRT_m), and Limiting (SRT_l) SRTs is illustrated in Figure 7.4, demonstrating the application of safety factors (SFs).

Operating variables affecting the theoretical SRT and the applied safety factor have been explored by Parkin and Owen and are illustrated in Tables 7.2 to 7.4 [15]. Table 7.2 illus-

trates the effect of operating temperature on required SRT and the resulting SF. As illustrated in Table 7.3, the mixing efficiency and the resulting effective reactor volume impact the process SRT, measured as the available safety factor applied to the theoretical (SRTe), limiting (SRTl), and minimum (SRTm) times. Critical temperature and mixing conditions producing washout conditions in Tables 7.2 and 7.3 are indicated by *safety factors of less than 1*. Finally, too high a solids residence time increases reactor volume and costs, but does not significantly improve the process efficiency, as indicated in Table 7.4.

As indicated in Table 7.4, three *theoretical* SRT values define the process limits. The minimum value to meet the effluent requirement is 11. Washout and limiting conditions (SRT 4.2 to 5.1) define the range where all the bacteria are removed faster than they can be replaced to the point where the system cannot sustain sludge growth. Clearly, if the system is to operate efficiently and reliably it must be operated at a SRT in excess of these critical values. It has been suggested that the design SRT be at least 10 times the minimum SRT (SRT_m) washout value for conventional digesters and 3 to 10 for high-rate systems [6].

A design SRT can be *approximated* from available municipal experience, using available data to adjust for specific industrial characteristics, or obtained directly through pilot studies for the specific waste. In either case, a *suitable SRT* must be selected, consistent with the feed and temperature range to assure a viable cell growth to sustain the process. In a conventional suspended growth system the HRT and SRT are equal, which could result in a large reactor and corresponding high capital costs. This problem can be overcome by using a system which allows independent solids and HRT control to maintain a high SRT with a relatively low HRT.

ANAEROBIC TREATMENT SYSTEMS

Anaerobic sludge treatment is a common basis for comparing any anaerobic treatment application because it is the

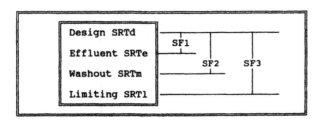

Figure 7.4 Applied Safety Factors

TABLE 7.2. **Effect of Temperature on Safety Factor (adapted from concepts in Reference [15]).**

Design basis: SRT_d = 20 days
S_o = 10 g/L COD
Removal, % = 85
Based on sludge constants indicated in Table 17.1

				Effective Safety Factor		
				$\dfrac{SRT_d}{SRT_l}$	$\dfrac{SRT_d}{SRT_m}$	$\dfrac{SRT_d}{SRT_e}$
Temp, °C	SRT_l	SRT_m	SRT_e			
25	6.0	9.7	47.5	3.4	2.1	0.4
30	5.0	6.8	20.6	4.0	2.9	1.0
35	4.2	5.1	11.0	4.7	3.9	1.8

TABLE 7.3. Effect of Mixing on Safety Factor (adapted from concepts in Reference [2,23]).

Temp, °C	% Vol Used	Real SRT$_r$	Effective Safety Factor		
			$\dfrac{SRT_r}{SRT_l}$	$\dfrac{SRT_r}{SRT_m}$	$\dfrac{SRT_r}{SRT_e}$
25	100	20	3.4	2.1	0.4
25	50	10	1.7	1.0	0.2
25	25	5	0.8	0.5	0.1
30	100	20	4.0	2.9	1.0
30	50	10	2.0	1.5	0.5
30	25	5	1.0	0.7	0.2
35	100	20	4.7	3.9	1.8
35	50	10	2.4	2.0	0.9
35	25	5	1.2	1.0	0.5

*Design basis: same as Table 7.2.

oldest applied system, with substantial available operating experience. An important consideration driving sludge digestion is final solids disposal costs. *Anaerobic waste treatment* design is driven by final regulated effluent criteria, with the system considered as a viable alternative to aerobic systems in special cases. Accordingly, suspended growth systems can be grouped into primarily sludge digestion or waste treatment processes, with special fixed film processes developed to optimize specific waste treatment applications.

ANAEROBIC SLUDGE DIGESTION

Anaerobic sludge digestion usually refers to municipal waste sludge treatment plants, detailed in many design texts [2,8,20]. Industrial and municipal sludge systems·are generally assumed to be similar and the same design guidelines applied, although this may not always be the case. In fact, the same process descriptions are often applied to anaerobic systems, regardless of the waste type. *Standard* and *High-rate* digesters configurations are frequently employed, with

enhanced *two-stage* and *contact* systems presenting some cost or process advantages and *phase separation* a more recent improvement. Municipal design texts devote a considerable amount of discussion on this subject, and the reader is referred to the cited texts for process and system details. Common sludge processes are summarized in this section.

Standard (Low) Rate Digestion System

This system is basically a feed-and-draw process, consisting of a tank into which the raw sludge is intermittently introduced and gradually digested. The digester contents become stratified, the stabilized sludge sinking to the bottom and periodically withdrawn. Above the settled sludge, the digesting process continues with more feed solids being treated. The supernatant layer is above the digesting sludge, above which exists a floating scum layer, both of which are periodically removed. Gas collects above the liquid layer, withdrawn to a gas handling system for process utilization. The digesting process is relatively slow, resulting in a low-capacity reactor. This system is sometimes added as a second stage thickener to a high rate digester, although this may not be practical if the feed is prethickened. When employed as a second stage, gas generated is withdrawn from both stages, with digested sludge removed from the second tank. This system is illustrated in Figure 7.5, and design criteria are summarized in Table 7.5.

High-Rate Digestion: Single Stage

The standard digester can be upgraded to a high-rate system, by employing reactor mixing and heating its contents to the 85 to 95°F mesophilic range. The faster digestion rate results in a reduced detention time, increased applied loading, and reduced reactor volumes. The digester volume can sometimes be reduced by sludge thickening. However,

TABLE 7.4. SRT vs. Removals.

Srt	Theoretical Removal, % @35°C
4.2	Limiting SRT
5.1	Washout SRT
6	50
7	67
8	75
9	80
10	83
11 <-SRT$_e$	85
15	90
20	92
25–30	94
35–40	95
45–50	96

Waste Properties
$S_o = 10{,}000$ mg/L COD
$Y = 0.04$
$k_d = 0.03$
$k = 6.67$
$K_c = 1800$

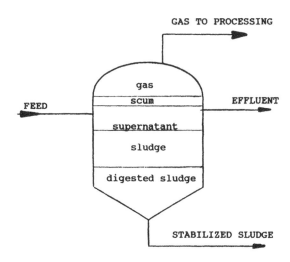

Figure 7.5 Standard rate digester.

TABLE 7.5. **Typical Standard Rate Design Criteria** (adapted from References [2,23]).

Often Unheated
30 to 60 Days detention time (heated)
0.6 to 1.6 kg VSS/m³/day
(0.04 to 0.10 lb VSS/ft³/day)
Intermittent feeding and withdrawal
Stratification

TABLE 7.6. **Typical High Rate Design Criteria** (adapted from References [2,23]).

Heated to 85°–95° degrees
10–20 days detention time
1.6–3.2 kg VSS/m³/day
(0.1–0.2 lb VSS/ft³/day)
Continuous or intermittent feeding
Homogeneity

this could inhibit biological activity or cause difficulties in mixing limit feed concentrations above 8 to 9% [23]. The process can be operated as a fill-and-draw or a continuous-feed system. Because the high rate process entails complete mixing, no supernatant layer is formed, and separate tanks are required if the sludge and supernatant are to be discharged separately. A single-stage digester is illustrated in Figure 7.6, and design criteria are summarized in Table 7.6.

High-Rate Digestion: Two Stages

Further municipal digester development led to the two-stage anaerobic digester system, where two stages are sequentially operated, with only the first stage mixed and heated. The second stage is stratified, the digested sludge settling to the bottom, the supernatant above the sludge, and the scum layer on the top. Gas is generated and removed from both stages, with digested sludge removed from the bottom of the second tank. Design criteria are a combination of those required for low- and high-rate digesters.

Phase Separation

Current developments include separate control of the acid- and methane-forming stages in an effort to reduce the overall

digester volume and increase process controllability. This system is actively being investigated in an effort to improve commercial municipal anaerobic treatment facilities.

ANAEROBIC WASTE TREATMENT SYSTEMS

In theory, any of the sludge digestion configurations discussed can be used for industrial waste anaerobic treatment, except that waste treatment could be a more formidable process design, and the effluent can seldom be directly discharged without further treatment. The major continuous systems applicable to industrial waste treatment include conventional low rate, anaerobic contact, and fixed bed technologies such as the upflow anaerobic sludge blanket, the anaerobic biofilter, and the anaerobic fluidized bed systems.

Dispersed Growth Anaerobic Systems

Theoretically, dispersed growth treatment can be applied to any influent concentration, the only requirement being that, like its aerobic counterpart system, a suitable microorganism inventory be maintained. Successful employment of a dispersed configuration depends on avoiding cell washout in an environment of slow methanogenic bacteria growth, and optimizing the cell concentration to minimize the required reactor volume. These requirements can be related to the reaction kinetics basics previously discussed and specific operating limitations:

(1) Microorganism cell growth rate is equal to the reciprocal of the solids retention time, as defined by Equation (7.4). The SRT is the principal parameter in designing a dispersed growth reactor.

(2) Because of the low cell yield in anaerobic systems, excess sludge generation is much lower than in a similar aerobic system. As an example, about 30 mg/L or less solids would be generated in a once-through system for a 1000 mg/L feed BOD, while more than 300 mg/L would be generated in an aerobic system. This is beneficial in minimizing by-product disposal, but could result in a reactor washout if the supernatant solids concentration cannot be controlled. The effect of low solids generation is compounded by process conditions promoting gasification and floc shear, favoring high solids entrainment and carryover. The result is that dispersed growth

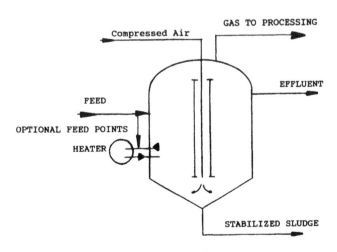

Figure 7.6 High rate digester.

systems can only be employed at high influent concentrations, usually well above that required to generate adequate methane to sustain the process.

(3) As with all biological systems the sludge retention time (SRT) and the hydraulic retention time (HRT) are critical design factors. The SRT/HRT ratio establishes the reactor volume, with a high ratio favored to stabilize the system, reduce the reactor size, and minimize sludge generation. However, in a once-through system this ratio is 1, resulting in a maximum, and in some cases prohibitive, reactor volume. The volume can be minimized by separately controlling SRT by including the capabilities to recycle sludge. Although this is the premise of the anaerobic contact system, the settling step required to incorporate recycling capability is more difficult than in the aerobic activated sludge system because of the gasification and sludge shearing conditions that promote solids carryover.

Although these conditions *do not* prohibit the use of dispersed systems for soluble organic wastewaters, they tend to drive the selection process toward newer contact technologies that accommodate biomass accumulation, encourage low solids carryover, and operate at a low F/M (high SRT) ratio.

In any of these processes influent pretreatment, nutrient addition, effluent post treatment, or further sludge processing prior to disposal may be required. Where effluent is directly discharged post treatment is usually necessary. At a minimum, effluent suspended solids reduction employing flotation, centrifugation, or sedimentation processes will be required.

Dispersed growth reactors should contain adequate mixing and be designed with provisions for future sludge recycle (if not part of the design) to allow for improved control. Gases from the system are discharged to a gas filter and then to methane utilization equipment, with excess gases usually discharged to a flare. Dispersed growth reactors can be operated as continuous flow-through or include recycle.

Anaerobic Lagoons (Sludge, Slurry, Soluble Substrates)

Anaerobic lagoons are once-through, low-rate, conventional treatment systems designed to degrade suspended solids and high strength wastes. They are anaerobic because they are deep and not aerated. Unfortunately, many times lagoons are anaerobic because of poor operation, not by design. They are easy to construct and operate, involving low capital and operating costs. Their inexpensive construction must be offset with large land area requirements, and their low operating costs with periodical sludge removal costs to restore the pond. Lagoons are commonly of earthen construction with a synthetic membrane liner, a membrane cover to collect the generated gases and maintain anoxic conditions, and with no means for sludge management.

Lagoons are usually designed for a minimum of 7 to 10 days retention, with 75 to 90% BOD removal possible, although performance is more an "occurrence" than a controllable objective. Data reported for two-stage lagoons treating recycled paper are summarized in Table 7.7 [7].

The major disadvantages of anaerobic lagoons are the inability to control performance and problems resulting from gases escaping from large enclosures that are difficult to maintain. Fugitive emissions present a potential air pollution or safety problem, or at least an odor nuisance. In addition, unless the influent is high strength the process operates at low SRT values, making the lagoons extremely susceptible to toxic upsets. These systems have limited industrial application because of stringent effluent discharge requirements, large land requirements, and the potential fugitive emissions associated with these systems.

Conventional Low Rate Anaerobic Treatment (Sludge, Slurry, Soluble Substrates)

Conventional anaerobic treatment systems are an upgrade of anaerobic lagoons employing closed vessels, heated and partially mixed to increase reactor effectiveness. Wastes can be added continuously or intermittently, functioning as a fill-and-draw or continuous flow-through system. They require large reactor volumes because sludge recycle is not employed, the SRT being equal to the HDT, operating in a manner similar to the standard rate digester previously discussed. Unless operational procedures or downstream equipment is included to separate the effluent from the biomass this process is limited to sludge concentration and stabilization. This system is illustrated in Figure 7.7. Performance for industrial facilities is indicated in Table 7.8 [12,13]. These systems are usually applied for small waste volumes with high influent concentrations to overcome the limitations discussed.

Anaerobic Contact System (Soluble Substrates)

The main elements of this system include a reactor, effluent degasification, solids separation, and solids recycle to maintain the desired SRT [4,7,13,19]. The system is similar

TABLE 7.7. Two-Stage Paper Plant Lagoons (adapted from Reference [7]).

Location:	Indiana	So. Carolina
Volume of pond, m³ (MG):	18,925 (5)	90,840 (24)
Influent, m³/d (MGD):	2612 (0.69)	15,140 (4)
Temperature, °C:	43–48	32–38
Influent BOD₅, mg/L:	1898	667–705
Effluent BOD₅, mg/L:		
Anaerobic basin effluent	297	251–318
Aerobic basin effluent	89	21–24
Overall BOD₅ removal, %:	95	97

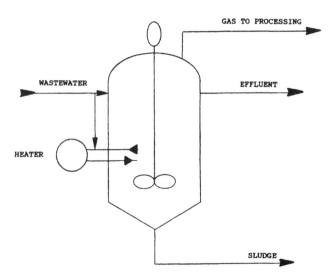

Figure 7.7 Conventional digester (adapted from Reference [19]).

Fixed-Bed Anaerobic Systems

An anaerobic filter utilizes a contact bed, allowing short retention times and high SRT resulting from biomass accumulating on the media [4,19]. Anaerobic beds are generally recommended for treating soluble organic wastes. Fixed beds allow higher biomass buildup due to entrapment in the media structure and therefore more effective substrate-to-mass loadings. However, they are prone to hydraulic short-circuiting, poor flow distribution, and plugging. Recycle is essential to assure good hydraulic distribution, minimize toxic effects, and minimize concentration gradients. They can be operated as upflow or downflow beds. The number of anaerobic filters treating industrial wastes is increasing, and the technology is rapidly being developed.

Upflow (or Downflow) Anaerobic Fixed-Bed Filter (Soluble Substrates)

Wastewater flows at a low velocity through an upflow anaerobic filter, allowing contact with attached biomass on an inert media such as plastic or rock bed [4,19]. The reaction chemistry is similar to that generally discussed for suspended growth systems. In many ways bed construction is similar to that of a trickling filter, sand bed, or activated carbon column. The basic elements consist of (1) an inlet waste distribution system, (2) media support, (3) inert packing material, (4) head space for methane collection and withdrawal, (5) effluent discharge nozzles, and (6) optional effluent recycle. Media selection requires consideration of adequate surface area to promote sufficient biomass growth and adequate void space to avoid plugging. Media maintenance includes provision for sludge wasting and media backwashing.

to an activated sludge system. HRT is decreased by increasing the system solids inventory and thereby separately controlling the SRT. Separate SRT control allows treatment of a wider range of influent concentrations, provided washout conditions are prevented and that the methane generated balances the energy requirements. Energy requirements include heating the influent to maintain the reactor temperature and replenish heat losses. High temperature wastes reduce the energy requirements, whereas the inherent lower reactor volumes minimize heat losses. Significantly, dilute wastes may not generate adequate methane to furnish the required system energy, requiring supplementary fuel, and thereby affecting process operating costs. Available industrial data are scarce; some reported performances are summarized in Table 7.9. The system is illustrated in Figure 7.8.

TABLE 7.8 **Anaerobic Conventional Data for Industrial Facilities (adapted from References [12,13]).**

Waste Type/Plant	Ref.	Parameter	Loading kg/m³/d	Loading lb/1000 cf/day	HRT days	Temp, °C	Stabilized kg/m³/d
Pea	12	VSS	11.2	700	3.5	55	9.3
Blancher	12	VSS	6.4	400	6.0	37	5.4
Winery	12	VSS	3.2	200		36	2.8
Butanol Fermentation	12	BOD₅	1.8	110	10.0		1.2
Rye	12	TSS	14.9	930	2.0	54	8.0 BOD₅
Corn	12	TSS	5.3	330	4.0	54	4.0 BOD₅
Whey Waste	12	TSS	2.4	150	29.0	54	1.7 BOD₅
Acid							
Acetic	12	BOD₅	21.9	1370	30.0	35	14.0
Butyric	12	BOD₅	13.3	830	30.0	35	14.6
							Removal, %
Sugar	13	BOD	0.6	39	4.0	37	60
Yeast	13	BOD	1.7	106	3.9	35	70

TABLE 7.9 Data for Anaerobic Contact Industrial Facilities
(adapted from References [7,12,13]).

Waste Type/Plant	Ref.	Influent, mg/L	Loading		HRT days	Temp, °C	Removal, %
			kg/m³/d	lb/1000 cf/day			
Starch	12	6280 BOD₅	1.8	110	3.3	23	88
Whisky	12	25,000 BOD₅	4.0	250	6.2	33	95
Cotton	12	1600 BOD₅	1.2	74	1.3	30	67
Citrus	12	4600 BOD₅	3.4	214	1.3	33	87
Brewery	12	3900 BOD₅	2.0	127	2.3	—	96
Starch	12	14,000 VSS	1.6	100	3.8	35	80
Wine	12	23,400 VSS	11.7	730	2.0	33	85
Yeast	12	11,900 VSS	6.0	372	2.0	33	65
Molasses	12	32,800 VSS	8.8	546	3.8	33	69
Meat	12	2000 BOD₅	1.8	110	1.3	33	95
Meat	12	1380 BOD₅	2.5	156	0.5	33	91
Meat	12	1430 BOD₅	2.6	164	0.5	35	95
Meat	12	1310 BOD₅	2.4	152	0.5	29	94
Meat	12	1110 BOD₅	2.1	131	0.5	24	91
Meat	13	— BOD₅	3.2	200	0.5	30	95
Meat	13	— BOD₅	2.5	156	0.55	35	90
Pulp Mill	7	3500 COD	2.5	156			67
Pulp Mill	7	30,000 COD	4.8	300			66
Pulp Mill	7	10,000 COD	4.3	268			65
Pulp Mill	7	6000 COD	4.2	262			85
Pulp Mill	7	4800 COD	2.7	169			77
Pulp Mill	7	7900 COD	6.0	374			40
Pulp Mill	7	6400 COD	4.0	250			—
Pulp Mill	7	20,000 COD	—	—			—
Pulp Mill	7	10,000 COD	3.0	187			49

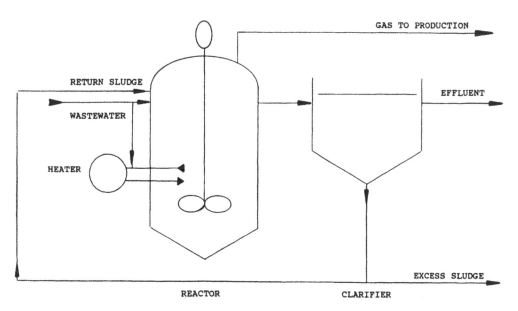

Figure 7.8 Anaerobic contact system (adapted from Reference [19]).

157

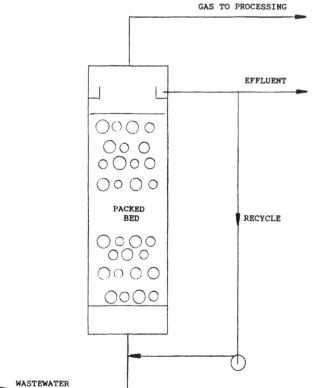

Figure 7.9 Anaerobic fixed bed filter (adapted from Reference [19]).

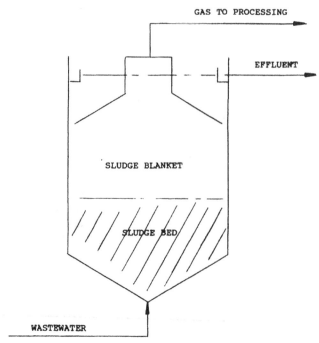

Figure 7.10 Anaerobic upflow sludge blanket (adapted from Reference [19]).

An advantage of a downflow process is that it can operate as a two-stage system, tolerating more inorganic sulfur because of separation of the more rapid sulfur reducing bacteria at the top and slower growing methanogenic bacteria at the bottom. The gas produced benefits the top section by stripping hydrogen sulfide [7].

The system is illustrated in Figure 7.9. Some reported performance data are summarized in Table 7.10.

Upflow Anaerobic Sludge Blanket (UASB)

This process resembles an upflow reactor except that it utilizes an *active sludge bed* at the bottom, instead of packing [7,19]. Bed characteristics are critical and include (1) main-

TABLE 7.10 Anaerobic Fixed Bed Industrial Facilities
(adapted from References [4,7,13]).

| Waste Type/Plant | Ref. | Influent, mg/L COD | Loading | | HRT days | Temp, °C | Removal, % |
			kg/m³/d	Media type			
Starch	4	8800	3.8	Graded rock	0.9		64
Guar gum	4	9140	16.0	Pall rings	1.0		60
Rum	4	95,000	8.9	Vinyl core	7–8		75
Chemical	4	12,000	9.6	Pall rings	1.5		80
Chemical	4	14,400	10.4	Pall rings	1.5		90
Soft drink	4	6000	9.6	Sand	0.25		77
Soy Process	4	9000	13.0	Sand	<1		—
Laboratory studies							
Synthetic	13		1.0			25	90
Food	13		1.6		3.5	35	86
Drug	13		3.5		2.0	35	98
Synthetic	13		2.5		4.0	35	92
Guar gum	13		7.4		1.0	37	60
Pulp mill	7	7900	12.7				70
Rum	7	75,000	12.8				50

taining a 1 to 2 meter (3 to 6 ft) depth, (2) a blanket containing 8 to 13% solids with a 60 to 90% volatile content, and (3) good settling particles in the range of 0.5 to 2.5 mm in size.

A secondary flocculent sludge layer formed on top of the blanket contains from 3000 to 10,000 mg/L solids. Wastewater flows into the bottom of the reactor, upward through the blanket and flocculent sludge layers, where the anaerobic process occurs. Entrained solids are removed by passing the gas through a settler, which separates the three phases (effluent, entrained solids, and gas) prior to the gas exiting to a collector. This system is illustrated in Figure 7.10. Some reported performance data are cited in Table 7.11.

Anaerobic (Expanded) Fluidized Bed

The process consists of fixed biomass growing on a suspended fluidized media [4,19]. The flow is evenly distributed through the bed bottom, fluidizing the bed. Bed volume consisting of media and biomass is constantly removed, treated to separate the biomass, and the media returned. Waste recycle is maintained to assure a controllable flow consistent with the system hydraulics and to assure that the influent variations are minimized.

The system components are similar to an upflow anaerobic filter, with provisions for bed fluidizing and continuous bed maintenance. Media investigated include PVC, sand, granular carbon, alumina oxide, and diatomaceous earth, essentially those used in sand filtration, ion exchange, and activated carbon adsorption. This system is illustrate in Figure 7.11. Some reported performance data are cited in Table 7.12.

PROCESS ENGINEERING DESIGN

The operating characteristics of an anaerobic system are summarized in Table 7.13. Anaerobic treatment is an extremely effective biological degradation process for both sludge and concentrated organic wastes, involving minimum net energy input by recovering methane rich generated gases. Anaerobic treatment systems possess three inherent qualities that affect process applicability and stability: (1) the system must be maintained oxygen free, (2) slow microorganism growth requires high feed concentrations and maintaining

TABLE 7.11 Anaerobic Upflow Sludge Blanket (USAB) Facilities (adapted from References [4,7,13]).

Waste Type/Plant	Ref.	Influent, mg/L COD	Loading kg/m³/d	Loading lb/1000 cf/day	HRT days	Temp, °C	Removal, %
Brewery	4	2500	14.1	880	0.20		86
Starch	4	22,000	11.0	686	1.96		85
Potato	4	4300	6.0	374	0.73		80
Sugar	4	17,000	13.3	830	1.0		94
Sugar beet	4	3000	14.5	905	0.20		85
Starch	4	7700	8.0	499	0.10		85
Potato	4	2260	8.3	518	0.27		80
Sugar beet	4	7500	12.0	749	0.63		86
Alcohol	4	5330	16.0	998	0.33		90
Sugar	13	Pilot	22.5	1404	0.25	30	94
Potato	13	Pilot	25–45	1560–2810	0.17	35	93
Beet sugar	13		16.0	998	0.17	35	88
Boxboard	7	6300	9	562			70
Corrugating	7	4500	20	1248			75
Tissue	7	1200	5	312			60
Corrugating	7	1100	6	374			70
Linerboard	7	2880	9	562			75
Box board	7	2500	10	624			65
Corrugating	7	3550	8.5	530			75
Corrugating	7	15,000	15	936			80
Corrugating	7	2500	8.5	530			70*
Fine paper	7	4000	7	437			75*
TMP/CIMP	7	7000	19	1186			35*
NSSC	7	16,000	20	1248			50*
Waste paper	7	2800	10	624			75*
NSSC/CIMP	7	12,000	11.9	743			60*
NSSC/TMP	7	25,000	18.5	1154			50*

*Design values operating. Data not available.

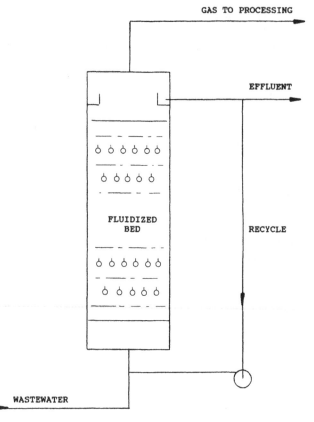

Figure 7.11 Anaerobic (expanded) fluidized bed (adapted from Reference [19]).

TABLE 7.13. Operating Characteristics.

	Waste Characteristics	
Variable	Operator Controllable	Critical
Waste generated	No	Yes
Composition	No	Yes
Concentration	No	Yes
Biodegradability	No	Yes
Toxicity	No	Yes
Operating Characteristics		
Flow rate	Minimal	No
SRT	Yes	Yes
HRT	Minimal	No
Temperature	Yes	Yes
Oxygen free	Yes	Yes
VSS level	Yes	Yes
Sludge recirculation	Contact system	Yes
Mixing	Yes	No
Gas production	No	Yes
Wasting	Yes	Yes
Nutrients	Yes	Yes
Alkalinity	Yes	Yes

REPORTED PERFORMANCE DATA

Performance for the various anaerobic treatment systems applied for waste treatment and sludge digestion is discussed in the Anaerobic Treatment Systems section and available operating data included. Data for anaerobic waste treatment application is sparse and not widely reported. In addition, some of the fixed-film processes have not been as widely applied as more conventional treatment methods; as a result full-scale operating data are limited and where available not fully detailed. Operating data for municipal sludge stabilization have been widely reported, with two-stage digester performance reported in the EPA *Treatability Manual* [17] and summarized in Table 7.14.

REQUIRED PROCESS DESIGN DATA

Specific design data required to design an anaerobic system depend on whether the system is applied for sludge

elevated reactor temperatures, and (3) the influent concentration must be well over 5000 mg/L BOD so that energy is available to sustain the process. Operating flexibility can be enhanced by providing sludge recycle or employing fixed beds to accumulate biomass, allowing for SRT control. Waste sludge stabilization is an easier application than anaerobic treatment of concentrated industrial wastewaters because upstream biological processes provide some degree of equalization minimizing sludge variations, and solids concentration is easier to achieve than "targeted" effluent quality.

TABLE 7.12. Anaerobic Fluidized Bed Industrial Facilities (adapted from References [4,7]).

Waste Type/Plant	Ref.	Influent, mg/L COD	Loading kg/m³/d	Loading lb/1000 cf/day	HRT, days	Temp, °C	Removal, %
Soft drink*	4	6900	9.6	599	0.25		77
Soy process*	4	9000	13.0	811	<1		—
Paper board	7	3000	35	2184			72.2

*Sand media employed.

TABLE 7.14. Two-Stage Sludge Anaerobic System.
(adapted from Reference [17]).

Sludge	
Influent total solids, %	2–7
Sludge total solids, %	2.5–12
Total solids reduction, %	33–58
Volatile solids reduction, %	33–50
Supernatant	
Suspended solids, mg/L	200–15,000
BOD$_5$, mg/L	500–10,000
COD, mg/L	1000–30,000
TKN, mg/L	300–1000
Total phosphorus, mg/L	50–1000

TABLE 7.15. Required Design Data—Sludge Digestion
(adapted from Reference [17]).

Critical pilot plant treatability data specific to the waste
(1) Design temperature
(2) k_d coefficient, 1/days
(3) k_{max} dS/dt/wt, 1/time
(4) K_s concentration at 1/2 max, mg/L
(5) Y, excess cell yield, mg/mg

Waste solids characteristics that should be obtained from laboratory studies but can be estimated
(6) Fraction of solids biodegradable
(7) Ultimate solids mg BOD per mg solids
(8) Ratio of SBOD to ultimate BOD
(9) Ratio MLVSS/MLSS

SELECTED operating characteristics
(10) Required final sludge concentration
(11) Mixed liquor Volatile Suspended solids
(12) MRCT, calculated from treatability data
(13) Summer ambient temperature
(14) Winter ambient temperature

Operating characteristics that should be obtained from pilot studies but can be estimated from treatability data
(15) Gas production

digestion or waste treatment, although most of the requirements are similar.

Sludge Digester

A complete sludge analysis is paramount! The characterization should establish the volatile suspended solid, soluble BOD, COD, nitrogen, alkalinity, and toxic content. Sludge samples should be taken, and analyzed, until characteristics and variability can be established. When the sludge is generated in a suspended growth biological system, the secondary treatment plant and related operations should be investigated for their affect on sludge properties. At a minimum the operating SVI should be monitored, and its value for collected samples should be recorded. Other tests discussed in Chapter II-7 defining dewatering and settling qualities may prove beneficial. The benefit of sludge concentration to digester performance should be established so that the advantages of thickening can be evaluated.

Digester design criteria can be obtained from batch laboratory testing. Resulting data can be tabulated or integrated into appropriate digestion models to formulate design criteria over the applicable operating range, establishing the required design data listed in Table 7.15. Important information not obtained from batch tests are the digested sludge dewatering characteristics, which is a significant property affecting downstream equipment capacity. This information is commonly included as part of an overall sludge management study, specifically designed to establish dewatering effectiveness and disposal limitations.

Anaerobic Waste Treatment

To design an anaerobic treatment system the Process Engineer must have adequate performance data to assure that (1) the waste is biodegradable and does not contain inhibitory components, (2) adequate gas is generated to economically justify the process, (3) effluent quantity can be defined, and (4) a specific operating HRT or SRT range, associated with the expected operating temperature, has been established.

Anaerobic treatment criteria can be obtained using pilot or laboratory testing techniques similar to those discussed for once-through aerated lagoons or activated sludge [14]. The major difference from aerobic systems is that anaerobic processes operate at elevated temperatures. As with aerobic systems, inhibitory effects of influent components must be established as well as the affects of residence time (or solids loading) and temperature to performance. Table 7.16 lists the required design data.

Scale-up from laboratory bench-top testing is common. Where bench scale data are used it is essential that upset conditions be evaluated to determine whether the data developed are effective through the wide range of influent conditions encountered.

WASTE EVALUATION

As with any biological system waste characteristics will determine whether the process will be effective and can be

TABLE 7.16. Required Design Data—Anaerobic
Waste Treatment.*

Selected operating characteristics
  (1) Clarifier underflow MLSS, %

Operating characteristics that should be obtained from pilot studies, but can be estimated from treatability data
  (2) Waste sludge generated
  (2) Nitrogen required per mg SBOD removed
  (3) Phosphorus required per mg SBOD removed
  (4) Gas production

Required criteria are as indicated in Table 7.15, with these additions.

sustained. Anaerobic systems require that the waste (1) be biodegradable, (2) concentrated to generate sufficient methane and sludge growth, (3) devoid of toxic substances, and (4) reasonable consistent to assure stable performance. Factors governing waste biodegradability and selection or rejection of individual plant wastes to a central anaerobic treatment plant are similar to those discussed for aerobic treatment in Chapter II-2 and will not be repeated here. Waste flow, concentration, and toxicity are special properties that can influence anaerobic process performance and therefore warrant special design attention.

Flow Rate

Except for contact stabilization systems, flow rate is the primary design and operating variable in anaerobic dispersed growths plants as significant as the reactor temperature, the reason being that for any flow-through system the HRT is equivalent to the SRT, thereby controlling cell growth. Therefore, if a high flow rate results in a HRT (SRT) less than the minimum SRT, cell growth rate cannot be maintained, and the system will "washout." The Process Engineer must be aware of this and select a reactor volume that will allow operating in the expected flow range. In a contact stabilization system the SRT is determined by the plant recycle capabilities, which must be selected to assure an operating range consistent with the feed variability.

Concentration

Although waste biodegradability determines whether biological treatment is a viable process, organic concentration establishes whether anaerobic treatment is viable. Because of slow microorganism growth rate, anaerobic reactors are operated at elevated temperatures of 35°C (95°F) or higher. Although this can be accomplished with supplementary fuel, this may not be practical. Accordingly, waste concentrations should be higher than 5000 mg/L (ultimate BOD) to generate adequate methane gas to provide required energy.

In addition, the substrate concentration must be adequate to generate an effective solids inventory. Although this is usually not a problem in sludge digestion, it could be a significant limitation in applying low rate processes to treat substrates with biodegradable ultimate BOD values less than 20,000 to 30,000 mg/L [26]. As the size of a low rate reactor increases, an anaerobic system will be prohibitively large unless a bed or contact anaerobic system is employed to reduce HRT, resulting reactor volume, and large heat loss surface area. AT very low influent concentrations, the advantages of anaerobic treatment diminished because either not enough energy is produced or not enough cells are generated to maintain the system and compensate for system losses.

Toxic Substances

As with any biological system, toxic substances retard the anaerobic reactions, resulting in reduced performance or total system washout. The criteria listed in Chapter II-2 for biological systems should be reviewed for anaerobic systems, with emphasis on substances such as dissolved solids, sulfides, and metals. Although the toxicity effects in anaerobic systems are comparable to aerobic systems, there are some significant differences. The most critical difference being that *cell growth rate in anaerobic systems is inherently slower, resulting in prolonging toxic or inhibitory effects.* In addition, anaerobic systems are adversely affected by oxidants, with the most obvious *being oxygen,* and other compounds such as ozone and peroxides. The concerns about toxics and their effects has resulted in extensive study of specific contaminants and the recovery capabilities of anaerobic systems.

McCarty investigated the toxic effects of specific compounds on anaerobic systems, concluding that three levels of activities are displayed by anaerobic bacteria [10]. Low level toxicity stimulates biological activity and the reaction rate increases. At increasing concentrations, the stimulating effects of the compound are reduced and the reaction rate decreases toward its original rate. As the concentration increases there is a crossover point, and the reaction rate starts to deteriorate rapidly with increasing toxin concentration.

Investigators have conducted considerable studies evaluating the effects of and recovery from toxins. Although absolute results may vary, some general conclusions can be summarized as follows [3,11,15]:

(1) Anaerobic reactors *can* recovery from toxic or inhibitory compound exposure.
(2) Higher biomass concentration and younger cell age are conducive to conditions favoring rapid recovery from dormant, nonproductive methane conditions.
(3) Recovery from toxic conditions improves with increasing temperature.
(4) The nature of the toxicity, the system configuration, and process variables greatly influences the recovery response.
(5) Transient (short-lived) toxicity is minimized by a system that immediately diminishes slug toxic injection effects by maintaining microorganisms vitality, removing the compounds as soon as possible without washing out the biomass. This suggests a system with a short HRT and a high SRT such as a fixed film or a once-through system.
(6) Chronic toxicity is minimized in a completely mixed suspended growth system because of instant dilution and acclimation.
(7) The length of the exposure, toxicant concentration, and SRT establishes recovery time. Shorter recovery times are required for low toxicant concentrations and expo-

sure time, higher SRT and biomass concentration, and younger cell age.

(8) The sooner the toxicant is removed from the reactor the more rapid the recovery and the higher the toxicant concentration it can tolerate.

(9) If the toxicant cannot be rapidly removed, dilution will dampen the chronic effects of the compound. Dilution is especially useful for system acclimation to continuous toxic compound injections.

Specific toxicants identified as detrimental to anaerobic treatment vary between investigators due to the configurations and range of conditions considered. However, there is some general agreement as to the *range* of specific compound concentrations that should alert the Process Engineer to a potential process problem. In such cases the Process Engineer should (1) evaluate acceptable process configurations, (2) carefully consider the operating sludge age range, or (3) remove the specific compound from the feed. Besides oxidants such as *oxygen,* some common compounds identified as toxic or inhibiting include [3,11,12,15]

Alkali and alkaline cations
Ammonia
Sulfides
Heavy metals
Specific organics

As discussed for aerobic reactors in Chapter II-2, specific component toxicity is probably better related to biomass concentration than influent. For that reason, the implied influent toxicity levels should be viewed as *alert* values, highly dependent on system configuration, biomass adsorption, and other related process considerations. Values presented are from the cited references.

Alkali and Alkaline Salt Toxicity

Toxicity resulting from alkali and alkaline salts are usually the result of the cation and not the anion. Alert levels for these cations are indicated in Table 7.17 [12,15]. The combined effects of these cations, together or with other compounds, are difficult to predict, in some cases being synergistic, antagonistic, or acting independent of the other

TABLE 7.17. Alkali and Alkaline Salt Toxic Limits (adapted from Reference [12]).

Cation, mg/L	Moderately Inhibitory	Strongly Inhibitory
Magnesium	1000–1500	3000
Calcium	2500–4500	8000
Potassium	2500–4500	12,000
Sodium	3500–5500	8000

TABLE 7.18. Ammonia Toxicity Limits (adapted from Reference [12]).

Ammonia Nitrogen, mg/L	
Below 1500	May be necessary as a nutrient
15–3000	Inhibitory
Above 3000	Toxic

compounds. An important consideration is not only the waste cations but sodium or calcium added with the alkalinity control chemicals. Attempts to control reactor pH could result in intolerable cation levels.

Ammonia Toxicity

Depending on the system pH, ammonia can be present as the dissolved ammonia gas or ammonium ion. At pH values of 8 or higher the equilibrium is shifted toward the ammonium ion, with potential effects indicated in Table 7.18 [12,15].

There are some contrary data suggesting various levels of ammonia tolerances, highly dependent on the reactor configuration and conditions [15]. An example is the investigation by Bhattacharya and Parkin in which studies at SRT values of 15, 25, and 40 indicated tolerances of "slug" dosages up to 8000 mg/L total ammonia nitrogen at a SRT of 15 (decreasing with increasing SRT), and tolerances of continuous dosages up to 5000 mg/L at a 40 SRT level [3].

Sulfide Toxicity

Process sulfides are either a result of direct waste components or conversion of sulfur compounds. Sulfides can remain dissolved in the reactor liquor, precipitate out, or form the hydrogen sulfide gas. Hydrogen sulfide formation is pH dependent, with reactor concentrations affected by the exiting gas stripping. Alert soluble sulfide levels are indicated in Table 7.19 [12,15].

Lee et al. [7] discussed the potential of sulfide toxicity by noting that inorganic sulfur compounds can effect an anaerobic reactor but that organic sulfur compounds (such as encountered in the pulp and paper industry) may not. This is because organic sulfur compounds do not decompose in an anaerobic reactor. In addition, they noted that inorganic sulfur compound toxicity increases from sulfate to thiosulfate, to sulfite, to sulfide.

TABLE 7.19. Sulfide Toxicity Limits (adapted from Reference [12]).

Soluble Sulfide, mg/L	
100–200	Minor effect
Above 200	Toxic

TABLE 7.20. Heavy Metal Toxicity Limits (adapted from Reference [15]).

Compound	mg/L
Soluble Cu	0.5
Total Cu	50–70
Soluble Cr VI	3
Total Cr VI	200–260
Total Cr III	180–420
Soluble Ni	2
Total Ni	30
Soluble Zn	1

*Strongly inhibitory.

Heavy Metal Toxicity

Dissolved heavy metals such as copper, zinc, nickel, and hexavalent chromium are toxic. They are unreactive if precipitated; in solution they can be tolerated in low concentrations. Some alert concentration are indicate in Table 7.20 [15].

Toxic Organic Compounds

Anaerobic treatment is effective for complex organics such as fats, proteins, and carbohydrates. However, many industrial organics may demonstrate toxic characteristics because of their inability to hydrolyze into the intermediate fatty acid. These include organic solvents, alcohols, long-chain fatty acids, and chlorinated compounds.

Parkin and Owen [15] listed the compounds shown in Table 7.21 as potentially inhibitory. The concentration of the organic, along with the acclimation period could significantly effect toxicity. Toxicity effects will be minimized if the organic can be diluted to tolerant levels in the reactor or into an acclimated system which can readily reduce the organic to the fatty acid intermediate. In both cases, the organics can be tolerated in continuously fed system more effectively than in a batch system. However, it must be assumed that as with the aerobic system, anaerobic systems acclimated to high concentrations of potentially toxic organic components are potentially unstable systems. Such systems should not be subjected to highly variable feed or operating conditions.

Waste Preparation

Sustaining an anaerobic process requires adequate nutrients and optimizing the process pH to encourage cell production, both of which must be considered as part of the upstream waste preparation.

Nutrient Addition

The basic nutrients required to sustain an anaerobic treatment system are nitrogen and phosphorus, which could be

TABLE 7.21. Organic Compounds Toxicity Limits (adapted from Reference [15]).

Compound	Potential Inhibitory Effects
Propanol	
Resorcinol	
Aniline	
Phenol	
Catechol	
Acrylic acid	
Ethyl acetate	
Acetaldehyde	
Vinyl acetate	*Increasing inhibitory potential based on measured millimolar concentrations required for reduction of 50% activity.*
2-Chloropropionic acid	
Crotonaldehyde	
3-Chlorol-1,2-Propandiol	
Acrylonitrile	
Ethyl benzene	
Lauric acid	
Formaldehyde	
1-Chloropropane	
Acrolein	
Nitrobenzene	
1-Chloropropene	

Inhibitory effects were also observed from organic compounds such as chloroform, ethylene dichloride, kerosene, and linear ABS (detergent).

deficient in industrial wastes. The quantity required is proportional to the sludge growth, which for design purposes nitrogen requirements can be estimated at 11% of the dry sludge produced and phosphorus at 2% [12]. The effect of nitrogen nutrients on the system pH, as discussed subsequently, will have to be evaluated in estimating feed requirements. In addition, as in aerobic systems other inorganics are required to maintain an anaerobic system. Some suggested nutrients include 1 to 5 ppm of iron and nickel and approximately 50 ppb of molybdenum, cobalt and selenium [7].

Alkalinity/pH Range

The pH range for sustaining an anaerobic treatment system is considered to be between 6 and 8 [10]. Methane production is directly related to pH, with evidence that methane production will drop below 6.5 to 6.8 and ceases completely below 6 and above 8.5, if the condition is prolonged [7]. Generally, a system is viable if an adequate bicarbonate alkalinity can be maintained to buffer the carbon dioxide released from the reactor. This is in agreement with conclusions reported by Parkin and Owen [15], who explored the McCarty interrelation of carbon dioxide concentration, pH, and bicarbonate alkalinity [10]. They demonstrated that a well-balanced anaerobic treatment system will operate at an optimum pH range of 6.5 to 7.6, at carbon dioxide concentrations of 25 to 40 vol %, and bicarbonate alkalinity concentrations of 1000 to 5000 mg/L $CaCO_3$.

In stable systems, an optimum pH range is maintained by a natural buffering capacity. The bicarbonate alkalinity being dependent on the reaction of carbon dioxide with a cation, as follows:

$$CO_2 + \text{cation} = HCO_3^-$$

This is usually accomplished with a breakdown of organic nitrogen present as part of the waste or sludge component or present as waste compounds such as soaps or organic salts, as follows [15]:

$$\text{Nitrogen compound} \rightarrow NH_3 + CO_2 \rightarrow NH_4 + HCO_3^-$$

If these carbonate forming compounds are not present, they must be added or adequate alkalinity will not be available.

Bicarbonate alkalinity in solution can be related to the digester gas carbon dioxide according to the relation:

$$H_2O + CO_2 \longleftrightarrow H_2CO_3 \longleftrightarrow H^+ + HCO_3^-$$

In turn, the system pH is hydrogen concentration related,

which can be stated in terms of the carbonic acid equilibrium constant:

$$[H^+] = K \frac{[H_2CO_3]}{[HCO_3^-]}$$

The balance among bicarbonate alkalinity, pH, and carbon dioxide concentration can be expressed by Equation (7.11) [15].

$$\text{Balk (mg/L } CaCO_3) = 0.00063 \times \frac{PCO_2, \text{ partial pressure (atm)}}{10^{-pH}} \quad (7.11)$$

A suggested correction of the constant 0.00063 to 0.00128 to account for ionic strength of the constituents has been reported [15]. In a balanced system the volatile acids formed are rapidly converted to methane, resulting in minimum volatile acid concentration, so that the total alkalinity in the system is essentially the bicarbonate alkalinity.

During unstable conditions the volatile acids formed are neutralized, reducing the bicarbonate alkalinity according to the stoichiometric relation:

$$HCO_3^- + H(Ac) \rightarrow CO_2 + H_2O + VAC^-$$

where H(Ac) is the unionized acetic acid, Ac the acetate ion.

Under unstable conditions the resulting bicarbonate alkalinity can be determined by analytically measuring the total alkalinity and the volatile acid content, the three components 2being expressed as Equation (7.12) [10].

$$\text{Bicarbonate alkalinity (mg/l } CaCO3) = \text{total alkalinity} - (0.85)(0.833)(\text{total volatile acid, mg/L}) \quad (7.12)$$

The 0.85 value corrects for the analytical limitation in that the titration analysis to pH 4 accounts for only 85% of the volatile acid, and 0.833 converts the mg/L of volatile acid to equivalent $CaCO_3$.

Acidity induced by influent characteristics or detrimental process conditions result in acid buildup, promoting unstable conditions, requiring alkalinity additions to correct the problem. Lime, sodium carbonate, and sodium hydroxide are the primary alkalinity control chemicals, with potassium hydroxide and ammonia sometimes considered. The specific chemical used depends on cost, availability, and convenience. [23]

(1) Lime has the advantages of being inexpensive and having an inherent buffering capacity, making control at the desired range relatively easy. Its great disadvantages include slow solution and reaction rate, increasing the inerts solids content, and carbon carbonate formation. The relatively slow solution and resulting reaction rate

of lime require a long reaction time, making excessive additions a possible result of operator "overreaction" to a seeming slow correction. Any resulting excessive lime additions can cause two problems. First, it increases the sludge solids content by increasing its ash content. This not only increases sludge volume but could increase incinerator slagging and fly ash production if final disposal is by thermal destruction. Next, if excess lime is accompanied with extreme conditions favoring calcium carbonate, excessive pressure (vacuum) reductions can occur.

(2) Sodium hydroxide is expensive but easier to handle than lime. It is purchased as a concentrated solution and diluted as needed. Storage, preparation, and handling capital costs are minimum, relative to other chemical alternatives. However, it readily ionizes, possessing no inherent buffering capacity. Unless carefully added, the reactor pH can rapidly rise to 10 or higher.

(3) Sodium bicarbonate has excellent buffering capacity, is relatively inexpensive, easily dissolves, readily reacts, and is ideal for pH control. In addition, it can increase bicarbonate alkalinity without reacting with carbon dioxide, eliminating the potential for a partial vacuum in the system.

(4) As discussed previously, care must be taken in adding any of these chemicals to avoid toxicity resulting from high alkali cation additions. The same concerns apply to the use of ammonia or potassium compounds.

PROCESS DESIGN VARIABLES

Anaerobic treatment design requires a sequential evaluation of process components and parameters to assure effective treatment and a stable process in the entire range of influent conditions encountered. Specific process parameters to be investigated include

(1) System configuration
(2) Reactor temperature
(3) HRT or SRT
(4) Reactor concentration
(5) Reactor volume
(6) Reactor mixing
(7) Sludge growth
(8) Gas production
(9) Energy considerations
(10) Fate of contaminants

System Configuration

Developing an anaerobic treatment system involves *three* basic steps, starting with assuring process applicability and finishing with reactor optimization

Step 1: Anaerobic Treatment Applicability

The waste characteristics must be appropriately analyzed to assure substrate biodegradability. Next, the feed organic concentration must be adequate to assure system stability, requiring that enough methane and cell growth be generated to sustain the system. A process energy and material balance must be completed to establish system stability. If these minimum conditions cannot be met, the advantages of anaerobic treatment may be negated, and *alternative treatments* should be considered.

Step 2: Can a Simple Dispersed Growth System Be Employed?

The confidence of applying simple low rate, high rate, or two stage digesters depends on whether the system is employed for sludge stabilization or treating concentrated wastes. In most industrial treatment facilities the sludge quantity generated is relatively small, appropriately lending itself to a conventional treatment system.

Anaerobic waste treatment applications are more complex than sludge digestion. Generally, the three conditions that govern whether simple conventional systems can be employed are the waste volume, concentration, and its physical state. First, applied waste volumes must generally limit the required reactor volume to less than 3,800 cubic meters (1 MGals) to avoid prohibitive equipment costs. Large waste volumes may be better suited to an anaerobic contact or bed system where the SRT can be controlled and the reacting volume significantly reduced. Next, as previously stated, the influent concentration must be adequate to assure an economical and stable process. Finally, the waste organic components must readily acclimate to a suspended growth system. Wastes containing high organic suspended solids could be expected to acclimate to the system in a manner similar to a sludge digester. When the components are soluble organics the acclimation of the waste to a low rate, contact stabilization or a contact bed system must be investigated.

Step 3: Reactor Optimization

If a low rate treatment system is not applicable or totally effective, alternative technologies must be evaluated. The applied technology selected should be based on incorporating the *least complex* system to achieve the required performance, applying proven technology. Some of the options are indicated in Table 7.22, indicating their relative complexity. The final application will be driven by process applicability, and economic considerations to minimize both capital and operating costs.

TABLE 7.22. Anaerobic Technology Classifications.

Experience	Complexity	
Low-rate system		Ambient temperature
		No mixing
		Small batch volumes
		Applicable to sludges
		Low substrate reductions
High-rate system		Higher temperatures
		Small batch volumes, unless continuous design
		Completely mixed
		Applicable to sludges, can be designed for soluble organics
		Improved substrate reductions
Two-stage		Stage one
		Ambient temperature
		No mixing
		Stage two
		Higher temperature
		Completely mixed
		Continuous operation possible
		Applicable to sludges, can be designed for soluble organics
		Improved substrate reductions
Anaerobic contact		Elevated temperatures
		Continuous operation
		Completely mixed
		Applicable to soluble organics
		Improved substrate reductions
Bed systems		Specialized designs
		Elevated temperatures
		Continuous operation
		Plug-flow configuration
		Applicable to soluble organics
		Improved substrate reductions
		Increased mass loadings

Reactor Temperature

Anaerobic treatment is applicable in the mesophilic range of 29 to 38°C (85 to 100°F), the common design range being from 30 to 38°C (86 to 100°F) and with 35°C (95°F) considered optimum [2,10]. Although the process can be promoted in the higher thermophilic range of 50 to 60°C (122 to 140°F), there is little evidence of the cost benefits of operating at elevated temperatures [7]. Within the commonly applied range, reduced temperature results in reduced cell growth, increasing the required minimum SRT to prevent "washout," the SRT required to achieve treatment efficiency, and the required reactor size. The effects of temperature on kinetic constants (b, K_d, K_i, k) commonly used to estimate the SRT are most frequently expressed using the Arrhenius relationship, simplified to Equation (7.13).

$$K2 / K1 = \Phi^{(T2-T1)} \tag{7.13}$$

where K is the reaction rate constant, T is the temperature in Kelvin, and Φ is the temperature constant.

The correction is applied to the specific constant as indicated in Table 7.1.

Solids Retention Time

Dispersed growth reactor design is based on the cell growth rate as expressed by the required effluent SRT (SRT_e) and related to the washout SRT_m by a safety factor. The relation between SRT to cell growth rate is discussed in detail in Chapter II-2, General Biological Oxidation section, and specifically in this chapter under the section entitled General Process Kinetics. The required SRT is temperature related but is usually in the range of 10 to 20 days for high rate (complete mix) sludge digestion at the applied temperatures indicated in Table 7.23 [2,12]. In single-stage, conventional sludge digesters the SRT can range from 30 to 90 days [2]. In industrial waste systems the required SRT can be estimated from the appropriate Monod relations defined by Equations (7.4) to (7.6), with the design SRT being 2 and 10 times the calculated minimum or "washout" value [2,12].

TABLE 7.23. **High Rate Sludge Digester**
(adapted from Reference [12]).

Suggested Solids Retention Times, Days		
Operating Temperature, °C (F)	Proposed Design	Minimum
18 (65)	28	11
24 (75)	20	8
29 (85)	14	6
35–41 (95–105)	10	4

In bed or sludge contact systems the effective SRT is a function of the reactor mechanics.

Reactor Concentration

The working mixed liquor volatile suspended solids (MLVSS) level is related to the reactor configuration.

(1) The *equilibrium* reactor concentration in a low rate completely mixed anaerobic system is not operator controllable, but the steady-state value is defined by Equation (7.14) [2,6].

$$X = \frac{Y \cdot (S_o - S_e)}{1 + k_d \cdot SRT} \qquad (7.14)$$

(2) In a batch reactor the initial solids level is a direct function of the *net* residual solids maintained. This is a function of the quantity of solids generated within a batch treatment cycle and the quantity wasted for sludge management.

(3) In a contact stabilization system the reactor solids level is a function of the separator performance and recycle capabilities. Separators can mechanically handle influent concentrations up to 5000 mg/L, thickening to a maximum 10,000 mg/L. As with an aerobic activated sludge system, sludge recycle depends on the required reactor concentration to obtain a required operating SRT, based on controlling predominant cell growth rate. In contact stabilization the solids concentration is defined by Equation (7.15).

$$X = \frac{Y \cdot (S_o - S_e)}{1 + k_d \cdot SRT} \cdot \frac{SRT}{HRT} \qquad (7.15)$$

The reactor concentration is commonly maintained at 3000 to 5000 mg/L, with provisions for a 2 to 4:1 recycle (recycle to influent flow) [2,12]. The recycle range must be adequate to allow the operator process control to (1) compensate for an expected increase in substrate loading, (2) increase the loading during periods of low biological activity, and (3) increase the inventory of fresh microorganisms during surges of toxic influent.

(4) In other contact systems, reactor concentration is not always directly controlled, but is an inherent property of the system's operating or physical properties.

Reactor Volume (Dispersed Growth)

Suspended growth reactor volume can be expressed in terms of the design SRT as defined by Equation (7.16).

$$Volume = wasting\ rate \cdot SRT/Xv \qquad (7.16a)$$

$$= wasting\ rate \cdot (1/cell\ growth)/Xv \qquad (7.16b)$$

$$Volume = \frac{[SRT] \cdot [Q \cdot (S_o - S_e) \cdot Y]}{X \cdot [1 + k_d \cdot SRT]}$$

All units are as defined for Equation (7.3), the volume expressed in cubic meters or gallons, depending on the selected flow units.

Therefore, the reactor volume is related to (1) the required performance as expressed by the influent and effluent concentrations, (2) the flow rate, (3) SRT, (4) the substrate conversion factor (Y), (4) the decay constant (k_d), and (5) the reactor solids concentration (X). Design and operating variables influencing the reactor volume are the reactor *MLVSS level* and the design flow rate, both of which must be carefully evaluated in establishing the system's operating range.

Mixing (Dispersed Growth)

Mixing controls influent dispersion and reactor volume consistency, maximized when a complete mix configuration is employed. As with most biological reactors, the primary considerations in evaluating mixing methods include (1) contact between the vessel solids and the feed, (2) reactor uniformity, preventing stratification and temperature gradients, (3) distribution of reactants and dilution of potential inhibitory substances, (4) effective use of the reactor volume, and (5) minimal scum formation or solids deposition. Because oxygen cannot be emitted into the system, devices common in aerobic treatment systems *are not suitable for anaerobic treatment.*

Mixing efficiency is gauged by the uniformity of the vessel contents, with the vessel being considered completely mixed if its solid concentration profile does not vary more that 10% [16]. Reactor mixing is accomplished by gas recirculation, pumping, or a mechanical device.

Gas Recirculation

Gas recirculation, as illustrated by Figures 7.12 and 7.13, is commonly practiced. This method has the advantages of (1) utilizing available gas product, (2) minimizing the possibility of oxygen (air) leakage, and (3) being designed as an integral part of the vessel. Digester gas is compressed and recirculated into the vessel contents through diffusers, lances, or confined

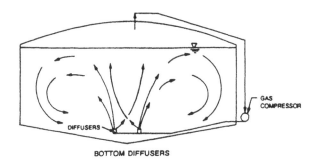

BOTTOM DIFFUSERS

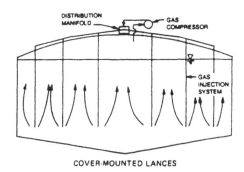

COVER-MOUNTED LANCES

Figure 7.12 Unconfined gas injection system (copied from reference [16]). Copyright © Water Environment Federation, reprinted with permission.

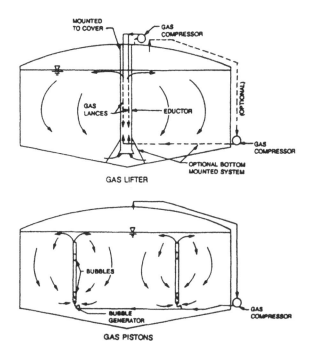

GAS LIFTER

GAS PISTONS

Figure 7.13 Confined gas injection system (copied from reference [16]). Copyright © Water Environment Federation, reprinted with permission.

tubes. Gas recirculation systems are generally classified as unconfined or confined. Unconfined systems inject the compressed gas at the vessel bottoms through one or more *lancers or diffusers,* moving the vessel contents and creating a circular mixing pattern. Confined systems move compressed gas through *confined tubes,* creating a venturi or piston effect, pushing reactor contents from the bottom of the tube, through the tube, and exiting at the top. Gas recirculation systems are suitable for fixed, floating or gas holder covered vessels.

Pump Recirculation

Pump recirculation (Figures 7.14 to 16) involves recirculating reactor sludge through a draft tube or eductor tube mixer. The resulting action disperses the feed, maintaining a uniform distribution. The tubes can be internal or external to the vessel; the pumping mechanism is external. Pump recirculation systems are suitable for fixed cover vessels.

Mechanical Mixing

Mechanical mixing (Figure 7.17 and 7.18) can include low-speed turbines and mixers, top or side mounted, utilized in baffled vessels. Mixing principals are discussed in Chapter I-5. Mechanical mixing is suitable for vessels employing either fixed or floating covers, provided mixers are always significantly submerged in the liquid, producing good bottom mixing.

Design Criteria

Mixing criteria, as summarized in an EPA technical evaluation report [16] for anaerobic reactors, are generally based on experience as opposed to finite calculated values. The performance is gauged on the basis of one of three criteria:

(1) *Unit power,* defined as the mixer motor power divided by the reactor volume, specified as W/m^3. Mechanical systems generally operate in the range of 5 to 8 watts per cubic meter (0.2 to 0.3 hp per 1000 ft^3).

Unit gas flow, defined as the digester gas delivered per reactor volume. Confined gas systems operate in the range of 5 to 7 m^3/min per 1000 m^3 (5 to 7 ft^3/min per 1000 ft^3). Unconfined gas systems operate in the range of 4.5 to 5 m^3/min per minute per 1000 m^3 (4.5 to 5 ft^3/min minute per 1000 ft^3).

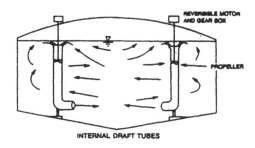

INTERNAL DRAFT TUBES

Figure 7.14 Mechanical pumping: internal draft tubes (copied with reference [16]). Copyright © Water Environmental Federation, reprinted with permission.

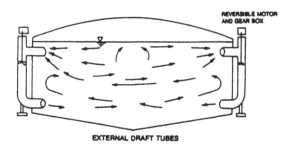

Figure 7.15 Mechanical pumping: external draft tubes (copied with reference [16]). Copyright © Water Environment Federation, reprinted with permission.

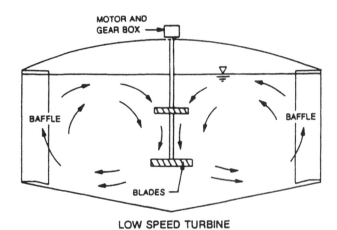

LOW SPEED TURBINE

Figure 7.17 Mechanical mixing: turbine (copied from reference [16]). Copyright © Water Environment Federation, reprinted with permission.

(2) *Velocity gradient,* a more theoretical estimate is based on a measure of power input directly applied to the liquid volume, defined by Equation (7.17).

$$G = [(P/V)/\mu]^{\frac{1}{2}} \qquad (7.17)$$

where P is the power input in W (lbf · ft/sec), V is the mixing chamber working volume in m³ (ft³), μ is the absolute viscosity of fluid in N · s/m² (lbf-s/ft²), and G is the velocity gradient, 1/s.
Anaerobic mixing operate at velocity gradients ranging from 50 to 80 1/s.

(3) *Turnover time* is defined as the vessel working volume divided by the internal flow rate. It is important to note that this is *not* the reactor residence time (volume divided by the feed rate) but the volume divided by the internal flow rate as influenced by the mixing device. The internal flow rate generated by a device is specific to the vendors proprietary design, affected by the vessel internal design. Anaerobic treatment reactors operate in the range of 20 to 30 min turnover time.

anaerobic systems. The sludge generated being 1/3 to 1/5 that of similar aerobic systems [7]. In a form similar to aerobic systems the volatile suspended solids (VSS) produced can be expressed using Equation 7.18 [15].

$$VSS = a\,F_r - b\,M_v \qquad (7.18)$$

where a is the grams of volatile biomass produced from each gram of COD removed, b is fraction of the biomass destroyed in endogenous metabolism, F_r is the quantity of COD (BOD, TOC) destroyed kg (lb), and M_v is the kg (lb) of volatile biomass in contact with the substrate.

Typical growth rate constants are indicated in Table 7.1.

McCarty proposed a modified form of the Monod cell growth relation to express sludge production in terms of

Sludge Growth

A significant difference between aerobic and anaerobic processes is the lower waste sludge quantities created by

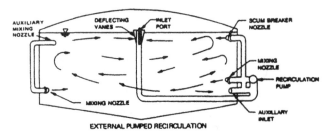

Figure 7.16 Mechanical pumping: external recirculation (copied from reference [16]). Copyright © Water Environmental Federation, reprinted with permission.

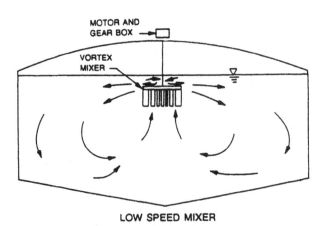

LOW SPEED MIXER

Figure 7.18 Mechanical mixing: mixer (copied from reference [16]). Copyright © Water Environment Federation, reprinted with permission.

the SRT and type waste stabilized, defined by Equation (7.19) [12].

$$A = \frac{a \cdot F}{1 + b \cdot (\text{SRT})} \qquad (7.19)$$

where A is the kg (lb) of volatile biological solids produced per day, F is the kg (lb) of ultimate BOD_L (COD) added per day, SRT is the solids retention time in days, and a and b are growth rate constants, related to the waste, Table 7.1.

Gas Production

The estimated total gas produced is critical in designing the gas handling equipment, and the methane production is crucial in evaluating the system operating economics. Stoichiometrically, the quantity of gases produced can be estimated using Equation (7.20) [15].

$$C_nH_aO_bN_c + [n-(a/4)-(b/2)+(3c/4)] \, H_2O = c \, NH_3$$

$$+ [(n/2)-(a/8)+(b/4)+(3c/8)] \, CO_2 \qquad (7.20)$$

$$+ [(n/2)+(a/8)-(b/4)-(3c/8)] \, CH_4$$

Typical gas compositions are indicated in Table 7.24 [20].

Methane production can be estimated using Equation 7.21 [12].

$$C = C_f \cdot (e \cdot F - 1.42 \, A) \qquad (7.21)$$

where C is the m^3 (ft^3) of methane produced per day at standard conditions (0°C, 1 atm), e is the fraction of BOD_L removed, from 0.80 to 0.95, F is the kg (lb) of BOD_L *added* per day, A is the kg (lb) of volatile biological solids produced per day, and Cf is equal to 0.35, the theoretical methane produced per kg of BOD_L, or if British units are used 5.62, 1.42 is a conversion factor to convert volatile biological solids to BOD_L.

At standard conditions approximately 0.35 m^3 of methane will be produced per kg of COD or ultimate BOD removed (5.62 scf/lb). The methane has a heating value of 8542 kg-calories per standard cubic meter of methane (960 BTU/scf) [2,15].

TABLE 7.24. **Typical Anaerobic Gas Analysis (adapted from Reference [23]).**

Component	Volume, %
Methane	42–75
Carbon dioxide	18–48
Hydrogen sulfide	0.01–1.5
Nitrogen	Trace–8
Hydrogen	Trace–3
BTU/ft³	459–791

Energy Considerations

Energy is required for sludge handling, maintaining the reactor temperature, pumping and transferring sludge, and reactor mixing. As a result total energy requirements are an important economic consideration in evaluating the feasibility of applying anaerobic treatment. Energy required for reactor mixing, sludge pumping, and sludge handling equipment are generally well defined, mechanically related to the force and work produced. In such services opportunities for capital and operating cost reduction depend on the equipment selected. However, the major energy use is for maintaining the reactor temperature. Energy cost optimization require implementing heat conservation, utilizing generated methane, and minimizing supplementary fuel requirements.

Methane collection and use is universally practiced because of the attractive operating cost reductions in large facilities, and in any size plant the gases must be converted to suitable exhaust products. The gases generated can be used for the many energy requirements mentioned, with any excess methane used for plant service. Heating efficiency can be estimated by conducting a system energy balance considering waste processing rate, reactor temperature, and methane generated.

Energy Balance

An anaerobic reactor (heat) energy balance involves heating the feed to the reactor temperature, and compensating for the heat losses from the system.

The *feed heat requirements* can be estimated using Equation (7.22).

$$H_r = W \cdot Cp \cdot (T_2 - T_1) \qquad (7.22)$$

where Hr is feed heat input, kg calorie (BTU) per hour, W is the design waste flow, kg/hr (lb/hr), Cp is the heat capacity, (taken as 1), T_2 is the required reactor temperature, °C (°F), and T_1 is the feed temperature, °C (°F).

The *system heat losses* can be estimated using Equation (7.23).

$$H_{loss} = U \cdot A \cdot (T_2 - T_1) \qquad (7.23)$$

where H_{loss} is the heat loss, kg-calories/hour (BTU/hour), A is the total heat transfer area, m^2 (ft^2), T_2 is the reactor temperature, °C (°F), T_1 is the surface temperature, °C (°F), and U is the overall heat transfer coefficient, kg cal/hr-m^2-C (BTU/hr-sf-F).

The total heat loss consists of the quantity transferred through the roof, walls, and floor, so that

$$H_l = h_{roof} + h_{walls} + h_{floor} \qquad (7.24)$$

$$= [(U \cdot A)_{roof} + (U \cdot A)_{walls} + (U \cdot A)_{floor}] \cdot (T_2 - T_1)$$

TABLE 7.25. Typical Anaerobic Heat Transfer Coefficients (adapted from Reference [23]).

	kg-cal/ m² · h · °C	BTU/ ft² · h · °F
Fixed covers (no insulation)		
1/4 in. steel	4.46	0.91
9 in., concrete	2.84	0.58
Floating covers, wood deck		
No insulation	1.62	0.33
12 In. concrete walls above ground		
Air space, 1 in. and		
4 in. brick	1.32	0.27
No insulation	4.21	0.86
Below ground 12 in. concrete walls or floors		
Dry earth	0.29	0.06
Moist earth	0.54	0.11

Typical heat transfer coefficients are given in Table 7.25 [23]. Estimates for total system heat losses range from 11.6 kg-calories per hour per cubic meter of digester volume (1.3 BTU/hr/cf) for locations in the southern part of the U.S. to 23.1 (2.6) for northern regions [2].

The *total system heat input* is the sum of the heat required to heat the reactor feed plus replacement of the system losses, as illustrated in Figure 7.19. The *available heat* sources are the heat content of the gases produced, supplementary heat, and/or the heat of reaction. The heat of reaction is normally neglected in evaluating the system heating requirements, except for autothermic system design. The *energy available* from the gas production is directly related to the methane generated.

Significantly, an anaerobic treatment system is economically viable if the heat available from the gas produced is adequate to balance the system requirements in the operating range. Any deficit in this balance will require a supplementary heat source. If supplementary fuel is necessary, the economics of the system must be carefully evaluated.

The Process Engineer should be aware that provisions must be made for an alternate heat supply for start-up purposes and to maintain the system during low treatment levels or upset conditions. The design must include evaluation of the following requirements:

(1) Heat available from the reactor gases for normal reactor temperature maintenance

(2) How surplus gas will be used

(3) Minimum supplementary heat source for anticipated normal system fluctuations

(4) Maximum supplementary heat source for start-up or emergency conditions

Finally, energy recovery may be as simple as an influent heater, or in large facilities a total heat recovery system using excess gas generated for plant heating, supplementary reboiler fuel, or power production. At a minimum, energy recovery will include a sludge heater. The sludge is usually heated in an external heater using hot water, as schematically illustrated in Figure 7.20.

The exchanger itself can be a water bath heat exchanger, a jacketed pipe heat exchanger, or a spiral heat exchanger. The water bath heat exchanger is directly linked to a boiler generating hot water, with the sludge passing through tubes immersed through a hot water (circulated) bath. Basically, the other two systems operate as tube-and-shell exchangers, with the hot water in the tubes and sludge in the shell.

The Process Engineer's primary responsibility is primarily in preparing the operating heat balance, with mechanical details of the heat exchanger equipment left to the equipment supplier and associated mechanical engineers. The design of this equipment is beyond the scope of this book. The reader is referred to the many heat transfer texts available for details on this subject.

Fate of Contaminants

Products from an anaerobic treatment system include treated effluent, waste sludge, and air emissions. Treated effluent or stabilized sludge are the system product. Either of these products may have to be further treated prior to discharge. Treated effluent may have to be further treated to meet discharge limits, stabilized sludge dewatered to reduced disposal costs, and further stabilized to meet RCRA requirements. Solids disposal volume, and corresponding costs, can be controlled by evaluating the various alternatives for sludge reduction, handling, and ultimate destruction.

Air emissions will depend on the anaerobic process stability and fugitive emission control. During normal operation the waste is stabilized to methane gas which is collected in a closed system and recovered. When the system is upset, a variety of odors can be generated and emitted form the system. With sulfur containing wastes, unstable conditions

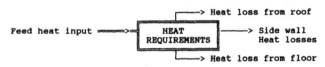

Figure 7.19 System heat balance.

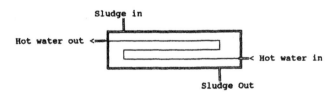

Figure 7.20 External sludge heater.

could generate sulfides that have a rotten egg-type odor and will result in immense problems.

GENERAL ENGINEERING CRITERIA

Figure 7.21 is symbolic of a contact anaerobic treatment system using a physical or sludge bed to treat industrial wastewater. System components are specific to the proprietary equipment employed. Process components for traditional anaerobic reactors include

Equalization (optional)   Chapter I-4
Waste feed system
Anaerobic reactors
Sludge settler (optional)
Sludge recycle (optional)
pH Control (optional)   Chapter I-8
Nutrient addition (optional)   Chapter I-5
Heating equipment
Gas-handling equipment

As indicated, many of these components are discussed in separate chapters, with specific criteria detailed. The mechanical design of anaerobic treatment systems components are frequently left to manufacturers of these specialized equipment because most of these systems are composed

of integrated components consisting of the reactors (or the reactor design details), mixing, sludge recycle, covers, heaters, and gas collection systems. Some pertinent process design criteria, common to all anaerobic systems, are summarized subsequently [8,20].

SPECIAL DESIGN CONSIDERATIONS

Reactor Design Considerations

Sludge digester construction criteria are relatively standard, employed for anaerobic treatment reactors operating in many municipal facilities. As with any biological system, the major process consideration is determining reactor volume and the number of reactors employed. Multiple units are recommended for large sludge capacity, especially for conventional systems to minimize total downtime resulting from process failures and to allow maximum operator flexibility.

Attention must be given to process conditions creating safety and health problems resulting from potential digester gas explosive and toxic conditions. A critical process *concern is preventing air leakage into the reactor, forming an explosive mixer with the digester gas.* For that reason digesters are operated at elevated pressures to prevent leakage into the vessels, although the vessel should be structural designed to allow for vacuum degasification. Because digesters are generally operated at a slight pressure, monitoring of poten-

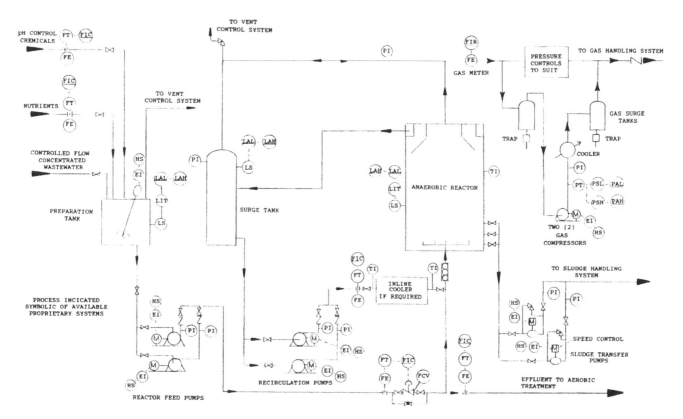

Figure 7.21 Anaerobic treatment preliminary concept flowsheet.

tial toxic components leaking from the reactor is an important consideration. In addition, the digester exhaust must be designed to prevent flash-back from accidental contact with a spark or flame.

Sludge digester shells are commonly constructed of reinforced concrete, as circular, square, rectangular, or egg-shaped configurations, and insulated for heat conservation. Cylindrical units, 6 to 35 m (20 to 115 ft) in diameter and 6 to 13 m (20 to 45 ft) in height, are frequently employed. A depth-to-diameter ratio of 0.3 to 0.7 allows for adequate mixing of the working volume, with the freeboard consistent with the cover provided. Floating cover freeboard can range from 45 to 60 cm (18 to 24 in.) above the maximum liquid level to the top of the wall, fixed roofs from 30 to 60 cm (12 to 24 in.), and fixed conical or domed roofs freeboard above the liquid level can be less than 30 cm (12 in.) [20,21]. Conical bottoms are sloped from 1:6 to 1:4. When mechanical sludge assisting devices are employed, slopes can be reduced to 1:12 [20,21].

Structural and construction design concerns include considering the extreme vessel temperature differences between process and ambient conditions, and corrosion conditions resulting from exhaust gas components such as hydrogen sulfide. The vessel must be designed with provisions for vacuum and pressure relief as well as flame arrestors and include ample manholes for entering, maintaining, and cleaning the tank. In addition, the vessel must contain port holes to observe digester conditions, multiports designed for representative sampling and supernatant removal at various tank levels, adequate bottom drain ports, and spare ports for future process considerations. All active processing, maintenance, and drain ports should be equipped with isolation valves for piping maintenance.

Waste treatment reactor specifics depend on the configuration selected, including fixed-film systems. Details of these systems depend on the supplier's proprietary design.

Digester Covers

Digester covers are commonly designated as floating or fixed, with floating covers being either the gas holding or sludge (liquid) holding type. The fixed cover provides a constant tank volume, allowing a more simplified reactor design, but provides less flexibility for changes in liquid volume or gas pressure. Floating liquid covers rests on the liquid surface, allowing a variation in liquid volume without a change in pressure. Floating gas holders float on a gas volume, allowing freeboard for *gas storage.*

Fixed roofs can be reinforced concrete constructed domes, conical or flat slabs, or steel domes. Floating covers are commonly steel constructed. A major consideration in dome design is preventing air leaking into the vessel to avoid explosive mixtures and because oxygen is toxic to the anaerobic process. A slight positive pressure is maintained in the reactor for that purpose. Floating roofs, although more expensive than flat roofs, allow process flexibility.

Gas Collection Systems

The gas collection system is an integral digester component providing temporary storage and a means to transfer and use the digester gases. Detailed design criteria are the responsibility of the digester manufacturer, in accordance with specified process design considerations, some of which could include [20]

(1) Digester gas collection design must carefully account for fluctuating downstream utilization demand, and the mechanical system designed to prevent gas leakage because of its explosive and odorous characteristics.

(2) Digester systems commonly operate at pressures less than 3.5 kPa (0.5 psi), making line and valve loss design evaluations critical.

(3) The gas collection system must be designed with proper sediment and drip traps to prevent liquid flow within the transfer lines, thereby deterring corrosion or freezing.

(4) Provision should be made for metering gas production and usage.

(5) The distance between gas combustion equipment and the digester tank is critical, with a *minimum* of 15 m (50 ft) recommended, consistent with local fire code and combustion regulation, as well as corporate and insurer safety standards. The gas collection piping should be segregated in a separate, properly ventilated structure.

(6) The corrosive nature of hydrogen sulfide must be considered in selection of gas-handling equipment materials of construction.

(7) Irregular gas generation may require some form of gas storage, either using floating digester covers or external storage tanks. The size of an external gas storage tank depends on the gas production variation rate and storage pressure. High-pressure vessels ranging from 138 to 690 kPa (20 to 100 psi) allow smaller volumes and prevent air leakage into the tank but are more expensive to construct.

Heating Systems

Influent heaters can consist of direct flame, direct steam injection, or external heat exchangers, with *external exchangers* being the most commonly employed. *External heat exchangers* can be configured as a water bath, a jacketed pipe, or a spiral heat exchanger. The most common heat source for digester exchangers is hot water generated from boilers utilizing digester gas.

A *water bath heat exchanger* is designed with tubes containing sludge immersed in a water bath, mounted on the side of a boiler, with hot water circulating from the boiler to the bath. A *jacket pipe heat exchanger* operates similar to a water bath exchanger, except that the hot water is circulated from the boiler through the heater jacket; the heater is separated from the boiler and located else where. A *spiral heat exchanger* consists of two concentric channels in a spiral pattern, one channel containing the sludge and the other the circulating hot water from the boiler system.

PROCESS SAFETY

A primary consideration in evaluating anaerobic system safety is controlling methane storage and transport and the release of relatively small amounts of carbon monoxide and hydrogen sulfide. As a result, the potential for air leaking into the reactor, or hydrogen sulfide into the workplace, must be considered in the design.

Methane is an explosive gas and must be handled as such. Methane and carbon dioxide components are not considered toxic. However, closed, unventilated housing conditions must be avoided to prevent conditions in which leaking digester gases depletes the oxygen supply.

Hydrogen sulfide presents a critical problem. First and foremost it is toxic! Its toxic effects have been evaluated for some time, with the maximum suggested exposure level generally at 20 mg/L [22,24]. Specific limits and occupational health requirements should be obtained from the local governing regulatory authority, the Corporate Health and Safety Officer, and the plant safety engineer. In addition, hydrogen sulfide is highly corrosive and potentially explosive.

When air is inadvertently injected into the system, usually in upset or start-up conditions, at least two dangerous conditions could develop.

(1) The system could be exposed to organic-air mixtures ranging from the lower explosive limits to the higher explosive limits, all of which represent a potential for dangerous explosive conditions.

(2) The potential for "pockets" of trapped gas that could be at the explosive limit range exists.

The Process Engineer must understand these concerns when evaluating the safety considerations of the system, which basically consist of three primary considerations:

(1) Preventing the release of hazardous gases to the working area or atmosphere

(2) Protecting the vessel from structural collapse

(3) Protecting a major disaster resulting from the system exploding

The system safety aspects must be evaluated with an understanding of a sequence of gas handling functions that is automatically triggered by defined pressure limits allowing the gas to be stored, recirculated in a mixing loop to the reactor, and excess quantities discharged to either heat recovery equipment or to a flare. Pressure regulators are utilized to maintain and direct the flows indicated, each leg of a system having a regulator at a preset conditions that is activated when specific conditions are reached. Pressure relief valves, strategically located in the system, assure that the excessive pressure is not reached in the operating line. In addition, vacuum valves assure that excessive gases are not vacated from the system

at rates producing a vacuum. Anaerobic systems also contain flame arresters and thermal shut off valves that prevent flame flashback from any external source, isolating the system from the source. Pressure and vacuum relief valves, as well as flame arrestors or applicable safety equipment, should be installed with independent operating backups.

Pressure and vacuum valves offer an immediate relief to the system but could present additional problems requiring immediate attention. One problem is the release of excess gases in the event of overpressure results in a possible safety or health problem to the working area and a possible environmental incident if released beyond the treatment facility. The other problem is that, when the vacuum relief system is activated, air is transmitted to the system, which could bring the system in the range of the explosive limits and thereby present a potential safety problem.

These problems can be controlled in several ways. First the system pressure (and vacuum) should be monitored at critical segments of the system and a primary alarm activated at preset levels well below the condition when the system relief system is activated. A secondary alarm should alert the operators that a condition activating the vacuum or pressure system has occurred. Obviously, this would require a reaction plan to either correct any problem or to shut-off the system. Another possibility is to connect the vacuum relief system to a nitrogen system that would inject nitrogen instead of oxygen into the system. Because the accumulation of hazardous or explosive gases in an enclosed vessel is not common in (aerobic) biological treatment, the potential dangers in aerobic systems could be overlooked. The Process Engineer should recognize that dangerous conditions could develop in the system, and incorporate suitable operating or safety equipment counter measures.

The Process Engineer must carefully evaluate the potential for hazardous conditions, incorporating prudent design measures to virtually eliminate the possibility for any safety or health problem. Part of the design should include

(1) Independent system HAZOP review to identify and remediate potential problems [1]

(2) Review of the design with plant safety, environmental, and insurer representatives

(3) Applying local codes relevant to storage of fuel gases

(4) Applying national codes such as those recommended by the National Firewriters Association and similar engineering codes

(5) Experience and common sense

PROCESS CONTROLS

Anaerobic treatment systems are as effective as they can be operated. The engineer can assist the operator in monitoring and controlling the plant by providing the following:

(1) Flow control (and recording) of influent, effluent, gas and sludge recycle flows

(2) Flow totalizing of sludge produced

(3) Sludge influent and effluent sampling

(4) Adequate means of sampling and withdrawing reactor supernatant at various tank levels

(5) Tank level monitoring

(6) A means of monitoring and maintaining tank pressure, with provisions for alarming the operator at the point of pressure buildup

(7) A means of monitoring and controlling reactor temperature and a means of alerting the operator when the reactor contents are outside of control limits

(8) A means of monitoring and controlling feed temperature

(9) Sludge heater temperature monitoring and control

(10) Waste gas monitoring and controls to each gas use, i.e., gas mixers, gas boiler, flare. This must include flow and pressure control, with adequate safety relief. The system must be designed with provisions to avoid volume or pressure buildup within the system.

(11) Running lights for all major equipment

(12) Inspection ports for all critical vessels

(13) Torque control for separator equipment (where applicable)

(14) Cover position indicators to track the floating cover level

(15) Sediment and condensate traps in the gas-piping system to collect moisture and remove pipe scale and particles

(16) Gas metering to measure gas production

(17) Gas pressure gauges

(18) Isolation valves to segregate and inspect equipment

(19) Provisions to flush sludge lines

(20) Gas analysis equipment for methane, carbon dioxide, and hydrogen sulfide

(21) Hydrogen sulfide removal capabilities where applicable

(22) Incorporating all required safety requirements

COMMON DIGESTER DESIGN DEFICIENCIES

Some design deficiencies common for municipal systems have been investigated by the EPA and are identified as a checklist of process and mechanical design considerations [18]. The cited reference should be reviewed for further details.

Digester Design

(1) The upstream treatment plant capacity is not adequate to treat the generated supernatant side streams.

(2) Sampling provisions are inadequate.

(3) Digester gas treatment for removal of impurities such as hydrogen sulfide prior to utilization is inadequate.

(4) Provision for digester cleaning is inadequate.

(5) Digester peak load capacity is insufficient.

(6) There is no provision for cleaning sludge heating lines.

(7) There is a lack of digester pH and temperature control.

(8) Relief valves are not winterized.

(9) Sludge thickening provisions to concentrate feed are not provided, reducing the digester capacity, especially for single-stage units.

(10) Supernatant withdrawal points are poorly located.

(11) Emergency chemical feed facilities to respond to upset conditions are not provided.

(12) Mixing is inadequate.

Gas Collection System

(1) Sampling provisions are inadequate.

(2) Gas line condensate traps are sparse and poorly located.

(3) High or low level pressure alarms are not provided.

(4) Condensate traps in the lines are not provided.

(5) Flame traps in the lines are not provided.

(6) Individual gas measurement devices are not provided for multiple digesters.

Sludge Withdrawal Design

(1) Sampling ports are inadequate.

(2) Sludge lines are not properly sized.

(3) Multiple sludge withdrawal and return points are not provided.

(4) Sludge metering devices are inadequate.

General

(1) No provisions are made for changing the supernatant withdrawal point within tank.

(2) Water seals on the digester pumps are inadequate (utilize mechanical seals).

(3) No provisions are made for supernatant disposal.

(4) Heating pipes are inadequately insulated.

(5) Heat exchanger capacity is inadequate.

(6) An inadequate number of sludge feed lines limits the system flexibility.

(7) No provision is made for flushing the sludge lines.

(8) Feed flow measurement or control devices are incorrect, incorrectly sized, or both.

CASE STUDY NUMBER 19

Develop an anaerobic treatment process design system for an industrial waste with the following characteristics:

Flow, MGD: 1
COD, mg/L: 3200

Applicable design criteria include:

Effluent COD, mg/L: 480
lb nitrogen required per pound solids formed: 0.12
lb phosphorus required per pound solids formed: 0.02
Fraction solids biodegradable: 0.12
Ultimate BOD solids: 1.42
Design temperature, °F: 95
Reactor depth, feet: 25

PROCESS CALCULATIONS

(1) Establish biokinetic coefficients: Laboratory data obtained indicate 85% removal from an influent containing 3200 mg/L COD with an SRT of 31 days at 35°C. The kinetic coefficients developed to define the treatability characteristics can be summarized as follows:

k_d: 0.015 (1/day)
k_{max}: 6.67 (g COD/g VSS-day)
K_s: 2235 (mg/L COD)
Y: 0.04 (g VSS/g COD)

(2) Performance requirements
85% removal minimum, equivalent to effluent of 480 mg/L COD

(3) Calculate MCRT

a. Guideline values: SRT minimum 10 days; SRT design 20 days
Laboratory results indicate SRT of 31 days at 35°C (95°F).

b. Calculated Effluent SRT

$$1/\text{SRT} = \frac{Y \cdot k \cdot S_e}{K_s + S_e} - k_d$$

$$= \frac{0.04 \cdot 6.67 \cdot 480}{2235 + 480} - 0.015$$

$$= 0.0322$$

$$\text{SRT} = 1/0.0322 = 31 \text{ days}$$

c. Washout SRT

$$1/\text{SRT} = \frac{Y \cdot k \cdot S_o}{K_s + S_o} - k_d$$

$$= \frac{0.04 \cdot 6.67 \cdot 3200}{2235 + 3200} - 0.015$$

$$= 0.142$$

$$\text{SRT} = 1/0.142 = 7 \text{ days}$$

d. Limiting SRT

$$\text{SRT} = 1 / (Y \cdot k - k_d)$$

$$\text{SRT} = 1/(0.04 \cdot 6.67 - 0.015) = 4 \text{ days}$$

e. Selected SRT

Assume a safety factor of 5

$$\text{SRT} = 5 \cdot 7 = 35 \text{ days}$$

Use 35 days SRT at 95°F.

(4) Calculate effluent

$$S_e = \frac{K_s \cdot (1 + k_d \cdot \text{SRT})}{\text{SRT} \cdot (Y \cdot k - k_d) - 1}$$

$$S_e = \frac{2235 \cdot [1 + 0.015 \cdot 35]}{35 \cdot (0.04 \cdot 6.67 - 0.015) - 1} = 436 \text{ mg/L COD}$$

(5) Calculate applied COD, lb/day

$$1 \text{ MGD} \cdot 8.34 \cdot 3200 \text{ mg/L feed} = 26,688 \text{ lb/day}$$

(6) Calculate COD utilized, lb/day

$$1 \text{ MGD} \cdot 8.34 \cdot (3200 - 436) \text{ mg/L} = 23,052 \text{ lb/day}$$

(7) Calculate effluent (untreated) feed COD, lb/day

$$1 \text{ MGD} \cdot 8.34 \cdot 436 \text{ mg/L} = 3,636 \text{ lb/day}$$

(8) Calculate solids produced, lb/day

$$P_r = \frac{[Q \cdot 8.34 \cdot (S_o - S_e) \cdot Y]}{(1 + k_d \cdot \text{SRT})}$$

$$P_r = \frac{1 \cdot 8.34 \cdot (3200 - 434) \cdot 0.04}{(1 + 0.015 \cdot 35)} = 605$$

Equivalent COD = 605 · 1.42 = 859 lb COD/day

(9) Calculate COD stabilized, lb/day

$$23052 - 859 = 22,193 \text{ lb COD stabilized/day}$$

(10) Calculate daily COD distribution, lb/day SCOD

Input SCOD: 26,688	Stabilized SCOD:	22,193
————————>	Solids SCOD:	859
	Effluent SCOD:	3,636
	Output SCOD:	26,688

CONTACT STABILIZATION

(11) Calculate inventory (MLVSS) solids, lb

$$\text{SRT} \cdot \text{Daily solids wasting} = \text{solids inventory}$$

$$35 \text{ SRT} \cdot 605 = 21{,}175 \text{ lbs solids in inventory}$$

(12) Select MLVSS, mg/L select 3500 mg/L

(13) Reactor volume

$$21{,}175/3500 = 6.05 \text{ million pounds in reactor}$$

$$6.05/8.34 = 0.73 \text{ million gallons}$$

CONVENTIONAL TREATMENT

(14) Calculate MCRT and HRT: MCRT = HRT = 35 days

(15) Calculate reactor storage quantities

$$35 \text{ days} \cdot 1 \text{ MGD} = 35 \text{ MG}$$

FINAL DESIGN

(16) Select working volume 0.73 MG, based on a contact stabilization configuration.

(17) Estimate methane production

5.62 SCF @ 32°F per lb COD stabilized at STP (32°F, 1 atm).

$$22{,}193 \cdot 5.62 \cdot [(460 + 95)/(460 + 32)]$$
$$= 141{,}000 \text{ cu ft/day @ 95°F.}$$

(18) Estimate total gas production

Assume methane 55 volume % of the total gas produced. Total gas produced: 141,000/0.55 = 256,000 ft^3/day.

(19) Alkalinity evaluation
The reactor alkalinity is dependent on the carbonic acid concentration, which is approximately equal to the dissolve carbon dioxide concentration. The alkalinity will be evaluated for the range of carbon dioxide concentrations commonly reported in the reactor *head space*, i.e., from 25 to 45%.

(20) Determine carbon dioxide partial pressure

For 25% CO_2 concentration in gas:
   P = 0.25 · 1 ATM = 0.25

For 45% CO_2 concentration in gas:
   P = 0.45 · 1 ATM = 0.45

(21) Determine alkalinity

Assume a pH of 7.2 to be maintained.

$CaCO_3$, mg/L = 0.00063 · $P^{CO2}/10^{-pH}$

For 25% CO_2 in gas:
$CaCO_3$, mg/L = 0.00063 · 0.25 / $10^{-7.2}$ ≡ 2500 mg/L

For 45% CO_2 in gas:
$CaCO_3$, mg/L = 0.00063 · 0.45 / $10^{-7.2}$ ≡ 4500 mg/L

Ideally, alkalinity should be maintained between 2,500 and 3,500 mg/L $CaCO_3$.

(22) Estimate mixing requirements

Reactor volume: 730,000 gal · 0.1337 = 100,000 ft^3
Mechanical mixing @ 0.3 hp/1000 ft^3
100,000/1000 · 0.3 = 30 hp *total*
Gas recirculation @ 5 cfm/1000 ft^3
100,000/1000 · 5 = 500 cfm recirculation *total*.

(23) Calculate nutrient requirements

0.12 lb nitrogen/lb solids formed
0.12 · 605 lb solids/day = 73 lb nitrogen per day.
0.02 lb phosphorus/lb solids formed
0.02 · 605 lb solids/day = 12 lb phosphorus per day.

(24) Complete digester design

Volume, MG:	0.73
Number of reactors:	2
Liquid depth, ft:	25
Projected surface area, sf:	4000, 2000 sf per reactor
Vessel diameter, ft:	50
Freeboard, ft:	2
Total depth, ft:	27

(25) Heat loss calculations
Assume a winter ambient condition of 50°F and a waste at 40°F, and a design ambient condition of 73°C and waste at 68°F.

	Approximate area, sf	H,BTU/h/sf/F	HEAT, BTU/hr/dF
Floor	4000	0.12	480
Roof	4000	0.50	2000
Side wall	8500	0.35	3000
			5500

5500 · (95 digester temp − 40 ambient) ≡ 300,000
                              ≡ 0.3 million BTU/hr (winter)

5500 · (95 digester temp − 73 ambient) ≡ 120,000
                              ≡ 0.1 million BTU/hr (summer)

(26) Feed heating requirements
Assume a 10°F dT for winter and 5°F for summer;
                              Tair − T liquid

1 MGD · 8.34 /24 hr daily operation
   · 1 BTU/dF · (95°F − 50°F) = 15.6 MBTU/hr (winter)

1 MGD · 8.34 /24 hr daily operation
   · 1 BTU/dF · (95°F − 68°F) = 9.4 MBTU/hr (design)

(27) Feed heating requirements

$$15.6 + 0.3 = 15.9 \text{ MBTU/hr (winter)}$$

$$9.4 + 0.1 = 9.5 \text{ MBTU/hr (summer)}$$

(28) Available heat

Methane generated: 141,000 ft^3 per day @ 95°F.
125,000 SCF per day @ STP (32°F, 1 atm).

Methane heating value: 960 BTU/SCF @ STP

$$125,000 \cdot 960 /1,000,000 = 120 \text{ MBTU/day}$$

$$120/24 = 5.0 \text{ MBTU/hr}$$

(29) Excess heat

$$5.0 - 15.9 = -10.9 \text{ MBTU/hr (winter)}$$

$$5.0 - 9.5 = -4.5 \text{ MBTU/hr (design)}$$

The waste does not generate enough methane to support anaerobic treatment, requiring supplementary heating to sustain the process at 95°F. A COD of approximately 10,000 mg/L would produce enough methane to sustain the reaction at 95°F without supplementary fuel.

CASE STUDY NUMBER 20

Design a *completely mixed, single-stage, anaerobic* treatment system for the sludge generated from the biological system described in Case Study 10, which is thickened to 2% (Case Study 25), having the following characteristics:

(1) 9658 gal/day of sludge
(2) 20,000 mg/L suspended solids
(3) 16,000 mg/L volatile suspended solids
(4) Sludge contains adequate nutrients

Applicable design criteria include

VSS reduction, %:	60
Ultimate BOD of solids:	1.42
Reactor temperature, °F:	95
Minimum waste temperature, °F:	40
Basin depth, ft:	25

PROCESS CALCULATIONS

(1) Establish biokinetic coefficients: Available data indicate that the sludge VSS can be reduced 60% when processed at 35°C with an SRT of 10 days. Kinetic coefficient applicable to the conditions tested are as follows:
k_d: 0.03 (1/day)
Y: 0.04

(2) Calculate applied COD, lb/day

$$\text{Feed: } 16,000 \text{ VSS} \cdot 1.42 \text{ COD/VSS} = 22,720 \text{ mg/L SCOD}$$

$$0.009658 \text{ MGD} \cdot 8.34 \cdot 22,720 \text{ mg/L feed} = 1830 \text{ lb/day}$$

(3) Calculate effluent COD, lb/day

$$16,000 \cdot 40\%/100 = 6400 \text{ mg/L effluent VSS}$$

$$6400 \text{ mg/L} \cdot 1.42 \text{ mg COD/mg VSS} = 9,088 \text{ mg/L COD.}$$

$$0.009658 \text{ MGD} \cdot 8.34 \cdot 9088 \text{ mg/L} = 732 \text{ lb/day}$$

(4) Calculate new solids produced, lb/day

$$P_x = \frac{[Q \cdot 8.34 \cdot (S_o - S_e) \cdot Y]}{(1 + k_d \cdot \text{SRT})}$$

$$P_x = \frac{0.009658 \cdot 8.34 \cdot (16,000 - 6,400) \cdot 0.04}{(1 + 0.03 \cdot 10)} = 24 \text{ VSS}$$

Equivalent COD $= 24 \cdot 1.42 = 34$ lb COD/day

(5) Calculate COD stabilized, lb/day

$$1830 - 34 - 732 = 1064 \text{ lb COD stabilized/day}$$

(6) Summarize daily COD distribution, lb/day SCOD

Input SCOD: 1830		Stabilized SCOD:	1,064
———>—		Solids SCOD:	34
		Effluent SCOD:	732
		Output SCOD:	1,830

CONVENTIONAL STABILIZATION

(7) Calculate MCRT and HRT: MCRT = HRT = 10 days
(8) Calculate reactor size

$$10 \text{ days} \cdot 9658 \text{ GPD} = 96,580 \text{ gal}$$

FINAL DESIGN

(9) Select working volume 100,000 gal
(10) Estimate methane production

5.62 SCF @ 32°F per lb COD stabilized.

$$5.62 \cdot 1064 \text{ COD stabilized} = 5,980 \text{ ft}^3 \text{ @ } 32°F$$

$$5980 \cdot [(460 + 95)/(460 + 32)] = 6746 \text{ ft}^3 \text{ @ } 95°F$$

(11) Estimate total gas production
Assume methane 55 volume % of the total gas produced.

Total gas produced: 6746/0.55 = 12,300 ft^3.

(12) Estimate mixing requirements
Reactor volume: 100,000 gal $\cdot$ 0.1337 = 13,400 ft^3
Mechanical mixing @ 0.3 hp/1000 ft^3
13,400/1000 $\cdot$ 0.3 = 4 hp
Gas recirculation @ 5 cfm/1000 ft^3
13,400/1000 $\cdot$ 5 = 70 ft^3/min recirculation.

(13) Complete digester design

Volume, gal:	100,000
Number of reactors:	1
Liquid depth, ft:	25
Projected surface area, sf:	535
Vessel diameter, ft:	26
Freeboard, ft:	2
Total depth, ft:	27

This digester design is similar to that developed for a first-stage of a two-stage system.

(14) Heat loss calculations

	Area, sf	H, BTU/h/sf/F	Heat, BTU/hr/dF
Floor	535	0.12	65
Roof	535	0.50	270
Side wall	2205	0.35	775
			1110

$1110 \cdot (95 \text{ digester temp} - 40 \text{ ambient}) = 61,000 \text{ BTU/hr}$

(15) Feed heating requirements

$9658 \text{ GPD} \cdot 8.34 / 24 \text{ hr daily operation} \cdot$
$1 \text{ BTU/dF} \cdot (95°F - 40°F) = 185,000 \text{ BTU/hr}$

(16) Total heating requirements

$61,000 + 185,000 \equiv 246,000 \text{ BTU/hr.}$

(17) Available heat

Methane generated: 5,980 ft³/day @ 32°F

Methane heating value: 960 BTU/cf at STP

$5980 \cdot 960 = 5,700,00 \text{ BTU/hr}$
$5,700,000/24 = 239,000 \text{ BTU/hr}$

Within the accuracy of this calculation the available heat is in balance with the operating requirements.

REFERENCES

1. American Institute of Chemical Engineers: *Guidelines for Hazard Evaluation Procedures,* AIChE, 1985.

2. Benefield, L.D., Randall, C.W.: *Biological Process Design for Wastewater Treatment,* Prentice-Hall, Inc., 1980.

3. Bhattacharya, S.K., Parkin, G.F.: "The Effect of Ammonia on Methane Fermentation," *Journal WPCF,* V 61, No 1, Pg 55, January, 1989.

4. Bowker, R.P.G.: "New Wastewater Treatment for Industrial Applications," *Environmental Progress,* V 2, No 4, Pg 235, November, 1983.

5. Chen, S.J., Li, C.T., Shieh, W.K.: "Anaerobic Fluidized Bed Treatment of an Industrial Wastewater," *Journal WPCF,* V 60, No 10, Pg 1826, October, 1988.

6. Lawrence, A.W. and McCarty, P.L.: "Unified Basis for Biological Treatment Design and Operation," *Journal Environmental Engineering,* Proceedings ASCE, V 96, Pg 757, 1970.

7. Lee, J.W. Jr., Peterson, D.L., Stickney, A.R.: "Anaerobic Treatment of Pulp and Paper Mill Wastewaters," *Environmental Progress,* V 8, No 2, Pg 73, May, 1989.

8. Medcalf & Eddy, Inc.: *Wastewater Engineering-Treatment, Disposal, Reuse,* McGraw-Hill, 1991, Third Edition.

9. McCarty, P.L.: "Anaerobic Wastewater Treatment Fundamentals, Part 1, Chemistry and Microbiology," *Public Works,* Pg 107, September, 1964.

10. McCarty, P.L.: "Anaerobic Wastewater Treatment Fundamentals, Part 2, Environmental Requirements and Control," *Public Works,* Pg 123, October, 1964.

11. McCarty, P.L.: "Anaerobic Wastewater Treatment Fundamentals, Part 3, Toxic Materials and Their Control," *Public Works,* Pg 91, November, 1964.

12. McCarty, P.L.: "Anaerobic Wastewater Treatment Fundamentals, Part 4, Process Design," *Public Works,* Pg 95, December, 1964.

13. Obayashi, A.W., Stensel, H.D., Kominek, E.: "Anaerobic Treatment of High Strength Wastes," *Chemical Engineering Progress,* Pg 68, April, 1981.

14. O'Connor, J.T. (Ed.): *"Environmental Engineering Unit Operations and Unit Processes Laboratory Manual,"* Association of Environmental Engineering Professors, July, 1972.

15. Parkin, G.F., Owen, W.F.: "Fundamentals of Anaerobic Digestion of Wastewater Sludges," *Journal of Environmental Engineering,* Proceedings ASCE, V 112, No 5, Pg 867, October, 1986.

16. U.S. Environmental Protection Agency: "EPA Design Information Report- Anaerobic Digester Mixing Systems," *Journal WPCF,* V 59, No 3, Pg 162, March, 1987.

17. U.S. Environmental Protection Agency: *Treatability Manual,* EPA-600/8-80-042a, 1980.

18. U.S. Environmental Protection Agency: *Handbook for Identification and Correction of Typical Design Deficiencies at Municipal Wastewater Treatment Facilities,* EPA-625/6-82-007, 1982.

19. U.S. Environmental Protection Agency: *Management of Industrial Pollutants by Anaerobic Processes,* EPA-600/2-83-119, November, 1983.

20. WEF Manual of Practice: *Design of Municipal Wastewater Treatment Plants,* Water Environment Federation, 1992.

21. WPCF Manual of Practice No 8: *Sewage Treatment Design,* Water Environment Federation, 1967.

22. WPCF Manual of Practice No 16: *Anaerobic Sludge Digestion,* Water Environment Federation, 1968.

23. U.S. Environmental Protection Agency: *Process Design Manual for Sludge Treatment and Disposal,* September, 1979.

24. Patty, F.A.: *Industrial Hygiene and Toxicology,* Interscience Publishers, John Wiley & Sons, 1963.

25. Canale, R.P.: *Biological Waste Treatment,* Intercience Publishers, John Wiley & Sons, 1973.

26. Grady, C.P.L. Jr., Daigger, G.T., and Lim, H.C.: *Biological Wastewater Treatment,* Marcel Dekker, Inc., 1999.

Sedimentation

Sedimentation is employed to remove settleable waste solids to achieve effluent quality, concentrate sludge, or both.

BASIC CONCEPTS

S EDIMENTATION is employed to remove suspended solids from a waste stream, either as a primary clarifier to reduce solids loading to the secondary treatment system, as the principal physical treatment device, or as a final clarifier to meet effluent discharge requirements. In industrial waste treatment it is frequently employed within a biological treatment system as a solids management device and to meet effluent quality. Thickening is a continuation of sedimentation, concentrating a sludges's solids content as a preliminary or sometimes sole dewatering step. In both cases, the principal separation mechanism is the density difference between the solids and water, creating a gravitational force "pushing" the solids downward. As illustrated in Figure 8.1, the specific gravity of the agglomerated mass relative to water establishes the direction and extent of the solids movement.

Settling operations fall into four basic classifications, depending on the solid characteristics and concentration: (type I) discrete, (type II) flocculent, (type III) zone, and (type IV) compression (thickening) settling [18].

The relationship among solids concentration, physical characteristics, and the settling process is illustrated in Figure 8.2, the major difference being that zone I is characteristic of discrete inert solids, retaining their form and resulting in unhindered settling. In the other processes free gravity settling is resisted by either a change in solids characteristics or interference resulting from solids concentration.

Settler design entails applying theoretical considerations, factored by many years of wastewater and water treatment operating experience. The best design data are those obtained from a large-scale prototype using fresh waste, tested on a continuous basis. However, time and cost restraints sometime drive the investigator into conducting simple laboratory tests to *verify settling is feasible*. Where laboratory tests are conducted, process design criteria are frequently established from interpreting batch data, comparing the results with similar operating units, a scale-up factor applied, and a conservative design attained. The mechanical factors impacting commercial, full-scale settlers are usually left to the manufacturer.

Laboratory test results neglect significant operating variables such as

(1) Sludge removal rate
(2) Blanket depth
(3) Short circuiting
(4) Solids scouring
(5) Surface wind velocity
(6) Weir design
(7) Inlet and outlet location

The effort expended in obtaining site-specific test data depends on the process application, and the importance of the settling unit in meeting regulatory compliance. Because full-scale testing is impractical and interpreting laboratory data difficult, design of clarifiers and thickeners that are not predominant treatment units is frequently based on available experience.

TYPE I DISCRETE SETTLING

Discrete settling involves removing *unhindered, nonflocculating,* and *unchanging solids* traveling at a constant velocity, the settling rate being a function of the liquid and solid properties. Wastes containing metal particles, grit, sand, and heavy inert materials are examples of discrete settling. An important process design consideration in defining the set-

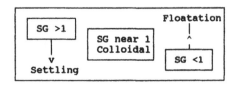

Figure 8.1 Specific gravity effects on settling.

tling characteristics is establishing the fluid flow regime, as either laminar or turbulent flow, as defined by the Reynold's number:

$$N = p \, d \, V_s / \mu \qquad (8.1)$$

Because type I solids settle at a constant velocity, not changing in physical characteristics, the settling velocity can be represented by Newton's law, which in the quiescent conditions required in a settler can be represented by Stoker's law.

For the special case where N is less than 0.3, Stoke's law is defined by Equation (8.2).

$$V_s = \frac{g \, (p_s - p) \, d^2}{18 \, \mu} \qquad (8.2)$$

where V_s is the settling velocity in m/s or fps, depending on the units selected for the other variables, p is the density of liquid, p_s is the density of solid, d is the particle diameter, g is the gravitational factor, and μ is the liquid viscosity.

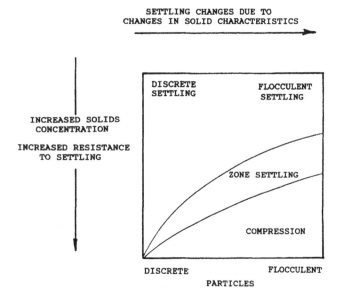

Figure 8.2 Types of settling (adapted from Reference [18]).

The clarifier overflow rate (V_o), defined by Equation (8.3), must be less than the critical solid settling rate for effective separation and to avoid excessive solids carryover.

$$V_o = Q/A \text{ (based on consistent units)} \qquad (8.3)$$

These relations conveniently explain the settling mechanisms and controlling variables, based on "ideal" solid properties. However, theoretical calculations are frequently replaced by laboratory testing when the settler is a *sole or critical treatment unit*.

Design Criteria

These units are often employed as primary treatment elements, used to reduce solids loadings, with the resulting waste discharged to secondary treatment systems. As a result, designs are frequently based on "an educated guess" and industry experience. Where laboratory testing is warranted, procedures detailed in waste treatment texts are employed to obtain process data [18].

Type I settling laboratory testing involves collection of data correlating particle size with settling velocity. The tank area is determined by selecting the effective settling velocity (V_c) based on the smallest particle size that is to be completely (100%) removed. This effective settling velocity (V_o), expressed as m/s (fps), is the basis for estimating the required tank area.

$$\text{Area} = Q/V_o \qquad (8.4)$$

where A is the area in m^2 (ft^2) and Q is the forward flow in m^3/s (ft^3/sec).

Theoretical estimates, however reached, must be adjusted for the variety of full-scale process conditions encountered. U.S. Environmental Protection Agency (EPA) design procedures suggest that the selected overflow rate be corrected by a factor of 1.25 to 1.75, whereas municipal design manuals suggest a 1.5 to 2 factor [12,18].

$$V_o/\text{correction factor} \qquad (8.5)$$

Scouring

Scouring is a major concern where high-level sludge inventories are subjected to excessive forward velocities, upsetting the solids, resuspending them, and potentially resulting in high carry over. Scouring velocities at which solids can be upset can be estimated from the Camp Equation (8.6) [18]:

$$V_s = 1.3 \, [(s - 1) \cdot d]^{0.5} \qquad (8.6)$$

where V_s is the critical liquid velocity at which particles start to scour, cm/s (fps), *s* is the specific gravity, and *d* is the diameter of the scoured particle, cm (mm).

Some general criteria cited as guidelines to prevent scouring include [18]

(1) The liquid forward velocity should be less than 9 to 15 times greater than the settling velocity of critical sized solids.
(2) Velocities in grit chambers should be limited to 1 fps (0.3 m/s).

TYPE II FLOCCULENT SETTLING

Flocculent settling involves removing *agglomerating* solids, caused by solids frequently in contact, or the presence of flocculating agents. The agglomerating conditions results in changes in solids physical characteristics and increasing settling velocities, which are depth (time) dependent. Flocculent settling is encountered in primary municipal treatment clarifiers and in waste facilities handling pulp, plastics, and special adhesive-type chemicals. In such cases, settling cannot be mathematically modeled because their characteristics change throughout the process.

Design Criteria

These units are often employed as primary units to reduce secondary treatment loading. As with type I settlers, design is frequently based on available data or industry experience, sometimes supplemented with laboratory *verification* testing. Specific design criteria can only be established by conducting extensive testing. The scope of testing depends on how critical the settler performance is to the overall treatment scheme.

Type II settling depends on flocculating solid size, retention time and vessel height. Typically, data are collected and developed into removal plots defining the settling characteristics of agglomerating solids, correlating settler depth, retention time, and total removal percentages. Experimentation techniques are detailed in waste treatment texts [18]. The tank area is established by selecting the effective overflow rate consistent with the required total removal efficiency, based on an "educated guess," industry operating experience, or from laboratory removal data. The selected settling velocity is the basis for estimating the required tank area, as defined by Equation (8.4). The theoretical area estimates must be adjusted for the variety of operational conditions encountered. EPA design procedures suggest that the selected overflow rate be corrected by a factor of 1.25 to 1.75, the correction applied as indicated by Equation (8.5) [12].

TYPE III ZONE SETTLING

Zone settling involves removing solids from highly *concentrated suspensions* with considerable solids contact, re-

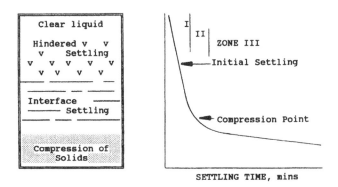

Figure 8.3 Type III settling zones (adapted from Reference [25]).

sulting in *hindered settling* and a distinct *mass zone* formed. As illustrated in Figure 8.3, three distinct zones could form: a top zone engaged in hindered settling at a relative constant rate, pushing clear water to the top, an interface zone in which settling is defined by the movement of a distinct blanket, and the bottom zone engaged in sludge thickening to its final concentration, caused by the compression force of the upper solid layers. To some degree, these three distinct zones could be encountered in both clarifying and thickening equipment, defined by fixed levels in continuous processes.

Mechanisms promoting type III settling depend on the slurry concentration as well as the solids flocculating and compressive solid properties. In any sedimentation process, these mechanisms result in continuous cocurrent clarification and solids transport, one of which controls performance. Type III settling has received considerable attention because it defines secondary clarifier activity critical to activated sludge treatment performance.

Theoretical relations have been developed to relate municipal sludge settling (V) to the initial concentration (X_1), its waste characteristics, and the measured sludge volume index (SVI) value. They are listed in Table 8.1 and include the Power [4] and Vesilind Exponential [15] Equations, as well those relating settling rate to the SVI [2,16]. The interested reader should review the cited references for the origin, application, and process limits of these correlations. They may be useful in correlating specific test or operating data, developing site-specific constants to define settling systems.

TABLE 8.1. Sludge Settling Relations.

Power Equation
$$V = a' \cdot X_i^{-n'}$$

Vesilind Exponential Model
$$V = a \cdot e^{-n \cdot x_i}$$

Dager and Roper Relation
$$V = 7.80 \, e^{-nC_i}$$
$$n = 0.148 - 0.00210 \, (SVI)$$

Clarification

Continuous clarification is commonly analyzed assuming ideal horizontal plug flow conditions, with available settling time equal to the travel time from the vessel inlet to the outlet. Under these conditions total solids removal depends on the collective terminal solids settling rate at the feed concentration.

$$V_d \geq Q_o/A = G_o \qquad (8.7)$$

As indicated by Equation (8.7), solids with terminal velocities (V_d) greater than the overflow rate (G_o) will be removed. If all solids do not have adequate terminal velocities, their removal efficiency will be proportional to their terminal velocity relative to the overflow rate. The overflow rate is equal to the flow volume divided by the horizontal area. In reality, ideal plug conditions are never fully achieved because of vessel turbulence, poor vessel distribution, short-circuiting, and solids scouring at the bottom thickening zone.

Solids Transport Capacity

Solids traveling from the clarification zone will be subjected to thickening forces resulting in a concentration gradient. The solids velocity decreasing with increasing concentration, solids and liquid being transported in opposite directions. The total solids transport *at any point* is a result of the combined effect of the solids settling forces and the discharge pump transport velocity, represented by Equation (8.8).

$$W = C \cdot V \cdot A + C \cdot U \cdot A \qquad (8.8)$$

At any point in the clarifier, W is the mass rate of solids being removed from the system, transported by the settling *velocity V* at concentration C through area A, plus the pump *velocity U* at concentration C through the area A. The velocity is expressed as m/hr, m^3/hr/m^2, or any other appropriate units. Solids transport can be represented by the total solids flux, defined by Equation (8.9).

$$W/A = G_t = C \cdot V + C \cdot U \qquad (8.9)$$

where G_t is the total solids removed per time per unit area.

General Design Considerations

Unlike type I and II settlers, type III settlers are almost always critical units in treatment systems, significantly affecting overall system performance and therefore warranting laboratory or field testing to define vital operating criteria. Settling tank process design involves estimating both clarification requirements and solids transport capacity, and designing the system on the basis of the controlling criteria.

In essence, this requires designing to eliminate the potential for process failure, which could result in massive effluent solids discharge. Sedimentation design procedures have been extensively studied, and various methods were proposed based on assumed ideal separation mechanisms and specific laboratory data requirements. Attempts to define zone III design procedures are complicated because zone I discrete settling, zone II flocculating settling, and zone IV compaction can be part of the process. Most proposed design procedures have evolved from municipal plant experience, based on final clarifiers containing flocculating sludge. In such cases, extensive effort has been expended to characterize the waste and the resulting settling process because of its critical effect to the activated sludge system performance.

Clarifier design techniques fall into one of three general categories, commonly referred to as (1) the Talmage and Fitch techniques [11], (2) the ISV method [20], and (3) the solids flux method [3,18]. A major consideration in employing these techniques is the required laboratory test data.

- The *Talmage and Fitch technique,* adopted from the work of Coe and Clevenger, utilizes a single batch test to evaluate both the clarification and thickening requirements. Because this method is based on the Kynch settling model, it reportedly has disadvantages limiting its application in industrial and municipal waste design [3].
- The *ISV method* is suggested by its proponents as an alternative to the solids flux method, *where the initial settling rate can be defined.* This method, as suggested by Wilson and Lee and others [2,6,10,20], proposes that the design overflow rate (Q/A) can be estimated using the initial settling velocity corrected by a safety factor, as indicated by Equation (8.10).

$$Q/A = \text{ISV/SF} \qquad (8.10)$$

The safety factor is related to the reflux ratio and waste characteristics. The reader is directed to the cited references for mathematical derivations exploring how identical results can be obtained using the solids flux on the ISV method.

- The *Solids flux method* assumes that there is a limiting solids loading, the clarifier "pinch point," representing the governing design criteria. This method has been generally accepted as applicable to most sedimentation applications involving high concentrated flocculating and compressible solids. However, extensive laboratory data are required to establish the relationship between concentration and settling velocity.

Although the ISV method can be employed with minimum data, adequate tests must be obtained to assure that the initial settling velocity is defined for the design conditions, and for the range of sludge quality that can be encountered. The ISV method has been interpreted by some as applicable with data obtained from a single batch settling test, thereby reducing test costs. Whether this can adequately define the sludge characteristics or is cost effective considering the capital investments is debatable. The total flux method will be discussed in detail because it allows the Process Engineer the ability to explore the operating range of a clarifier as well as the potential for process failure.

The total flux theory assumes that assumptions included in ideal settling techniques are not valid. Instead it is assumed that the solids have flocculating characteristics, are not all the same size, are not all the same shape, and are not rigid. Studies conducted on secondary sludge concluded that it did not meet the ideal settling characteristics because of the following reasons [3]:

(1) Subsidence rate is dependent on concentration, sludge depth, and mixing of the underlying layers.

(2) The settling rate of activated sludge has been shown to be an exponential function of the concentration, as indicated in Table 8.1.

The greatest deviation from ideal settling theory is not only the characteristics of individual particles but also that flocculating particles form a structured suspension that is subject to forces other than hydraulic transport.

Total Flux Theory [3,4]

The total flux theory is based on the settler solids capacity, described by Equation (8.9). In this equation the $C \cdot V$ term is the gravity solids flux, the magnitude of which depends on the floc settling characteristics; $C \cdot U$ is the system transport and withdrawal rate, the magnitude of which depends on the underflow pumping rate and underflow concentration. C represents the clarifier solids concentration, ranging from the influent to the underflow concentration.

Graphically, the limiting flux equation can be described as a combined or total flux curve represented by Figure 8.4, a composite of the floc settling characteristic as depicted in Figure 8.4(a), and the transport characteristics depicted in Figure 8.4(b).

Some basic clarifier characteristics are evident from the total flux curve, Figure 8.4(c).

(1) The minimum flux (G_L) is defined by a horizontal line drawn tangent to the x-axis at the minimis point of the total flux curve.

(2) A line from this minimis point to the transport line represents the clarifier's solids settling capacity ($C \cdot V$). The line from the transport line to the x-axis represents the underflow transport capacity $C \cdot U$, whereas the x intercept X_m represents the maximum underflow concentration.

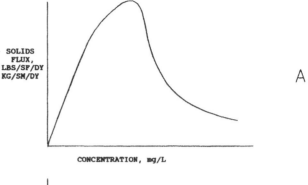

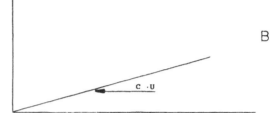

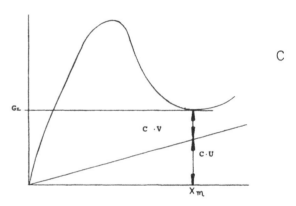

Figure 8.4a Type III: settling characteristics (adapted from concepts in Reference [3]). (b) Type III: transport characteristics (adapted from concepts in Reference [3]). (c) Type III: total flux curve (adapted from concepts in Reference [3]).

(3) Increasing the underflow pumping capacity moves the total curve upward, including the minimis point, and thereby increasing the allowable minimis point (G_L).

The potential for clarifier process failure can be evaluated from an analysis of the *total flux curve.*

Potential for Process Failure

Clarification failure is defined as a massive solids discharge resulting from thickening overloading. The potential for process failure can be best illustrated using the total flux graphical method, as illustrated in Figure 8.5. The detailed evaluation is illustrated in Figures 8.5 to 8.8 [3,4].

Referring to Figure 8.5: a line from the desired underflow concentration C_u, tangent to the equilibrium line, represents the transport line (C_u, G_L). The slope of the line represents

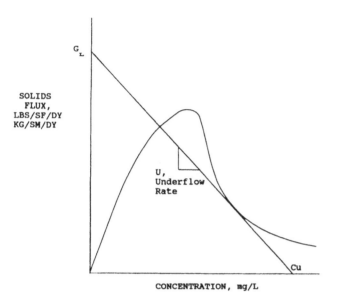

Figure 8.5 Total flux curve, transport line (adapted from concepts in Reference [3]).

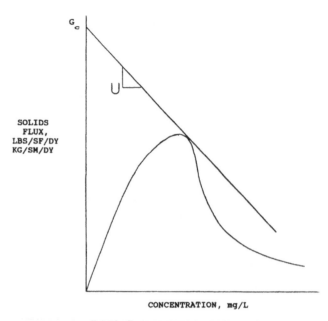

Figure 8.7 Total flux curve, total allowable solids flux (adapted from concepts in Reference [3]).

the underflow rate (U). The intercept of the transport line with the y-axis is the limiting total flux value (thickening), G_L.

Referring to Figure 8.6: the slope of the line from point "0–0" to the tangent point, defined by the equilibrium line and the transport line, represents the settling velocity, V. A perpendicular line from the tangent line to the x-axis represents C_L, the limiting feed concentration. A line from the tangent point to the y-axis, parallel to the x-axis, represents the maximum solids flux, G_b. The vertical distance from the G_L to G_b on the y-axis is the solids transport flux $C \cdot U$, whereas the vertical distance from G_b to the x-axis is the

solids settling flux $C_i \cdot V$, the sum being the total flux as defined by Equation (8.9).

Referring to Figure 8.7: a line tangent to the equilibria curve at a slope U will intercept the y-axis at the maximum solids flux (G_c) for the system.

As the underflow (maximum) concentration increases, the allowable total solids transport flux decreases, as indicated in Figure 8.8.

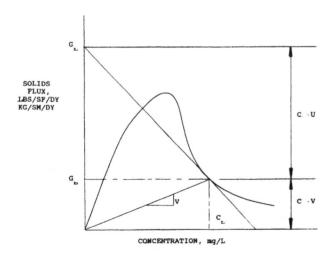

Figure 8.6 Total flux curve, settling line (adapted from Reference [3]).

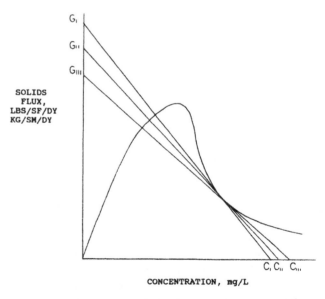

Figure 8.8 Total flux curve, total solids flux variation (adapted from concepts in Reference [3]).

GRAVITY THICKENERS

Historically, the Coe and Clevenger equation or equivalent graphical methods have been used to analyze critical thickener designs [5,11]. In addition, data obtained for clarifier solids transport analysis can be extended to estimate allowable thickener loadings. However, because thickeners are not as critical as clarifiers to upstream biological processes, their design is seldom subjected to detail analysis but "selected" to improve downstream dewatering performance, using an appropriate solids flux. In fact, if the final sludge is difficult to thicken there is probably a trail of malfunctions from the reactor to the clarifier, making thickening performance irrelevant. Where performance is critical, thickeners are best designed from continuous data utilizing prototypes of the same height as commercial operating units.

CLARIFICATION AND THICKENING SYSTEMS

Settlers or thickeners are designed on the basis of three general considerations:

(1) Adequate process capacity to achieve required solids removal, sludge concentration, and effluent quality
(2) An adequate transport device to continuous remove sludge
(3) Internal mechanical components to reduce turbulence and stabilize the process, minimizing effluent leakage

The system can contain internal or external chemical coagulation and flocculation capabilities to condition the solids and improve settling. Settling equipment can be classified as either *clarifiers* or *thickeners* and can be *rectangular* or *circular.*

CLARIFIERS

Either *rectangular* or *circular* configurations can be employed for clarification, the principle differences being process and construction considerations. As the required clarifier area increases rectangular clarifiers are easier and more economical to construct. When multiple units are required, the construction alternative becomes one of using multiple circular vessels with overall minimum perimeter lengths or simpler constructed rectangular units using common walls. For industrial facilities total area is usually not a major consideration, single units are commonly used, two units included only for process flexibility, and more than two are seldom required. Available area in a vicinity next to the secondary treatment, biological or chemical, is usually the major concern.

Circular clarifier diameters range from 3 to 90 m (10 to 300 ft) and 3 to 5 m (10 to 16 ft) deep. Rectangular length-to-width ratios range from 1.5:1 to 15:1, with a minimum 3 m (10 ft) length and depths from 2 to 4 m (7 to 12 ft) [18]. System hydraulics and the resulting flow pattern are a major process concern. Circular settler flow patterns are radial, whereas those of rectangular units are plug flow.

Theoretically, plug flow configurations minimize short circuiting or influent leakage from the system. Circular clarifier may require more sophisticated internal flow control to achieve rectangular configuration results, but this is not a major problem.

Clarifier design must incorporate some basic physical features to achieve performance criteria. This includes

(1) Adequate area to achieve required performance
(2) Adequate depth to allow agglomeration, sufficient concentration, and some solids storage
(3) An adequate inlet design to distribute flow and dissipate energy, thereby minimizing system turbulence
(4) Internal mechanisms to minimize short-circuiting and feed leakage to the effluent
(5) A discharge system to collect exit flow in a quiescent manner to minimize system turbulence
(6) Sludge collection and discharge mechanisms
(7) A scum removal system

These requirements apply to circular or rectangular units. *Mechanical details* for both rectangular and circular are discussed in the General Engineering section and will not be repeated here. Each manufacturer may have specific proprietary designs to accommodate these considerations.

Selection of either a circular or rectangular clarifier is commonly based on their relatively advantages and disadvantages, and the costs. Circular clarifiers have the advantages of lower maintenance, simpler sludge collection mechanisms, and shorter detention times and the disadvantages of a higher potential for short circuiting, uneven sludge collection, and lower allowable weir loadings. They are used in activated sludge systems and water treatment plants and for chemical sludges [18].

Rectangular clarifiers have the advantages of simpler construction for large area or multiple units, less chance of short-circuiting, good sludge distribution at the collection device, and higher allowable weir loadings and the disadvantages of being less effective for high solids loadings and frequently requiring longer detention times. They are commonly applied to activated sludge systems [18].

Flocculation capabilities to improve settling is common for all clarification systems. This is an important concern if the sludge solids have been subjected to shearing forces prior to clarification, such as from activated sludge pumping systems, or the feed innately contains significant submicron particles. Table 8.2 details typical flocculating methods [9].

As illustrated in Table 8.2, these systems can be categorized as (1) separate flocculation units, (2) integral flocculation employing center feeds, or (3) flocculation resulting from contact between the waste solids and the sludge blanket.

TABLE 8.2. Clarifier Flocculation Methods (adapted from Reference [9]).

	Center Well Residence, Minutes	Sludge Blanket, Meters	Feet
Separate flocculation	Special design as needed		
Conventional circular:			
Low sludge blanket	3–5	<0.3	1
High sludge blanket	3–5	1–2	3–7
Center well	20–30	<0.3	1
Upflow settler	<2	4	13

More details of coagulation and flocculation are discussed in Chapter I-7.

THICKENERS

Thickeners resemble conventional circular clarifiers with the exception of having a greater bottom slope to accommodate heavier sludge characteristics and the system mechanically designed for heavier solids loading. The internal mechanism normally consists of two scrapper arms mounted to a hoist mechanism. Gravity thickeners are commonly circular, up to 24 m (80 ft) in diameter and with steep floor slopes (2:12 or 3:12) [19]. Drive mechanisms inherently are heavier than clarifiers and designed to accommodate the type of sludge handled.

PROCESS ENGINEERING DESIGN

General process limitations and defining operating characteristics of settlers are summarized in Table 8.3. Clarifier (or thickener) control is primarily dependent on upstream

TABLE 8.3. Operating Characteristics.

Variable	Operator Controllable	Critical
Waste Characteristics		
Waste generated	No	Yes
Composition	No	Yes
Concentration	No	Yes
Solid properties	No	Yes
Settleability	No	Yes
Operating Characteristics		
Flow rate	Minimal	Yes
Overflow rate	Minimal	Yes
Solids transport	No	Yes
Recycle	Yes	No
Sludge blanket depth	Yes	Yes
Sludge wasting	Yes	No
Chemical addition	Yes	Yes
Anaerobic conditions	No	Yes

process conditions affecting sludge characteristics. When poor settling does occur, the only viable operator control is chemical addition to stabilize the problem, followed by inspection and correction of the upstream process conditions generating the sludge. One major settling variable difficult to observe is solids transport capacity, which could affect clarifier performance even at a minimal overflow rate. Solids transport capacity is a product of total solids feed rate, affected by sludge concentration and sludge recycle rate. A review of common clarifier and thickener operating deficiencies detailed in the General Engineering section demonstrates the effects of poor design.

REPORTED PERFORMANCE DATA

Sedimentation systems are sometimes designed using municipal plant criteria, assuming that equivalent effluent quality can be achieved by "proper" operation. As a frame of reference, results of approximately 40 industrial facilities reported in the EPA *Treatability Manual* are summarized in Tables 8.4 through 8.9 [14]. The data reported are for a variety of industries such as paper mills, petrochemicals, iron and steel, hospitals, leather, textile, organic chemicals, pharmaceutical, rubber, and resin, where sedimentation was used as primary or principal treatment, with and without chemical treatment. These data are cited to demonstrate the range of performance possible, with no assurance that the facilities reported were properly designed or operated, that these results are the best that can be achieved, or that these results can be achieved where influent characteristics differ.

REQUIRED PROCESS DESIGN DATA

Preliminary criteria for critical settling and thickening processes can be obtained using laboratory scale equipment, settling columns, or field prototypes. Although bench scale tests are commonly used for obtaining design data, settling column tests, using 6- to 8-ft test units, are recommended to obtain more extensive settling characteristics. In addition, tests should be repeated with different waste samples to observe any variation in waste characteristics. Tests should be conducted using fresh sludge and not waste samples that have been stored. Commonly employed test procedures are detailed in the literature [18]. Regardless of the manner of obtaining data, the required design criteria are listed in Table 8.10.

WASTE EVALUATION

The waste characteristics significantly affecting settling are sludge concentration, solids characteristics, and flow rate. Waste stream concentration establishes the settler's function as a clarifier or thickener. Generally, with a combination of clarification and thickening the maximum achievable concentration is 10%; that of a standard clarifier alone is 1%. A clari-

TABLE 8.4. Sedimentation, with No Chemicals (adapted from Reference [13]).

	BOD$_5$	COD	TOC	TSS	O/G	TTL P
			Effluent, mg/L			
Minimum	980	4	1	<1	2.7	
Maximum	6670	25,300	940	5700	522	
Median	1150	18	12	29	9	
Mean	2500	1620	63	212	70	13.9
			Removal, %			
Minimum	0	0	0	0	0	
Maximum	69	>99	>99	>99	99	
Median	25	91	29	97	40	
Mean	33	71	40	82	47	3

fier feed from biological systems is seldom greater than 5000 mg/L or a thickener feed greater than 10,000 mg/L; chemical sludge performance can vary, with thickeners frequently employed in place of clarifiers. Concentration affects settling with slurries above 1000 mg/L potentially encountering hindered settling because of increased solids contact. Solid characteristics affecting settling are density, diameter, and their flocculating or nonflocculating nature.

Flow rate is an important clarifier (or thickener) design and operating variable, being a major consideration in vessel sizing and operating stability. Wide flow variations subject the settler to increased turbulence, short-circuiting, and increased solids carry over. In addition, increased flows result in increased overflow rates, decreased residence time, and reduced settling efficiency. A positive means of distributing flow during both normal and upset conditions is critical for multiple units.

Clarifier loading is a major consideration in developing the treatment plant hydraulic profile. A selected design flow rate must be evaluated for flow variation and peak flow impact. Modern industrial complexes commonly segregate industrial waste water from storm water sewers so that dry and storm water conditions are not always applicable. In industrial systems wastewater variations are almost always production related. However, the same general criteria recommended for municipal design should be evaluated [18,19,21]:

(1) Select an appropriate overflow rate consistent with the waste characteristics, applying an appropriate safety factor.

(2) Establish area requirements for average flow conditions.

(3) Estimate the overflow rate and weir rates for peak production and establish maximum anticipated variation.

(4) Make a reasonable area adjustment for peak conditions.

(5) If settling tank influent variation exceeds a factor of *two*, conduct a detailed process evaluation considering potential effects on the settling equipment and corrective measures that could be implemented at the generating source.

PRETREATMENT

Settler pretreatment requirements are generally related to equipment protection or conditioning to assure a settleable floc. Primary consideration is screening large solids which could damage the drive mechanism or the motor. In addition, pretreatment should be considered to reduce oily and floating materials producing excessive clarifier scum.

PROCESS DESIGN VARIABLES

Clarifier and thickener design involves both process and mechanical considerations to develop a vessel configuration

TABLE 8.5. Sedimentation with Alum (adapted from Reference [13]).

	BOD$_5$	COD	TOC	TSS	O/G	TTL P
			Effluent, mg/L			
Minimum	3.6	212	72	28		2.3
Maximum	2900	7600	1500	122		43
Median	33	416	105	50		
Mean	1040	2410	437	55.8	11	22.7
			Removal, %			
Minimum	0	4	5	0	0	12
Maximum	82	71	80	99		15
Median	16	61	63	79		
Mean	<47	45	53	<67	99	14

TABLE 8.6. Sedimentation with Alum and Polymer (adapted from Reference [13]).

	BOD$_5$	COD	TOC	TSS	O/G	TTL P
	Effluent, mg/L					
Minimum	57	125	40	46	4	
Maximum	3800	30,000	4800	6000	880	
Median	2800	10,000	2850	1370	80.5	
Mean	2150	12,100	2640	2200	261	1.6
	Removal, %					
Minimum	7	38	37	0	48	
Maximum	65	80	71	99	99	
Median	25	69	58	67	80	
Mean	35	62	56	58	77	77

to achieve high separation efficiency and low internal solids disturbance. The major process evaluation involves selecting an appropriate overflow rate, solids loading rate, and estimating a suitable vessel area for the controlling condition. In so doing, the Process Engineer must carefully balance theoretical considerations with available operating experience, interpreting laboratory testing to include the range of expected sludge characteristics. Mechanically, consideration must be given to internal vessel design to assure favorable feed distribution and dissipating inlet energy feed energy, as well as the corresponding effluent discharge hydraulics, to assure minimal short-circuiting, vessel turbulence currents, and disturbance of the settled (stored) sludge.

The Process Engineer must be aware that industrial waste sludges are not necessarily similar to that generated in municipal systems, even when the sludge is generated from a biological treatment system. In some cases the sludge may be a direct result of a manufacturing process or from a chemical treatment process, not from a biological system. A safety factor, commonly referred to as a scale-up factor, is recommended by the EPA for all settling vessels to account for the physical differences between laboratory test units and operating units. This is an excellent barometer of the state-of-the-art of clarifier and thickener design, based more on the data available from numerous existing operating clari-

fiers and thickeners than application of any specific design method.

One final word, the engineer will encounter numerous studies where theoretical theories are proposed for poor and unpredictable settling in existing municipal plants. As discussed in the Biological Treatment chapter, poor sludge quality is frequently an upstream process problem, and may not be easily solved at the clarifier. Waste treatment plant operators are aware poor settling sludge will not produce a suitable effluent quality regardless of the clarifier area (or polymer dosage) considered, and must be remedied as part of a entire system upgrade.

Clarifier or thickener process design requires selecting and evaluating the following process parameters:

(1) Chemical treatment
(2) Depth
(3) Weir loading
(4) Hydraulic loading
(5) Solids loading
(6) Nonideal conditions
(7) Special criteria for thickeners
(8) Fate of the contaminants

Chemical Treatment

Coagulating agents such as alum, lime, and polymers are sometimes used to agglomerate solids to promote floccula-

TABLE 8.7. Sedimentation with Lime (adapted from Reference [13]).

	COD	TOC	TSS	O/G
	Effluent, mg/L			
Minimum	8	9	4	1
Maximum	37	<20	150	1.5
Median	23.8	<12	12.5	
Mean	23.1	13.7	44.7	1.25
	Removal, %			
Minimum	0	>5	0	0
Maximum	50	37	96	66
Median	14	18	71	
Mean	32	>20	61	33

TABLE 8.8. Sedimentation with Polymer (adapted from Reference [13]).

	BOD$_5$	COD	TOC	TSS	O/G
	Effluent, mg/L				
Mean	4700	8000	1600	39	22
	Removal, %				
Mean	2	71	82	>99	98

TABLE 8.9. Thickener with No Chemicals
(adapted from Reference [13]).

	Loading, lb/day/ft²		
	kg/day/m²	lb/day/ft²	Bottom, %
Primary	98–146	(20–30)	8–10
WAS	24–29	(5–6)	2.5–3
Trickling filter	39–49	(8–10)	7–9
Limed tertiary	(293)	60	12–15
Primary and WAS	29–49	(6–10)	4–7
Primary and TF	49–59	(10–12)	7–9
Limed primary	98–122	(20–25)	7–12

tion and settling. The required chemical dosage depends on the sludge type and the sludge process history. In many cases pH control is required to optimize the coagulating process. Chapter I-9 discusses coagulation and flocculation and the related chemical systems. Treatment chemistry is specific to the waste characteristics and established by bench-top testing.

Depth

There are no hard and fast criteria governing clarifier depth. The design basis is frequently predicated on recommended minimum depths based on available experience, factored by cost and construction restraints. Recommended depths are commonly related to vessel shape, vessel diameter, sludge characteristics, type of service, and sludge blanket depth. Since clarifier depth will directly affect vessel cost, it must be balanced against other possible process improvements. Common factors affecting depth include [9,18,19,23]

(1) Depths required for circular or rectangular units are generally assumed to be the same, although rectangular units can be shallower if the sludge is withdrawn at the effluent end.

(2) Studies suggest that clarification depth begins at the sludge blanket top, from the sludge top to the weir, implying similar performances with low blanket shallow tanks and high blanket deep tanks [1,19]. This is not universally accepted. The general consensus being that

TABLE 8.10. Required Design Data.

Critical pilot plant treatability data specific to the waste
  (1) Design temperature
  (2) Type of settling
  (3) Settling vs concentration relation

    Selected operating characteristics
  (4) Overflow rate
  (5) Solids loading rate
  (6) Polymer dosage requirements
  (7) Achievable effluent quality
  (8) Solids concentration

effluent quality improves with increased depth as well as increased depth between the sludge and the weir [9,18,19,23]. However, depth may not be an independent parameter affecting effluent quality but interrelated with detention time, overflow rate and the clarifier physical characteristics [9,19,22].

(3) Blanket depth serves as a source of sludge inventory and a buffer for process variations. A 0.6- to 0.9-m (2 to 3 ft) blanket allowance should be included in the design.

Generally, municipal facilities clarifier depths range from 4 to 5 m (12 to 16 ft) often related to the tank diameter, as illustrated in Table 8.11 [17–19]. Overflow rate corrections of 0.17 m/h (100 gpd/ft²) for each 0.3 m (1 ft) of depth less than the recommended minimum stated in Table 8.11 are suggested. Table 8.12 details EPA-recommended depths based on type of municipal system service [12]. Some investigatory data suggest a significant effluent quality improvement at depths greater than 5 m, up to 5.5 m (18 ft) [9].

Weir Loading

Weir loadings are frequently arbitrarily chosen based on the concern for effluent quality, driven by a desire to minimize solids carryover. Conservative loadings frequently imposed by regulatory requirements for municipal systems range from 125 m³/m-day (10,000 gpd/ft) maximum for small treatment plants of less than 160 m³/hr (1 mgd) to 190 m³/m/day (15,000 gpd/ft) for larger plants [19]. The EPA guidelines (Ten States Standards) for municipal secondary sedimentation suggests a 190 m³/m-day loading (15,000 gpd/ft) based on average design flow rates, with a maximum rate of 250 m³/m-day (20,000 gpd/ft) if excessive density currents or solids disturbance can be avoided in the weir location [12].

Weir configuration and location are selected to prevent excessive solids migration, although the method of achieving this criteria can vary, and is frequently driven by construction costs. Generally, single baffled peripheral weirs are employed, with double lauders installed where increased length is required. The weir location is specific to the clarifier configuration, its internal design features, and manufacturer's proprietary designs. There is some evidence that weir placement can significantly affects solids carryover, with inboard placement better than locating them at the wall [9].

Hydraulic Loading

Commonly applied municipal design loadings are summarized in Table 8.13 [19]. Table 8.14 enumerates EPA recommendations for municipal design, based on process service and flow conditions [12]. These loading can serve as a guide for similar industrial applications as a check of the design

TABLE 8.11. Clarifier Depths per Diameter (adapted from References [17,19]).

Tank Diameter					
Meters	<12	12–21	21–30	30–43	>43
Feet	40	40–70	70–100	100–140	140
Proposed SWD					
Meters	3.4	3.7	4.0	4.3	4.6
Feet	11	12	13	14	15
Minimum SWD					
Meters	3.0	3.4	3.7	4.0	4.3
Feet	10	11	12	13	14

estimates obtained from laboratory studies and design procedures based on a total flux analysis.

Solids Loading

Suggested municipal maximum solids loadings range from 4 to 6 kg/m²-hr (20-30 lb/d/ft²), with EPA-suggested criteria ranging from 4 to 6 kg/m²-h (20-30 lb/day/ft²) for average flow rates, and 10 kg/m²-h (50 lb/day/ft²) for peak flow rates [12,19]. Floor slope affects sludge removal and the potential for solids entrainment, influencing circular tank design overflow rates. If a flat floor is employed the hydraulic loading is corrected by 8 m³/day/m² (200 gpd/ft³). A 1:12 slope is considered good engineering practice [19]. These loadings can serve as a guide for similar industrial applications as a check of the design estimates obtained from laboratory studies and design procedures based on a total flux analysis.

Special Criteria for Thickeners

Thickener design generally follow the principles discussed for clarification, with design parameters adjusted for thickening performance. Common parameters cited for municipal sludge characteristics are summarized in Table 8.15 [24].

Nonideal Conditions

General clarifier criteria are based on near ideal hydraulic conditions within the vessel. Ideal conditions cannot be

achieved by solely selecting a suitable area based on hydraulic loadings, adjustments must be made for expected solids loading. In addition depth and weir design must be selected to minimize vessel turbulence and solids entrainment. Actually, clarify performance will always be less than predicted from the ideal models, with settling performance being influenced by internal vessel factors such as [12]

(1) Short-circuiting caused by wind, inlet and outlet turbulence, and temperature (density) differences in the basin

(2) High concentration solids settling in mass, further impacted by agglomeration of flocculating solids

(3) Scouring and resuspension of solids

(4) Other forces that could cause internal turbulence

Estimated design parameters are adjusted for these deviations from ideal hydraulics by a scale-up factor, as indicated in Table 8.16 [12].

FATE OF CONTAMINANTS

Air emissions from processing primary or secondary treatment biological sludge, where any volatiles have already been stripped, are generally not a problem unless anaerobic conditions occur. The clarifier products are treated effluent and waste sludge. Waste sludge classification depends on what influent components are partitioned to the sludge, which could be significant. If any of these components are classified as hazardous or toxic, final disposal could be expensive, and final disposal sites limited. Primary sludge generated in industrial facilities will contain the gamut of chemicals processed in the manufacturing facilities.

TABLE 8.12. Clarifier Depths per Service (adapted from Reference [12]).

	Depth	
	Meters	Feet
Primary clarifier	3–4	10–12
Primary clarifier		
Plus activated sludge	4–5	12–15
Secondary clarifiers		
Trickling filter	3–4	10–12
Any form activated sludge system	4–5	12–15

TABLE 8.13. Common Municipal Loadings (adapted from Reference [19]).

Forward Feed	Range	Design
Average flow	16–29 (400–700)	23 (560)
Peak flow	41–65 (1000–1600)	50 (1230)*
		51 (1240)**

*Circular.
**Rectangular.
Based on *Forward feed rate*, m³/day/m² (GPD/sq ft)

TABLE 8.14. EPA Municipal Design Loadings (adapted from Reference [12]).

| | Waste Flow Basis | | Corresponding Depth m (ft) |
| | m³/day/m² | (GPD/sf) | |
	Average	Peak	
Primary clarifier	33–49 (800–1200)	81–122 (2000–3000)	3–4 (10–12)
Primary with activated sludge	24–33 (600–800)	49–61 (1200–1500)	3–4 (10–12)
Secondary clarifiers			
Trickling filter	16–24 (400–600)	41–81 (1000–2000)	3–4 (10–12)
Conventional	16–33 (400–800)	41–49 (1000–1200)	4–5 (12–15)
Extended	8–16 (200–400)	33 (800)	4–5 (12–15)

GENERAL ENGINEERING CRITERIA

Clarifier or *Thickener* devices and auxiliary components are commonly designed and purchased separately, to be installed in a steel or concrete shell designed and constructed by others. Vendor supplied design details frequently includes fabrication and installation details. Figures 8.9 and 8.10 illustrates a typical clarifier and thickener preliminary concept flowsheet.

COMPONENTS OF CIRCULAR CLARIFIERS [18]

- Diameter: 3 to greater than 100 m (10 to 300 ft)
- Primary SWD: 2.5 to 4 m (8 to 13 ft)
- Secondary clarifier and thickener SWD: 3 to 5 m (10 to 16 ft).

The major components of a circular clarifier include the *vessel*, the *drive*, the *sludge collection mechanism*, and the *scum removal system*. The vessel includes an *influent pipe* with a *stilling well* to direct the incoming flow and an *effluent weir* around the vessel circumference to effectively distribute the discharge to a *lauder*, directing the collected effluent to

an *effluent pipe*. Feed methods are designed to dissipate the inlet energy over as large a volume as possible, minimizing disturbance at any isolated location. Common configurations are illustrated in Figures 8.11 to 8.13. Internal baffling is common in feed and effluent discharge systems to further minimize short circuiting and turbulence. Weir designs, as illustrated in Figures 8.11 to 8.13, are an extremely important consideration in clarifier design. The same considerations for minimizing entrance turbulence are employed in the effluent discharge design. As illustrated in the configurations, these include either a peripheral center or rim discharge.

The *drive* consists of the *motor, gear reducer,* and the *turn table.* The *sludge collection mechanism* consists of a *rake arm, scraper blades,* and *adjustable squeegees.* Sludge collectors can be classified as either mechanical scrapers or hydraulic suction types. Scrapper mechanisms can be designed as plow or spiral scrappers. Hydraulic suction mechanisms utilize a hydraulic lift to "vacuum" the sludge from across the tank bottom radius. The sludge removal mechanism, along with the floor slope and design, are an integral part of the manufacturer's proprietary design. Skimming devices are employed to prevent floating materials

TABLE 8.15. Common Municipal Thickener Data (adapted from Reference [24]).

	Primary	Primary + WAS	WAS	TF
Influent, %	1–7	0.5–2	0.2–1.5	1–4
Thickening, %	5–10	4–6	2–4	2–6
Hydraulic:				
m³/m²/day	24–33	4–10	2–4	2–6
gal/ft²/day	590–811	98–246	49–98	49–147
Solids				
kg/m³/day	90–144	25–80	10–35	35–50
lb/ft²/day	18–29	5–16	2–7	7–10
Solids capture, %	85–98	85–92	60–85	80–92
Overflow TSS, mg/L	300–1000	300–1000	200–1000	200–1000

TABLE 8.16. Clarifier Scale-Up Factors
(adapted from Reference [12]).

Sizing Parameter	Scale-up Factor
Area	1.25–1.75
Volume	1.5–2

from being discharged with the effluent. The *scum removal system* consists of a *surface skimmer, scum baffle,* and a *scum box,* the scum box having the capability to discharge the collected scum from the system for disposal. Typical system configurations are as illustrated, whereas alternate component designs are listed in Table 8.17 [18].

COMPONENTS OF RECTANGULAR CLARIFIERS [18]

- Length to width ratio: 1.5:1 to 15:1; 3:1 recommended minimum
- Minimum length: 3 m (10 ft)
- Minimum depth: 2 m (7 ft)

The components of a rectangular clarifier are basically similar to a circular, modified to suit the configuration. *Inlet, Outlet,* and *Sludge collection* systems are illustrated in Figures 8.14 and 8.15, and detailed in Table 8.18 [18].

HYDRAULIC FLOW CONSIDERATIONS

When multiple parallel clarifiers are included in the design, hydraulically balancing the flows to the individual units is critical, especially when a unit is removed from service. Flow distribution in circular tanks is commonly accomplished with separate piping or channels, while for rectangu-

lar tanks flows from common channels are controlled with gate valves. Equal flow distribution can be achieved by a variety of methods utilizing proper piping design techniques. Some commonly employed schemes include the following:

(1) Utilizing a *symmetric piping arrangement* and providing manual valves to fine tune the flows, or adjust the flows when units are out of service, as illustrated in Figure 8.16.
(2) Control devices can be utilized for each unit, but this is expensive. A control pressure device at each leg will result in equal flows, as will individually controlled flow devices, as illustrated in Figure 8.17.
(3) Equal flow distribution can be accomplished by weirs splitting the flows, as illustrated in Figure 8.18.

SITE SPECIFIC CRITERIA

Any proposed design should be checked by a civil engineer early in the project for available land, soil conditions, groundwater conditions, and orientation. First consideration in clarification design is the available land space, based on the entire treatment system. The clarification equipment should be designed for gravity flow, and equal flow through multiple units. The available space may dictate whether rectangular or circular tanks are utilized, and where available land is limited common wall rectangular construction may be required. *In no circumstance* should the equipment design be compromised to "fit into" an available area.

Any proposed clarifier and thickening design should be checked for construction restrictions early in the process design. *Site-specific* soil conditions should be evaluated to assure that the proposed facilities can be structurally supported, considering the vessel sizes, the liquid operating

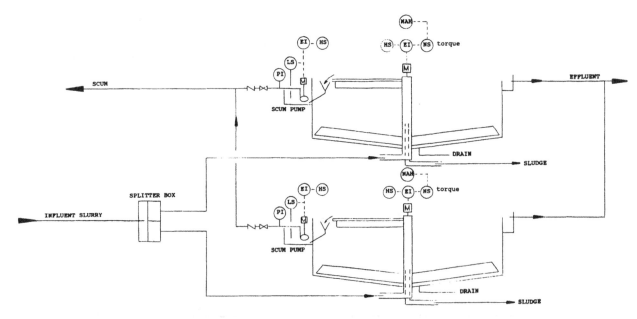

Figure 8.9 Clarifier preliminary concept flowsheet.

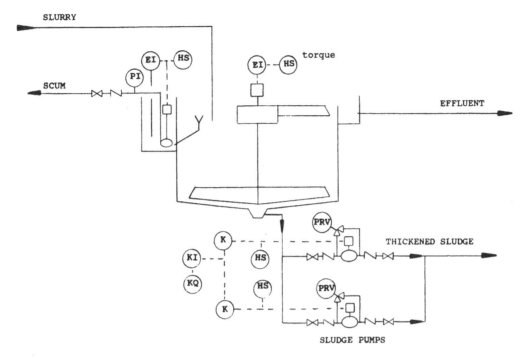

Figure 8.10 Thickener preliminary concept flowsheet.

levels, empty conditions, and the combined structural forces of multicompartment systems. Potential total or select vessel settling should be considered. In fact, the soil conditions could dictate the number of units, size of each unit, the feasibility of common wall design, the system hydraulics, and the vessel depths.

The proposed site groundwater table can dramatically influence the cost of sedimentation equipment. Specifically, water tables can result in

(1) Expensive construction costs, especially in site preparation
(2) More expensive designs to compensate for potential buoyancy forces
(3) Lower allowable vessel depths
(4) Costly lining design to prevent seepage or to comply with RCRA groundwater prevention requirements

The primary effect of wind direction and velocity is wave motion creating a disturbance within the tank and the effluent flow over the weir. Such action is minimized by adequate freeboard provision but in extreme cases may require clari-fier covers. The Process Engineer should take into consideration the vessel orientation to minimize the effects of the site's prevailing wind patterns.

CONSTRUCTION MATERIALS AND RELATED CRITERIA

Clarifier walls can be constructed of reinforced concrete or structural steel, depending on the configuration, size, and cost. Tank floors are commonly concrete constructed. Internal mechanisms are commonly fabricated of cast steel. Outlet and inlet baffles can be constructed of concrete, fiberglass, or steel; with serious attention given to corrosion resistance. The weir is designed as a separate plate, capable of being adjusted for vertical or horizontal alignment to maintain a required working depth. Effluent weirs are usually designed as 90 degree "V" notches, located on 10, 15, or 30 cm (4, 6, or 12 in.) centers, at least 5 cm (2 in.) deep. Scum baffles are usually placed 20 cm (8 in) away from the weir plate and are 30 cm (12 in.) deep. The weir system includes the plate and the trough; constructed of concrete, steel, or fiberglass [18].

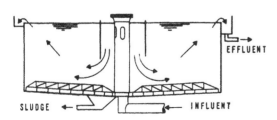

Figure 8.11 Center feed, scraper removal (copied from Reference [12]).

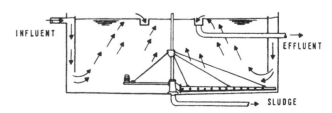

Figure 8.12 Rim feed, suction sludge removal (copied from Reference [12]).

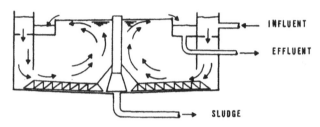

Figure 8.13 Rim feed, rim takeoff (copied from Reference [12]).

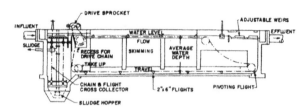

Figure 8.14 Chain and flight collector (copied from Reference [12]).

COMMON CLARIFIER AND THICKENER DESIGN DEFICIENCIES

Some design deficiencies common for municipal systems have been investigated by the EPA and are identified for the reader as a checklist of process and mechanical design considerations. Reference is made to the cited literature for a more detailed discussion of these potential problems. The most common deficiencies cited include [14]

Primary Clarifiers

(1) Rectangular clarifier geometry employed is incorrect. Clarifier length to width should be at least 3 to 1. Minimum length should be at least 3 m (10 ft). Side water depth should be at least 2 m (7 ft), and up to 4 m (14 ft) for larger clarifiers.

TABLE 8.17. Circular Clarifier Component Alternatives (adapted from Reference [18]).

Shape
Circular
Square
Hexagonal
Octagonal

Tank Depth
Shallow
Deep

Feed Point
Center
Peripheral
Flocculating, center feed

Outlet Weirs
Single
Dual
Multiple

Sludge Collectors
Plow scraper
Spiral scraper
Suction mechanism

Skimmers
Beach and scraper
Full radius beach and scraper
Positive scum removal device
Submersible pump

(2) The sludge removal piping systems is inadequate. A common sludge pipe is used for two or more clarifiers resulting in unequal sludge removal. Separate removal pipes (or isolation valves) are not provided for each clarifier.

(3) The effluent weir is not at the proper level.

(4) Short-circuiting results from improper baffling.

(5) High effluent solids results from short-circuiting.

(6) Septic conditions results from excessive hydraulic loading and insufficient sludge withdrawal capacity.

(7) Poor inlet design results in high inlet velocity and poor flow distribution.

(8) Sludge removal mechanism has an inadequate torque.

(9) No provision is included for influent grit removal, resulting in heavy scraper wear.

(10) No flushing or cleanout connections are provided for critical sludge line locations, such as elbows, tees, etc.

(11) Sludge pumps are too far from the clarifiers.

(12) No screen or cutter mechanisms are included upstream of the waste pumps to protect the pump internals. Pump internals are difficult to access; there is no hatch or access port.

(13) Only one clarifier is installed, making repair and part replacement difficult without complete shutdown.

(14) Provisions for influent sampling are inadequate. Separate sampling points are not provided for multiple clarifiers.

(15) No site-glass or equivalent device is included to observe sludge pumps lines.

(16) The clarifier sludge piping system is inadequate. Multi-clarifier pumps and piping are not interconnected, and no spare pumps are provided.

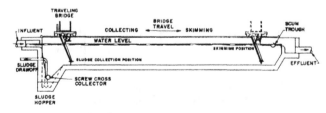

Figure 8.15 Traveling bridge collector (copied from Reference [12]).

TABLE 8.18. Rectangular Clarifier Component Alternatives (adapted from Reference [18]).

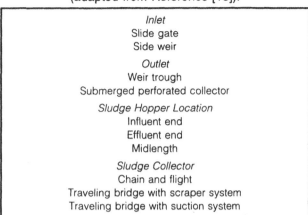

Inlet
Slide gate
Side weir

Outlet
Weir trough
Submerged perforated collector

Sludge Hopper Location
Influent end
Effluent end
Midlength

Sludge Collector
Chain and flight
Traveling bridge with scraper system
Traveling bridge with suction system

(17) Materials of construction are not adequate for the service.

(18) No provision was made to measure sludge withdrawal or recycle.

(19) Poor sludge piping design increases solids plugging problems. Solids collect and plug in the line because small lines are employed, and the system is designed with long lines containing sharp turns.

(20) Grease accumulation results from excess floating materials. No pretreatment is provided for their removal.

(21) Scum removal equipment is not placed near the clarifier outlet, and therefore ineffective.

(22) No provision is made for scum wasting, resulting in recycle and accumulation.

(23) Poor scum pump selection results in excessive wear from concentrated and large solids.

(24) Removal mechanism does not include provisions for freeze protection.

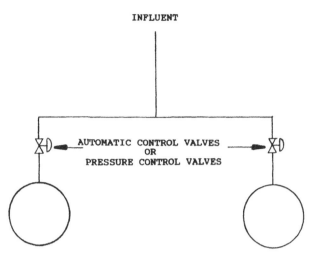

Figure 8.17 Clarifier pressure or flow device flow distribution.

Secondary Clarification

(1) No provision was made for chemical addition or flocculation.

(2) The removal mechanism selected is not adequate. The mechanism is not suited for the clarifier geometry (circular or rectangular), type sludge, or both. Positive withdrawal mechanism should be selected for lighter secondary sludges, plow type for primary heavier sludges.

(3) The side-water depth is incorrect relative to the clarifier diameter.

(4) No provisions are made for *safe* access to the weirs for maintenance and sampling.

(5) Clarifier piping design does not permit tank isolation, or provisions to adjust flows to multiple clarifiers, reducing operating flexibility and complicating repairs.

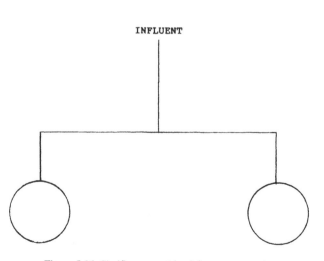

Figure 8.16 Clarifier symmetric piping arrangements.

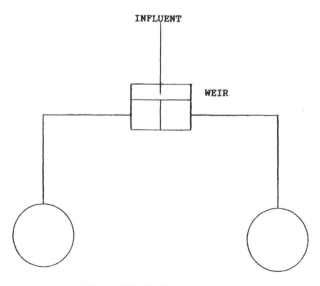

Figure 8.18 Clarifier weir splitting.

(6) The clarifier is undersized because too high an over-flow rate was selected, and the selected influent design flow did not include in-plant side streams or consider peak flows.

(7) No provision was made for adequate flow distribution to multiple clarifiers.

(8) No provision was made to minimize short-circuiting. Inlet flow distribution, minimizing turbulence, or the short-circuiting path was not considered.

(9) Inlet and outlet head loss estimates did not adequately considered the clarifier hydraulics.

(10) The design overflow rate is too high as a result of improper scale-down of laboratory results.

(11) The weir is placed too close to the troughs creates local velocities and turbulence. Weirs should be placed over the entire effluent end (rectangular) or perimeter (circular).

(12) The rectangular clarifier geometry is incorrect. The clarifier length-to-width ratio should be at least 3 to 1. The minimum length should be at least 3 m (10 ft) and width to depth ratio at 1:1 to 2:1.

(13) Scum removal provisions are inadequate.

(14) Scum lines frequently clog as a result of long pipe lengths and a lack of clean-outs or provisions for flushing the pumps or lines.

(15) Scum cannc be removed from the collection tanks once superna ant is removed. Tanks are not sloped or provided with water flush to facilitate removal.

(16) Clogged sludge lines cannot be dismantled to remove obstacles because of bolted flanges. No provision for flushing and rodding the lines compounds the problem.

(17) Clarifiers are not winterized to minimize heat losses, the resulting temperature reductions affects the upstream biological treatment. Below ground tanks and covers were not considered in severe climates, exposed piping are not heat traced and insulated.

Thickeners

(1) Hydraulic loading was not considered in sizing the thickener, especially where dilution water is added to the thickener.

(2) Short-circuiting through the tank results in heavy solids carryover. Influent and effluent baffling was not used.

(3) No provisions were made for chemical addition.

(4) Skimming provisions to remove scum from the treatment plant recycle were inadequate.

(5) The side water depth was inadequate, less than 3 to 4 m (10 to 12 ft) was installed.

(6) Freeze protection measures provided were inadequate (as discussed for final clarifiers).

CASE STUDY NUMBER 21

Develop a clarifier design for a one million gallon per day waste containing 360 mg/L of solids generated in Case Study 6. Whether type I, II, or III settling was dominant was difficult to establish from laboratory studies. However, the slurry was concentrated to 1% and an 18 ppm effluent obtained at a 2 centimeters per minute settling rate.

PROCESS CALCULATIONS

(1) Calculate the clarifier area based on hydraulic loading requirements

$$2 \text{ cm/min} = 3.93 \text{ ft/hr} = 707 \text{ gal/day/sf}$$

$$1,000,000 \text{ GPD} / 707 \text{ GPD/ sf} = 1414 \text{ ft}^2$$

(2) Establish a rectangular clarifier physical dimensions
Number of units: One
Length to width ratio: 5
Width equals 17 ft
Length equals 85 ft

(3) Establish a circular clarifier physical dimensions
Number of units: One
Diameter equals 42 ft

(4) Select side water depth: 11 ft

(5) Retention time

$$\text{Hours} = \frac{(1414 \text{ ft}^2 \cdot 11 \text{ ft deep})}{1,000,000 \cdot 0.1337 \cdot 1/24} \cong 3 \text{ hr}$$

CASE STUDY NUMBER 22

Develop a clarifier design for the activated sludge system designed in Case Study 10, for a forward waste flow of 1 MGD, a design MLVSS concentration of 2000 mg/L and a maximum MLVSS concentration of 3000 mg/L. Base the design on an underflow concentration of 10,000 mg/L, but explore the performance at an underflow range between 8000 and 13,000 mg/L. Apply a 1.5 safety factor to both the theoretical hydraulic and solids loading. Data from a series of laboratory settling tests relating initial settling velocity (ISV) to influent concentration are summarized as follows:

Influent, mg/L	1500	2000	2500	3000	4000
ISV, m/hr	4.00	2.70	1.4	0.82	0.41
ISV, ft/hr	13.12	8.86	4.59	2.69	1.35
Overvflow, g/sf/d	2356	1590	825	483	242

Influent, mg/L	4500	5000	5500	9000
ISV, m/hr	0.32	0.22	0.17	0.04
ISV, ft/hr	1.05	0.72	0.56	0.13
Overflow, g/sf/d	189	130	100	24

PROCESS CALCULATIONS

(1) Develop an initial settling velocity curve and obtain enough points to develop a solids flux curve

MLSS, mg/L	m/hr	kg/m²/hr	MLSS, mg/L	m/hr	kg/m²/hr
1000	4.80	4.800	7000	0.08	0.560
1500	4.00	6.000	8000	0.06	0.480
2000	2.70	5.400	9000	0.04	0.360
2500	1.40	3.500	10,000	0.03	0.300
3000	0.82	2.460	11,000	0.023	0.253
3500	0.55	1.925	12,000	0.019	0.228
4000	0.41	1.640	13,000	0.014	0.182
5000	0.22	1.100			
6000	0.12	0.720			

Figure 8.19 represents the correlation of the laboratory settling data.

(2) Develop a solids influx curve and establish limiting flux for critical MLSS concentrations

Points are selected from Figure 8.19 to develop a solids flux curve, Figure 8.20, covering the expected operating range. Figure 8.20 solids loadings corresponding to expected underflow concentrations are as follows:

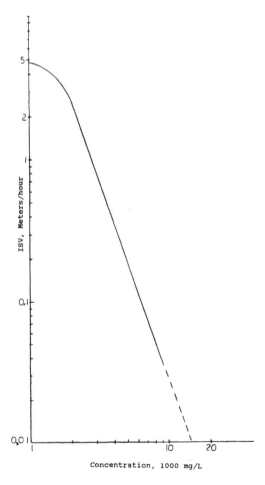

Figure 8.19 Solids settling curve (for calculation).

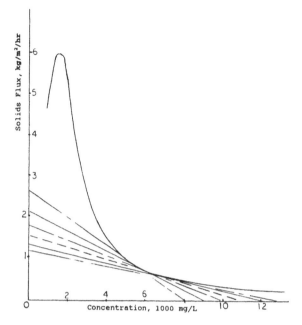

Figure 8.20 Total flux curve (for calculation).

	Underflow			Underflow	
mg/L	kg/m²/h	lb/ft²/hr	mg/L	kg/m²/hr	lb/ft²/hr
6000	4.60	0.94	10,000	1.85	0.38
7000	3.40	0.70	11,000	1.56	0.32
8000	2.70	0.55	12,000	1.35	0.28
9000	2.18	0.45	13,000	1.20	0.25

(3) Develop critical values for design conditions

    Flow rate: 1 MGD
Design MLVSS: 2000 mg/L
  Design MLSS: 2000/0.8 = 2500 mg/L

Overflow rate corresponding to 2500 mg/L is 1.4 m/hr
$$1.4 \cdot 589 = 825 \text{ GPD/sf}$$

Hydraulic loading related area:
$$1{,}000{,}000 \text{ GPD}/ 825 = 1213 \text{ ft}^2$$

The table values are calculated using the following relations:

(1) Recycle ratio = MLSS / (underflow − MLSS)
(2) Solid, lb/day = 1 MGD $\cdot$ (1 + σ) $\cdot$ 8.34 $\cdot$ MLSS
(3) Area

$$\text{Area} = \frac{\text{lb solids/day/24}}{\text{limiting flux (lb/hr/ft}^2)}$$

Underflow, mg/L	8000	9000	10,000	11,000	12,000	13,000
Recycle ratio	0.455	0.385	0.333	0.294	0.263	0.238
Solids, lb/day	30,337	28,877	27,793	26,980	26,334	25,812
Limit flux, lb/hr/ft²	0.55	0.45	0.38	0.32	0.28	0.25
Area, ft²	2300	2700	3100	3500	3900	4300

(4) Develop critical values for maximum conditions

Flow rate: 1 MGD
Design MLVSS: 3000
Design MLSS: 3000/0.8 = 3750

Overflow rate corresponding to 3750 mg/L:
$$0.40 \text{ m/h} \cdot 589 = 236 \text{ GPD/ft}^2$$

Hydraulic loading related area:
$$1,000,000 \text{ GPD}/236 = 4200 \text{ ft}^2$$

Underflow, mg/L	6000	7000	8000	9000	10,000
Recycle ratio	1.67	1.15	0.88	0.71	0.60
Solids, lb/day	83,504	67,241	58,797	53,480	50,040
Limit flux, lb/day/ft²	0.94	0.70	0.55	0.45	0.38
Area, ft²	3700	4000	4500	5000	5500

(5) Select clarifier area

The selected design condition is based on the solid flux requirements related to an underflow of 8000 to 9000 mg/L and an aeration basin MLSS concentration of 3750 mg/L, resulting in a theoretical area requirement of 4800 ft². An area of 4800 ft² allows an underflow concentration above 10,000 mg/L at MLSS design conditions of 2500 mg/L, leaving the operator considerable flexibility in operating the activated sludge system.

$$\text{Actual area} = 1.5 \cdot 4800 = 7200 \text{ ft}^2$$

Within the accuracy of the estimate, this is adequate for the maximum hydraulic area requirements,

$$= 1.5 \cdot 4200 = 6300 \text{ ft}^2.$$

(6) Establish a circular clarifier physical dimensions

Selected area: 7200 ft²
Number of units: Two
Diameter equals 68 ft
Calculated installed area, 3600 ft² per *unit*
Calculated installed area, 7200 *total* square feet
Overflow rate, 1,000,000 GPD/7200 = 139 GPD/sf

(7) Retention time

$$\text{Hours} = \frac{(7200 \text{ ft}^2 \cdot 11 \text{ ft deep})}{1,000,000 \cdot 0.1337 \cdot 1/24} = 14 \text{ hr}$$

DISCUSSION

This calculation illustrates the effect of required aeration basin MLVSS, the inert concentration in increasing the resulting MLSS, and the balance between hydraulic and solids transport loadings in estimating the installed area. What should be obvious is that equal results can be obtained in smaller clarifiers by reducing the achievable underflow concentration, and increasing the required recycle rate. In the final analysis, an economical judgment will be made between clarifier operating flexibility and the economics of recycle pumping and clarifier cost.

CASE STUDY NUMBER 23

Develop a *thickener* design for the waste sludge system described in Case Study 10. Assume that the sludge can be further thickened to 2% using solids loading data developed in Case Study 23. The design should be based on the maximum waste sludge generation of 1611 lb/day at a 1% concentration and a maximum flow rate of 19,317 gal/day.

PROCESS CALCULATIONS

(1) Allowable solids loading
Based on the settling data generated in the Case Study 23 a limiting flux of 1.20 kg/m²/hr or 0.25 lb/ft²/hr will be assumed. This is in the range of 10 to 35 kg/m²/day commonly employed for waste activated sludge.

(2) Calculate the thickener area based on solids loading requirements
$$[1611 \text{ lb/day}/24]/0.25 \text{ lb/hr/ft}^2 = 269 \text{ ft}^2$$
Applying a 1.5 safety factor:
$$1.5 \cdot 269 = 404 \text{ ft}^2.$$

(3) Establish a circular clarifier physical dimensions

Selected area: 404 ft²
Number of units: One
Diameter equals 22.6 ft, select 23 ft
Installed area = 415 ft²
Solids loading rate, 1611/415 = 4 lb/day/ft²
$$= 0.17 \text{ lb/hr/ft}^2.$$

(4) Select side water depth: 11 ft

(5) Retention time

$$\text{Hours} = \frac{(415 \cdot 11)}{19,317 \cdot 0.1337 \cdot 1/24} = 42 \text{ hr}$$

DISCUSSION

The long retention time will be an operating problem causing potentially anaerobic conditions. In fact, the quantity of sludge generated does not warrant the size and cost of the thickener estimated. The sludge should be discharged directly to the dewatering equipment, increasing the dewatering performance requirements; or an aerobic sludge holding tank of 20,000 gallons or less employed, decanting the separated water once or twice a day prior to dewatering or disposing the sludge.

REFERENCES

1. Crosby, R.M.: Evaluation of Hydraulic Characteristics of Activated Sludge Secondary Clarifiers," U.S. Environmental Protection Agency, EPA-68-03-2782, 1984.

2. Daigger, G.T., Roper, R.E. Jr.: "The Relationship Between SVI and Activated Sludge Settling Characteristics," *Journal WPCF,* V 57, No 8, Pg 859, August, 1985.

3. Dick, R.I.: "Role of Activated Sludge Final Settling Tanks," *Journal San Eng Div.,* Proceedings ASC, V 96, No SA2, Pg 423, April, 1970.

4. Dick, R.I., Young, K.W.: "Analysis of Thickening Performance of Final Settling Tanks," 27th Purdue Industrial Waste Conference, May 25, 1972.

5. Fitch, B.: "Batch Tests Predict Thickener Performance, *Chemical Engineering,* Pg 83, August 23, 1971.

6. Keinath, T.M.: "Diagram for Designing and Operating Secondary Clarifiers According to the Thickening Criteria," *Journal WPCF,* V 62, No 3, Pg 254, May/June, 1990.

7. Kynch, G.J.: "A Theory of Sedimentation," *Trans Faraday Soc,* V 48, Pg 166, 1952.

8. Laquidara, V.D., Keinath, T.M.: "Mechanism of Clarification Failure," *Journal WPCF,* V 55, No 1, Pg 54, January, 1983.

9. Parker, D.S.: "Assessment of Secondary Clarification Design Concepts," *Journal WPCF,* V 55, No 4, Pg 349, April, 1983.

10. Riddel, M.D.R., Lee, J.S., Wilson, T.E.: "Method for Estimating the Capacity of an Activated Sludge Plant," *Journal WPCF,* V 55, No 4, Pg 360, April, 1983.

11. Talmage, W.P., Fitch, E.B.: "Determining Thickener Unit Area," *Industrial and Engineering Chemistry,* V 47, No 1, Pg 38, 1955.

12. U.S. Environmental Protection Agency: *Process Design Manual for Suspended Solids Removal,* EPA-625/1-75-003a, January, 1975.

13. U.S. Environmental Protection Agency: *Treatability Manual,* EPA-600/8-80-042a, 1980.

14. U.S. Environmental Protection Agency: *Handbook for Identification and Correction of Typical Design Deficiencies at Municipal Wastewater Treatment Facilities,* EPA-625/6-82-007, 1982.

15. Vesiland, P.A.: The Influence of Stirring in the Thickening of Biological Sludge, Ph.D Thesis, University of North Carolina, Chapel Hill, 1968.

16. Wahlberg, E.J., Keinath, T.M.: "Development of Settling Flux Curves Using SVI," *Journal WPCF,* V 60, No 12, Pg 2095, December, 1988.

17. WEF Manual of Practice No 8: *Wastewater Treatment Plant Design,* Water Pollution Control Federation, 1982.

18. WEF Manual of Practice No 8: *Clarifier Design,* Water Pollution Control Federation, 1985.

19. WEF Manual of Practice: *Design of Municipal Wastewater Treatment Plants,* Water Environment Federation, 1992.

20. Wilson, T.E., Lee, J.S.: "Comparison of Final Clarifier Design Techniques," *Journal WPCF,* V 54, No 10, Pg 1376, October, 1982.

21. Medcalf & Eddy, Inc.: *Wastewater Engineering-Treatment, Disposal, Reuse,* McGraw-Hill, 1991, Third Edition.

22. Dietz, J.D., Keinath, T.M.: "Experimental Evaluation of the Significance of Overflow Rate and Detention Time," *Journal WPCF,* V 56, No 4, Pg 344, April, 1984.

23. Tekippe, R.J., Bender, J.H.: "Activated Sludge Clarifiers-Design Requirements and Research Priorities," *Journal WPCF,* V 59, No 10, Pg 865, October, 1987.

24. Qasim, S.R.: *Wastewater Treatment Plants; Planning, Design, and Operation,* Technomic Publishing Co., Inc, 1994.

Index

Printed and bound by CPI Group (UK) Ltd, Croydon, CR0 4YY

23/10/2024

01778249-0019